T0282538

CAMBRIDGE LIBRARY COLLECTION

Books of enduring scholarly value

Darwin

Two hundred years after his birth and 150 years after the publication of 'On the Origin of Species', Charles Darwin and his theories are still the focus of worldwide attention. This series offers not only works by Darwin, but also the writings of his mentors in Cambridge and elsewhere, and a survey of the impassioned scientific, philosophical and theological debates sparked by his 'dangerous idea'.

Palaeontology

Richard Owen (1804–1892) was a contemporary of Darwin, and like him, attended the University of Edinburgh medical school but left without completing his training. His career as an outstanding palaeontologist began when he was cataloguing the Hunterian Collection of human and animal anatomical specimens which had passed to the Royal College of Surgeons in London. His public lectures on anatomy were attended by Darwin, and he was entrusted with the classification and description of the fossil vertebrates sent back by Darwin from the Beagle voyage. He was responsible for coining many of the terms now used in anatomy and evolutionary biology, including the word 'dinosaur'. Palaeontology (published in 1860) defines, describes and classifies all the fossil animal forms then known, and discusses the origin of species, commenting on the theories of Buffon, Lamarck, the then anonymous author of Vestiges of Creation, Wallace and Darwin.

Cambridge University Press has long been a pioneer in the reissuing of out-of-print titles from its own backlist, producing digital reprints of books that are still sought after by scholars and students but could not be reprinted economically using traditional technology. The Cambridge Library Collection extends this activity to a wider range of books which are still of importance to researchers and professionals, either for the source material they contain, or as landmarks in the history of their academic discipline.

Drawing from the world-renowned collections in the Cambridge University Library, and guided by the advice of experts in each subject area, Cambridge University Press is using state-of-the-art scanning machines in its own Printing House to capture the content of each book selected for inclusion. The files are processed to give a consistently clear, crisp image, and the books finished to the high quality standard for which the Press is recognised around the world. The latest print-on-demand technology ensures that the books will remain available indefinitely, and that orders for single or multiple copies can quickly be supplied.

The Cambridge Library Collection will bring back to life books of enduring scholarly value across a wide range of disciplines in the humanities and social sciences and in science and technology.

Palaeontology

*A Systematic Summary of Extinct Animals
and their Geological Relations*

RICHARD OWEN

CAMBRIDGE UNIVERSITY PRESS

Cambridge New York Melbourne Madrid Cape Town Singapore São Paolo Delhi

Published in the United States of America by Cambridge University Press, New York

www.cambridge.org
Information on this title: www.cambridge.org/9781108001335

© in this compilation Cambridge University Press 2009

This edition first published 1860
This digitally printed version 2009

ISBN 978-1-108-00133-5

This book reproduces the text of the original edition. The content and language reflect the beliefs, practices and terminology of their time, and have not been updated.

PALÆONTOLOGY.

PALÆONTOLOGY

OR

A SYSTEMATIC SUMMARY OF EXTINCT ANIMALS

AND

THEIR GEOLOGICAL RELATIONS.

BY

RICHARD OWEN, F.R.S.

Superintendent of the Natural History Departments in the British Museum,
Fullerian Professor of Physiology in the Royal Institution of Great Britain,
Foreign Associate of the Institute of France, etc.

EDINBURGH:

ADAM AND CHARLES BLACK.

MDCCCLX.

CONTENTS, OR SYSTEMATIC INDEX.

———◆———

Cuts 2 to 22 inclusive are from Original Drawings by
S. P. WOODWARD, F.G.S.

TABLE OF STRATA AND ORDER OF APPEARANCE OF ANIMAL LIFE UPON THE EARTH.

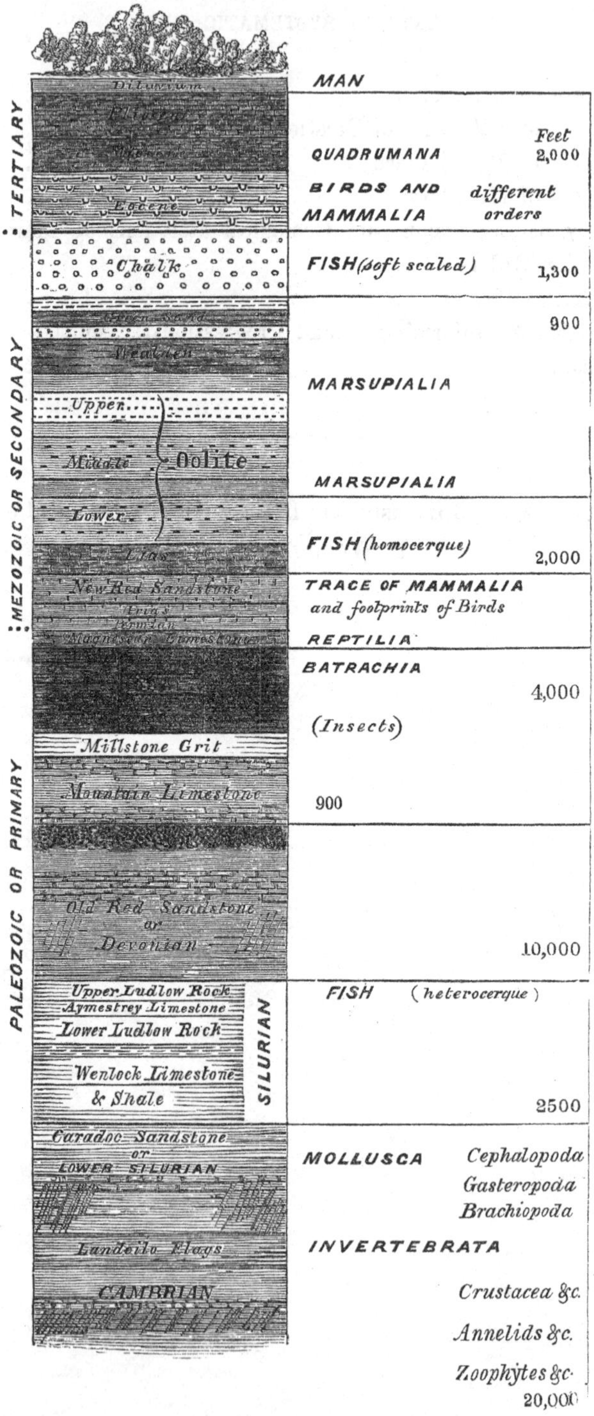

Strata	Animal Life	Feet
TERTIARY	MAN	
Diluvium / Pliocene	QUADRUMANA	2,000
Eocene	BIRDS AND MAMMALIA	different orders
Chalk	FISH (soft scaled)	1,300
MEZOZOIC OR SECONDARY		900
Weald	MARSUPIALIA	
Upper / Middle / Lower — Oolite	MARSUPIALIA	
Lias	FISH (homocerque)	2,000
New Red Sandstone, Trias, Permian, Magnesian Limestone	TRACE OF MAMMALIA and footprints of Birds	
	REPTILIA	
	BATRACHIA	4,000
	(Insects)	
PALEOZOIC OR PRIMARY		
Millstone Grit		
Mountain Limestone	900	
Old Red Sandstone or Devonian		10,000
Upper Ludlow Rock, Aymestrey Limestone, Lower Ludlow Rock (SILURIAN)	FISH (heterocerque)	
Wenlock Limestone & Shale (SILURIAN)		2500
Caradoc Sandstone or LOWER SILURIAN	MOLLUSCA — Cephalopoda, Gasteropoda, Brachiopoda	
Landeilo Flags	INVERTEBRATA	
CAMBRIAN	Crustacea &c.	
	Annelids &c.	
	Zoophytes &c.	20,000

Fig. 1.

PALÆONTOLOGY.

PALÆONTOLOGY* is the science which treats of the evidences in the earth's strata of organic beings, which mainly consist of petrified or fossil remains of plants and animals, belonging, for the most part, to species that are extinct.

The endeavour to interpret such evidences has led to comparisons of the forms and structures of existing plants and animals, which have greatly and rapidly advanced the science of comparative anatomy, especially as applied to the animal kingdom, and more particularly to the hard and enduring parts of the animal frame, such as corals, shells, spines, crusts, scales, scutes, bones, and teeth.

In applying the results of these comparisons to the restoration of extinct species, physiology has benefited by the study of the relations of structure to function requisite to obtain an idea of the food and habits of such species. It has thus been enriched by the well-defined law of " correlation of structures."

Zoology has gained an immense accession of subjects through the determination of the nature and affinities of extinct animals, and its best aims have been proportionally advanced. Much further and truer insight has been carried into the natural arrangement and subdivision of the classes of animals since palæontology expanded our survey of them.

The knowledge of the type or fundamental pattern of certain systems of organs, e.g., the framework of the Verte-

* From παλαιός, ancient; ὄντα, beings; λόγος, a discourse.

brata and the teeth of the Mammalia, has been advanced by the more frequent and closer adherence to such type discovered in extinct animals, and thus the highest aim of the zoologist has been greatly promoted by palæontology.

But no collateral science has profited so much by palæontology as that which teaches the structure and mode of formation of the earth's crust, with the relative position, time, order, and mode of formation of its constituent stratified and unstratified parts. Geology, indeed, seems to have left her old handmaiden mineralogy, to rest almost wholly upon her young and vigorous offspring, the science of organic remains.

By this science the law of the geographical distribution of animals, as deduced from existing species, is shown to have been in force during periods of time long antecedent to human history, or to any evidence of human existence ; and yet, in relation to the whole known period of life-phenomena upon this planet, to have been a comparatively recent result of geological forces determining the present configuration and position of continents. Hereby palæontology throws light upon a most interesting branch of geographical science, that, viz., which relates to former configurations of the earth's surface, and to other dispositions of land and sea than prevail at the present day.

Finally, palæontology has yielded the most important facts to the highest range of knowledge to which the human intellect aspires. It teaches that the globe allotted to man has revolved in its orbit through a period of time so vast, that the mind, in the endeavour to realize it, is strained by an effort like that by which it strives to conceive the space dividing the solar system from the most distant nebulæ.

Palæontology has shown that, from the inconceivably remote period of the deposition of the Cambrian rocks, the earth has been vivified by the sun's light and heat, has been fertilized by refreshing showers, and washed by tidal waves ; that the ocean not only moved in orderly oscillations regulated, as now, by sun and moon, but was rippled and agitated by winds and storms; that the atmosphere, besides these movements, was healthily influenced by clouds and vapours, rising, condensing, and falling in ceaseless circulation. With these conditions of life, palæontology demonstrates that life has been enjoyed during the same countless thousands of years; and that with life, from the beginning, there has been death. The earliest testimony of the living thing, whether coral, crust, or shell, in the oldest fossiliferous rock, is at the same time proof that it died. At no period does it appear that the gift of life has been monopolized by contemporary individuals through a stagnant sameness of untold time, but it has been handed down from generation to generation, and successively enjoyed by the countless thousands that constitute the species. Palæontology further teaches, that not only the individual, but the species perishes ; that as death is balanced by generation, so extinction has been concomitant with the creative power which has continued to provide a succession of species; and furthermore, that, as regards the various forms of life which this planet has supported, there has been " an advance and progress in the main." Thus we learn, that the creative force has not deserted the earth during any of the epochs of geological time that have succeeded to the first manifestation of such force; and that, in respect to no one class of animals, has the operation of creative force been limited to one geological epoch; and perhaps the most important and significant result of palæontological research has been the establishment of the axiom of *the continuous operation of the ordained becoming of living things.*

The present survey of the evidences of organic beings in the earth's crust commences with the lowest or most simple forms, and embraces chiefly the remains of the animal kingdom.

A reference to the subjoined "Table of Strata" (fig. 1) will indicate the relative position of the geological formations cited. The numerals opposite the right hand give the approximative depth or vertical thickness of the strata.

Organisms, or living things, are those which possess such an internal cellular or cellulo-vascular structure as can receive fluid matter from without, alter its nature, and add it to the alterative structure. Such fluid matter is called "nutritive," and the actions which make it so are called "assimilation" and "intus-susception." These actions are classed as "vital," because, as long as they are continued, the "organism" is said "to live."

When the organism can also move, when it receives the nutritive matter by a mouth, inhales oxygen and exhales carbonic acid, and developes tissues the proximate principles of which are quaternary compounds of carbon, hydrogen, oxygen, and nitrogen, it is called an "animal." When the organism is rooted, has neither mouth nor stomach, exhales oxygen, and has tissues composed of "cellulose" or of binary or ternary compounds, it is called a "plant." But the two divisions of organisms called "plants" and "animals" are specialized members of the great natural group of living things; and there are numerous beings, mostly of minute size and retaining the form of nucleated cells, which manifest the common organic characters, but without the distinctive superadditions of true plants or animals. Such organisms are called "Protozoa," and include the sponges or *Amorphozoa*, the *Foraminifera* or Rhizopods, the *Polycystineæ*, the *Diatomaceæ*, *Desmidiæ*, *Gregarinæ*, and most of the so-called *Polygastria* of Ehrenberg, or infusorial animalcules of older authors.

PROTOZOA.

Class I.—AMORPHOZOA.

Fossil sponges take an important place among the organic remains of the former world, not only on account of their great variety of form and structure, but still more because of the extraordinary abundance of individuals in certain strata. In England they specially characterize the chalk formation,— extensive beds of silicified sponges occur in the upper greensand, and in some beds of the oolite and carboniferous limestone. In Germany a member of the Oxford oolite is called the "spongitenkalk," from its numerous fossils of the present class.

Existing sponges are divided into *horny*, *flinty*, and *limy*, or "ceratose," "silicious," and "calcareous," according to the substance of their hard sustaining parts, which parts are commonly in the shape of fine needles, or *spicula*, of very varied forms, but in many species of sufficient constancy to characterize such species. The soft organic substance called "sarcode" appears to be structureless, and is diffluent; it is uncontractile and impassive, but consists of an aggregate of more or less radiated corpuscles, in some of which the trace of a nucleus may be discerned. The larger orifices on the surface of a sponge are termed "oscula," and are those out of which the currents of water flow: these enter by more numerous and minute "pores."

The calcareous sponges abound in the oolitic and cretaceous strata, attaining their maximum of development in the chalk; they are now almost extinct, or are represented by other families with calcareous spicula. The horny sponges appear to be more abundant now than in the ancient seas, but their remains are only recognisable in those instances where they were charged with silicious spicula.

M. d'Orbigny enumerates 36 genera and 427 species of fossil sponges; and this is probably only a small proportion of the actual number in museums, as the difficulty of determining the limits of the species is very great, and many remain undescribed.

Palæospongia and *Acanthospongia* occur in the lower Silurian; and *Stromatopora*, with its concentrically laminated masses, attains a large size in the Wenlock limestone. *Steganodictyum*, *Sparsispongia*, and species of *Scyphia*, are found

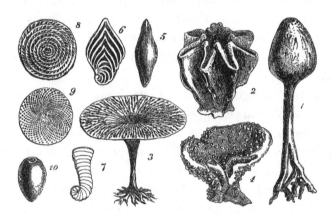

Fig. 2.

Amorphozoa; Rhizopoda.

1. Siphonia pyriformis, Goldf.; *Greensand*, Blackdown.
2. Guettardia Thiolati, D'Arch.; *U. Chalk*, Biarritz.
3. Ventriculites radiatus, Mant.; *U. Chalk*, Sussex.
4. Manon osculiferum, Phil.; *U. Chalk*, Yorkshire.
5. Fusulina cylindrica, Fisch.; *Carboniferous*, Russia.
6. Flabellina rugosa, D'Orb.; *Chalk*, Europe.
7. Lituola nautiloidea, Lam.; *Chalk*, Europe.
8. Nummulites nummularia, Brug.; *Eocene*, Old World.
9. Orbitoides media, D'Arch.; *U. Chalk*, France.
10. Ovulites margaritula, Lam.; *Chalk*, Europe.

in the Devonian; and *Bothroconis*, *Mamillopora*, and *Tragos*, in the Permian or magnesian limestone. Several genera are common to the trias and oolites, and several more are peculiar to the latter strata. The Oxfordian sponges belong chiefly to the genera *Eudea, Hippalimus, Cribrispongia, Stellispongia,* and *Cupulispongia.* Their fibrous skeleton appears to have been entirely calcareous, and often very solid; their form is cup-shaped, or mammillated, or incrusting, and many have a sieve-like appearance, from the regular distribution of the excurrent orifices (*oscula*) over their surface.

The greensand of Faringdon in Berkshire is a stratum prolific in sponges, chiefly cup-shaped and calcareous, of the genera *Scyphia* and *Chenendopora ;* or mammillated, like *Cnemidium* and *Verticillopora.* The Kentish rag is full of sponges, which are most apparent on the water-worn sides of fissures. Some beds are so full of silicious spicula as to irritate the hands of the quarrymen working those beds. The greensand of Blackdown is famous for the number and perfect preservation of its pear-shaped *Siphoniæ* (fig. 2, 1) ; whilst those of Warminster are ornamented with three or more lobes. The latter locality is the richest in England for large cup-shaped and branching sponges (*Polypothecia*), which are all silicified : the long stems of these sponges have been mistaken for bones. The sponges, chiefly *Siphoniæ,* of the upper greensand of Farnham are infiltrated with phosphate of lime, and have been used in agriculture.

The sponges of the chalk belong to several distinct families. *Choanites* resembles the *Siphonia,* but is sessile, and exhibits in section, or in weathered specimens, a spiral tube winding round the central cavity. It is the commonest sponge in the Brighton brooch-pebbles. Others are irregularly cup-shaped and calcareous ; and many of the Wiltshire flints have a nucleus of branching sponge (*S. clavellata*). The chalk flints, arranged in regular layers, or built up in columns of "Paramoudræ," all contain traces of sponge structure, and their origin is in some measure connected with the periodic growth of large crops of sponges. Frequently the crust or outer

surface only of the sponge has been silicified, while the centre
has decayed, leaving a botryoidal or stalactitic cavity. The
cup-shaped sponges are almost always more or less enveloped
with flint, which invests the stem and lines the interior, leaving
the rim exposed. The sponges of the Yorkshire chalk are of
a different character : some are elongated and radiciform,
others horizontally expanded, but they contain comparatively
little silica; while those belonging to the genus *Manon*
(fig. 2, 4), having prominent " oscula," are superficially silicified, and will bear immersion and cleaning with hydrochloric
acid. The largest group of chalk sponges, typified by *Ventriculites* (fig. 2, 3), have the form of a cup or funnel, slender or
expanded, or folded into star-like shape (*Guettardia*, fig. 2, 2),
with processes from the angles to give them firmer attachment. Some have a tortuous or labyrinthic outline, and
others are branched or compound, like *Brachiolites*. Curious
sections of these may be obtained from specimens enveloped
with flint or pyrites. The burrowing-sponge, *Cliona*, is commonly found in shells of the tertiaries and chalk. The great
cretaceous *Exogyræ* of the United States are frequently mined
by them ; and flint casts of *Belemnites* and *Inocerami* are
often covered by their ramifying cells and fibres. Thin sections of chalk flints, when polished and examined with the
microscope, sometimes exhibit minute spherical bodies (*Spini-
ferites*) covered with radiating and multicuspid spines. From
their close resemblance to the little fresh-water organism
Xanthidium, they long bore that name ; but they are certainly
marine bodies, and probably the spores of sponges.

Class II.—RHIZOPODA.

The organisms of this class are small and for the most
part of microscopic minuteness, of a simple gelatinous
structure, commonly protected by a shell. The most simple

rhizopods, called *Amœba*, present a globular form when contracted, but can extend portions of their substance ("sarcode") like roots, and use them to draw along the rest of the mass, like the feet or tentacles of polyps, whence the name of the class. These root-like processes can also attach themselves to foreign particles, and draw them into the "sarcode," or substance of the body, where the soluble organic part, so "intus-suscepted," may be assimilated, the insoluble part being extruded. A solid hyaline corpuscle or nucleus is commonly discernible in the interior of the *Amœba*, sometimes accompanied by one or more clear contractile vesicles. When the productions of the sarcode are numerous, filiform, and seemingly constant, radiating from all parts of the body, the rhizopod presents the characters of *Actinophrys*. When the tentacles are produced from only one extremity of the body we have the genus *Pamphagus*. When such a rhizopod is enclosed in a membranous sac it is a *Difflugia* ; if the sac be discoid with a slit on the flat surface for the protrusion of the tentacles, it is an *Arcella*. In other rhizopods the sac is calcified, or becomes a "shell," which is sometimes simple, but usually consists of an aggregate of chambers, inter-communicating by minute apertures, whence the name *Foraminifera* given to the testaceous rhizopods. These chambers grow by successive gemmation from a primordial segment, sometimes in a straight line, more commonly in a spiral curve ; and each segment so developed has its own shelly envelope. As, however, they are organically connected, the whole seems to form a "chambered" or "polythalamous" shell. The last-formed segment is usually distinguished by the very long, slender, pellucid, colourless, contractile filaments which have suggested the name "Rhizopods" for the class. But, in the *Foraminifera*, both the outer wall and the septa of the compound shell are perforated by minute apertures, through which either connecting or projecting filaments of the soft organic tissue can pass.

The several segments or jelly-filled chambers are essentially
repetitions of each other ; and there is no proof that the inner
and earlier segments derive their nourishment from the outer
and last-formed one. A Foraminifer may therefore be regarded
either as a series of individuals, organically united, or as a
single aggregate being, compounded according to the law of
vegetative repetition.

The minute chambered shells of *Foraminifera* enter
largely into the composition of all the sedimentary strata, and
are so abundant in many common and familiar materials, like
the chalk, as to justify the expression of Buffon, that the very
dust had been alive. The deep-sea soundings of the Atlantic
Telegraph ·Company have shown that the bed of that great
ocean, at a depth approaching, or even exceeding, two miles,
is composed of little else than the calcareous shells of a
Globigerina and a few other Rhizopods, with the silicious
shields of the allied *Polycystineœ*. The composition of the
chalk is extremely similar : when the finer portion, amounting
to half or even less, has been washed away, the remaining
sediment consists almost entirely of foraminated shells, some
perfect, others in various stages of disintegration. They have
also been found in other marine formations, which are soft
enough to be washed, down to the Lower Silurian ; and in the
hard limestones and marbles they can be detected in polished
sections, and in thin slices laid on glass. The greater part of
these shells are microscopic, but some of the large extinct
Foraminifera, called, from resembling a piece of money,
" Nummulites," are two inches in diameter.

The generic divisions in use for these shells were mostly
invented by M. d'Orbigny, but on artificial grounds, viz.,
exclusively upon the plan of growth, or mode of increase in
the number of chambers. The structure and anamorphoses
of these complex atoms have been recently investigated by
Messrs. Williamson and Carpenter, and especially by Mr.

W. K. Parker. The following are the primary groups of *Foraminifera* in the system of d'Orbigny :—

1. *Monostega.*—Body consisting of a single segment : shell of one chamber.
2. *Stichostega.*—Body composed of segments disposed in a single line : shell consisting of a linear series of chambers.
3. *Helicostega.*—Body consisting of a spiral series of segments : shell made up of a number of convolutions.
4. *Entomostega.*—Body consisting of alternate segments spirally arranged : shell chambers disposed on two alternating axes forming a spiral.
5. *Enallostega.*—Body composed of alternate segments not forming a spiral : chambers arranged on two or three axes which do not form a spiral.
6. *Agathistega.*—Body consisting of segments wound round an axis : chambers arranged in a similar manner, each investing half the entire circumference.

A somewhat different arrangement has been adopted by Schultze, who divides the *Polythalamia* into three sections, viz.—

1. *Helicoidea,* including those forms in which the several chambers of the shell are arranged in a convolute series, and answering to the last four orders of d'Orbigny.
2. *Rhabdoidea,* in which they are placed in a direct line (*Stichostega,* d'Orb.) ; and
3. *Soroidea,* where they are disposed in an irregular manner (*Acervulina*).

Lagena is a genus of *Monostega,* or single-chambered Foraminifera, with a flask-shaped shell, sometimes presenting a beautiful fluted exterior. *Entosolenia* is like a

Lagena, with the tubular neck inverted into the cavity of the shell.

Among the many-chambered Foraminifers the modifications of form seem endless. *Nodosaria* resembles a cylindrical beaded rod : *Cristellaria* begins by being spiral and afterwards becomes straight : most species are wholly spiral : in some, as *Nummulites,* the convolutions are on the same plane : in many the spiral turns obliquely round an axis, and gives the shell a trochoid form.

Upwards of six hundred and fifty-seven fossil species, belonging to seventy-three genera, have been described : they commence in the palæozoic age, increase in number and variety with each successive stratum, and attain their maximum in the present seas. Most of the fossil genera, and even some of the species, pass through many formations ; indeed, if correctly observed, the existing forms are the oldest known living organisms. *Dentalina communis, Orbitolites complanatus, Rosalina italica,* and *Rotalina globulosa,* all living species, are said to be found in the chalk; *Rotalina umbilicata* ranges to the gault ; and *Webbina rugosa* is common to the upper lias, the chalk, and present sea. It has, however, been observed, that fossil Rhizopods, set free by the disintegration of rocks, are mingled with the recent shells on every beach ; and Mr. M'Andrew has obtained them in this condition from great depths of the mid-channel.

The earliest important form is the *Fusulina* (fig. 2, 5), which forms layers many inches, or even feet, in thickness in the carboniferous limestone of Russia. The recent genera *Dentalina* and *Textularia* are found in the magnesian limestone ; *Nodosaria, Cristellaria,* and *Rotalia,* in the lias. *Flabellina* (fig. 2, 6) is peculiar to the chalk ; *Orbitoides* (fig. 2, 9) to the chalk and tertiary series ; *Ovulites* (fig. 2, 10) is peculiar to the eocene tertiaries; *Operculina, Orbitolites,* and *Alveolina* appear first in the tertiary, and are still living. The *Lituola* (fig. 2, 7) occurs in the chalk and chalk flints, and has been described as a species of " Spirolinite." Many of the cretaceous *Foraminifera* contain a brown colouring matter, which remains after the shell has been dissolved with weak acid, and has been regarded as the remains of the organic substance which once filled all the cells.

The " calcaire grossier," which is employed at Paris as a building-stone, contains *Foraminifera* in such abundance that one may say the capital of France is almost constructed of those minute and complex shells.

But it is in the middle eocene, or " nummulitic period," that the Rhizopods attained their greatest size, and played their most important part. Wherever limestones or calcareous sands of this period are met with, these Foraminifers abound, and literally form strata which in the aggregate become mountain masses. These " nummulitic limestones" are found in Southern Europe, in Northern Africa, and in India ; they also occur in Jamaica. The commonest form is the *Nummulite* (fig. 2, 8), which occurs in the building-stone of the Great Pyramid. The *Nummulites* were evidently sedentary organisms ; and, in the large thin species, one side is moulded to the inequalities of the sea-bed on which it grew.

Polycystineæ.—The tertiary marls of Barbadoes afforded to Ehrenberg an extensive series of novel and extraordinary microscopic organisms, composed of silica, but foraminated like the shells of the Rhizopods. The same forms, and others similar to them, have been met with in the deep-sea mud of the Gulf of the Erebus and Terror, and more recently in the mud of the North Atlantic soundings. They are quite distinct in form and character from most of the silicious-shielded *Diatomaceæ*, but some of them resemble the *Coscinodiscus* and *Actinocyclus*. No less than 282 forms, grouped in 44 provisional genera, have been described.

CLASS III.—INFUSORIA.

(*Polygastria*, Ehrenberg.)

Numerous genera and multitudes of so-called species of free and locomotive microscopic organisms, which, because they do not present the distinctive characters of plants or animals, have been by turns referred to one or other kingdom, possess shells of flint, and consequently enter largely into the domain of fossil evidences of former life. The silicious shells of these infusory organisms present under the microscope the most definite as well as beautiful characters of form and sculpture, which are as recognisable and distinctive as those of the calcareous shells of Mollusca. The plates of the incomparable works and memoirs of Ehrenberg abound with exact figures of the delicate sheaths, shells, and shields of the loricated Infusoria of past and present æras of life, the deposits of which, by reason of their pure, flinty, atomic constitution, were known in the arts long before science had detected their nature and vital origin. In 1836 portions of the stone called "tripoli" or "polierschiefer" (polishing-slate of lapidaries) were microscopically examined by Ehrenberg, who discovered it to be wholly composed of the silicious shells of Infusoria, and chiefly of an extinct species called *Gaillonella distans*. At Bilin, in Bohemia, there is a single stratum of polierschiefer, not less than fourteen feet thick, forming the upper layer of a hill, in every cubic inch of which there are forty-one thousand millions of the above-named organic unit. This mineral likewise contains shells of *Navicula, Bacillaria, Actinocyclus*, and other silicious organisms. The lower part of the stratum consists of the shells compacted together without any visible cement ; in the upper masses the shells are cemented together, and filled by amorphous silicious matter formed out of dissolved shells. At Egea, in Bohemia, there is a stratum of two miles in length, and averaging twenty-eight feet in thickness, of which the uppermost ten feet are composed wholly of the silicious shells of Infusoria, including the beautiful *Campylodiscus ;* the remaining eighteen feet consist of the shells mixed with a pulverulent substance. Corresponding deposits of the silicious cases of Infusoria have since been discovered in many other parts of the world, some including fresh-water species, others marine species of Infusoria.

The conditions of such depositions will be readily understood by examining the sedimentary deposits of bogs and of stagnant or slow-flowing sheets of water. In warm latitudes and seasons, such water swarms with infusorial life, and the indestructible cases of the loricated kinds are found in great quantities in the sedimentary deposits. Beneath peat bogs they have been found to form strata of many feet in thickness, and co-extensive with the turbary, forming a silicious marl of pure whiteness. A quantity of pulverulent matter is deposited upon the shores of the lake near Uranea, in Sweden, which, from its extreme fineness, resembles flour : this has long been known to the poorer inhabitants under the name. of " bergmehl," or mountain-meal, and is used by them, mixed up with flour, as an article of food. It consists in great part of silicious shells of Infusoria, with a little organic matter. With regard to the source of fossil infusorial remains in sea-water, the following evidence is given in the *United States Coast Survey,* 1856 :—

Soundings of the gulf-stream near Key Siscayne, Florida, varying in depths from 147 fathoms to 205 fathoms, give a light greenish-grey mud composed chiefly of Foraminifers, Diatoms, Polycystins, and Geolites, in a profusion only surpassed by the fossil polycystinous strata of Barbadoes. The Foraminifers compose the largest part of these muds, including *Textularia Americana, Marginula Bachei,* and other forms, particularly many species of the *Plicatilia* of Ehrenberg, which had been supposed to live only in shallower haunts. The silicious shells of Diatoms abound in the residue, after the calcareous Foraminifers have been dissolved by acid. The inorganic portion of the soundings is chiefly quartz sand, and its proportion is quite small.

Such manifestations of life, with its mineral results, have been detected from the earliest sedimentary deposits to the present time ; but as regards the Infusoria, they are given on the grandest scale in formations of the tertiary age. The town of Richmond, in Virginia, United States, is built on barren silicious strata of marine origin and tertiary age. The strata are twenty feet in thickness, composed chiefly of infusorial flint-shells, including the well-known and beautiful microscopic objects, *Actinocyclus* and *Coscinodiscus*.

Most of the infusorial formations, as the polishing-slates at Cassel, Planitz, and Bilin, are astounding monuments of the operation of microscopic organisms at former periods of the history of this planet. The minute size, elementary structure, tenacity of life, and marvellous reproductive power of the Infusoria have enabled them to survive as species those destroying causes which have exterminated contemporaneous higher forms of organism. Species of *Bacillaria* still exist which were in being at the period of the deposition of the chalk. Existing species of *Diatomaceœ* have been detected as low down as the oolite. The discovery by Ehrenberg of more than twenty species of silicious-shelled Infusoria, fossil, in the chalk and chalk-marls, which are identical in species with some now living in the bed of the Baltic, is an instructive addition to the obscure history of the introduction of species of living things in this planet, and must add greatly to the interest of the infusorial class in the eyes of the geologist and philosopher. " For these organisms," writes Ehrenberg, " constitute a chain which, though in the individual link it be microscopic, yet in the mass is a mighty one, connecting the life-phenomena of distant ages of the earth, and proving that the dawn of the organic nature co-existent with us reaches further back in the history of the earth than had hitherto been suspected." "The microscopic organisms are very inferior in individual energy to lions and elephants, but in their united influences they are far more important than all these animals." If it be ever permitted to man to penetrate the mystery which enshrouds the origin of organic force in the wide-spread mud-beds of fresh and salt waters, it will be, most probably, by experiment and observation on the atoms which manifest the simplest conditions of life.

ANIMALIA.

INVERTEBRATA.

Remains of invertebrate animals occur in strata of every age, from the partially metamorphic and crystalline rocks of the Cambrian system to the deposits formed by the floods of last winter and the tides of yesterday. They are found in every country, from the highest latitude attained by Arctic voyagers to the extremities of the southern continents, and at the greatest elevation hitherto climbed in the Andes or Himalaya. If some classes—e. g., Tunicata, Acalephæ—seem not to be represented in stratified deposits, they are such as, either from the soluble or perishable tissues composing the entire frame, could not be expected to be fossilized under any conceivable circumstances ; or from the same cause, are only not so recognisable at one of their metagenetic phases. Evidence of compound Hydrozoa—i. e., of the polypes which Ellis called "Corallines"—and especially of the genus Campanularia, would show that the acalephal type and grade of organization had been manifested at the period of the formation of the strata containing such fossil Polypi. With the above seeming

C

exceptions, every class of invertebrate animal is represented by fossil remains.

They consist of corals and shells, of the petrified skeletons of star-fishes and sea-urchins, of the hard coverings of crabs and insects, of the tracks and shelly habitations of worms, and of impressions of surfaces, and casts of cavities, of organisms, retained by the matrix after the animals had perished.

The condition in which invertebrate fossils occur depends on the nature of the matrix and other accidental circumstances ; for while some are scarcely altered in composition, or even in colour, others are silicified or infiltrated with carbonate of lime. Some may be cleared by the action of acid or exposure to the weather, and some require the chisel of the mason or the mill of the lapidary for the proper exhibition of their structure.

Multitudes of recent species are fossilized in the newer tertiaries whose history can be made out perfectly from living specimens ; but the number of these diminishes gradually in each older stratum, while the proportion of extinct forms is ever on the increase. No living *species* more highly organized than a Rhizopod is found in the secondary rocks. Recent *genera* extend further back in time ; indeed a few may be recognised in strata of palæozoic age, shedding a light on the probable affinities and conditions of their associates. Many of the smaller groups of genera, called *families*, disappear in the secondary, and still more in the palæozoic period, and are to a limited extent replaced by groups which no longer exist. But as to the larger groups of *Protozoa* and of true invertebrate animals, it may be affirmed that every known fossil belongs to some one or other of the existing classes, and that the organic remains of the most ancient fossiliferous strata do not indicate or suggest that any earlier and different class of beings remains to be discovered, or has been irretrievably lost, in the universal metamorphism of the oldest rocks.

PROVINCE I.—RADIATA.*

Sub-Province POLYPI.

A polype is a small soft-bodied aquatic animal which generally presents a cylindrical oval or oblong body, with an aperture at one of its extremities surrounded by a crown of radiating filaments or "tentacles." This aperture leads to the digestive cavity, which, in most Polypes, is without intestine or vent. A very large proportion of these animals has organs of support, called "polyparies" or corals, of various form and substance, but for the most part consisting of carbonate of lime ; and, as a general rule, locomotion is lost with the development of the polypary, which usually attaches the polype to some foreign body. The organization of the soft tissues is in general simple ; the faculties of the Polypes are very limited ; and the vital phenomena, save those of irritability and contractility, are inconspicuous. Nevertheless, the influence of the combined powers of some of the species, in adding to and modifying the crust of the earth, is neither slight nor of limited extent.

CLASS I.—HYDROZOA.

Char.—Polypary, when present, flexible, external; for the most part developing cells for the polypes according to regular patterns.

FAMILY I.—GRAPTOLITIDÆ.

To this class may probably belong the organic remains called "Graptolites," which are exclusively and characteristically Silurian fossils. A certain knowledge of their affinities

* For the characteristic organization of the provinces, classes, orders, and families of *Invertebrata*, reference may be made to the writer's " Lectures on Invertebrata," 8vo, Longmans, 1855.

would require examination of the soft parts ; and the family
has long been extinct. Indications of the flexible consistency
of the polypary, and M. Barrande's statement of the existence
of a cylindrical canal in its axis, which he conjectures to have
contained the common connecting tissue of the polypes, have
weighed with the writer in placing the Graptolites provisionally in the present class of Polypi. The axis of the polypary is
sometimes straight (fig. 3, ₃), sometimes spiral (fig. 3, ₆). The
ordinary form, as given by the *Graptolites priodon* (fig. 3, ₃),

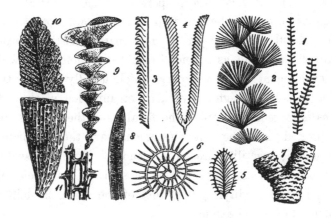

Fig. 3.

Hydrozoa; Anthozoa; Bryozoa.

1. Protovirgularia dichotoma, M'C.; *Silurian*, Dumfries.
2. Oldhamia antiqua, Forbes; *Cambrian*, Wicklow.
3. Graptolites priodon, Brun.; *Silurian*, Britain.
4. Didymograpsus Murchisoni, Beck; *L. Silurian*, Wales.
5. Diplograpsus folium, His.; *L. Silurian*, Britain.
6. Rastrites peregrinus, Barr; *Silurian*, Bohemia.
7. Cœnites juniperinus, Eichw.; *U. Silurian*, Dudley.
8. Ptilodictya lanceolata, Lonsd.; *U. Silurian*, Tortworth.
9. Archimedipora Archimedea, Lesuer.; *Carboniferous*, Kentucky.
10. Ptilopora pluma, M'C.; *Carboniferous*, Ireland.
11. Fenestrella membranacea, Ph.; *Carboniferous*, Britain.

is serrated on one side only, and is found abundantly in the
Cambrian or older Silurian beds of Scotland and Wales ; it
occurs also in the Ludlow rocks. The double Graptolites

(*Diplograpsus*, fig. 3, 5, and *Didymograpsus*, fig. 3, 4) are Cambrian forms. *Rastrites* (fig. 3, 6) had the polypes only in one side, and they are less crowded: it characterises Barrande's division E of the Lower Silurian beds of Bohemia, and has not yet been found in Britain. The Graptolites occur in argillaceous strata, especially in the mud-stones of Wales and Cumberland, and in the alum-slates of Sweden. These beds remind one of the mud bottoms in which the Virgularia and other long and slender graptolite forms of " Pennatulidæ" flourish in forest-like crowds. The primeval Graptolite may have presented a more generalized polype structure than is now met with in the specially differentiated Sertularians and sea-pens.

CLASS II.—ANTHOZOA.

In this class of Polypes the tentacles are hollow, and, in most, with pectinated margins. The polypary is usually internal, and forms the bodies more properly called " corals and " madrepores."

Asteroida.—Great doubt attaches to some of the fossils referred to this class of Polypi. The terms " Gorgonia" and " Alcyonium" have been applied to objects not well understood, and usually proving to be *Bryozoa* and sponges. The Lower Silurian fossil called *Pyritonema* consists of a fasciculus of silicious fibres, and has been supposed to be related to the glass zoophyte (*Hyalonema*). The miocene deposits of Piedmont contain a species of the Mediterranean genus *Corallium*, an *Antipathes*, and an *Isis* (or *Isisina*, d'Orb.), which is also found in Malta. The London clay contains one coral (*Graphularia*), referred to the *Pennatulidæ*, and two *Gorgonidæ* (*Mopsea* and *Websteria*).

Actinoida.—The lamelliferous or stony corals are (next to the *Testacea*) the largest and most important class of invertebrate fossils. They attained a great development in

the earliest seas, and were perhaps more widely diffused and
individually abundant in the Silurian age than at any subsequent period. " Reef-building" corals are now confined to
warm seas, and are wanting even on great tracts of tropical
coast. The *Oculina* is the only large coral now found in the
north. But in palæozoic times the representatives of the
modern Astræas and Caryophyllias extended as far northward
as Arctic voyagers have penetrated ; and at a much later
period they formed reefs of considerable thickness and extent
in the area of the coralline oolite. The Silurian limestone of
Wenlock Edge is itself a coral reef thirty miles in length ;
and the Plymouth limestone and carboniferous limestone have
frequently the aspect of coral-banks skirting the older regions
of Cambrian slate and Devonian " killas." The structure of
coral-banks may be studied in the lofty limestone cliffs of
Cheddar, and in the wave-worn shores of Lough Erne, as well
as in the upheaved coral islands of the southern seas. In the
fields about Steeple-Ashton, every stone turned up by the
plough is a coral ; and our inland quarries and chalk-pits
afford to the palæontologist materials for the study of a class
almost wholly wanting on the present sea-shores of Europe.
The history of the British fossil corals, as given by Milne
Edwards and Haime in the "Monographs of the Palæontographical Society," exhibits, equally with that of the fossil
shells by other authors, a transition from a state very different
from that which now subsists in our part of the world, and
a gradual approximation to the present order of things.

In the palæozoic strata the corals belong chiefly to two
extinct orders ; those of the secondary period more resemble
living corals of warmer climates than ours ; and the few tertiary genera and species resemble those of Southern Europe
and our own coast.

The distinction between one large group of the palæozoic
Actinoida (*Cyathophyllidæ*) and more modern corals consists

in the quadripartite character of their plaited cups or stars, whereas the lamellæ (or *septa*) of the other families are developed in multiples of 6. A remarkable exception exists in the *Holocystis* (fig. 5, 8), an Astrea-like coral with quadripartite stars, which is found in the lower greensand. The old-rock corals are also remarkable for the manner in which they are partitioned off by horizontal " tabulæ" (fig. 4, 3), like the septa of the *Nautilus* and *Spondylus*. This character obtains not

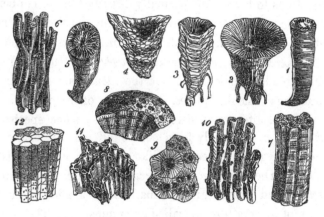

Fig. 4.

Palæozoic Corals (Anthozoa).

1. Amplexus Sowerbyi, Ph.; *Carboniferous*, Ireland.
2. Cyathophyllum turbinatum, Lin.; *U. Silurian*, Wenlock.
3. Cyathophyllum subturbinatum (section); *U. Silurian*, Wenlock.
4. Cystiphyllum Siluriense, Lonsd.; *U. Silurian*, Wenlock.
5. Zaphrentis Phillipsi, M. Edw.; *Carboniferous*, Somerset.
6. Lithodendron irregulare, Ph.; *Carboniferous*, Europe.
7. Lithostrotion striatum, Flem.; *Carboniferous*, Europe.
8. Acervularia luxurians, Eich.; *U. Silurian*, Europe.
9. Heliolites interstincta, Wahl.; *U. Silurian*, Europe.
10. Syringopora ramulosa, Goldf.; *Carboniferous*, Europe.
11. Halysites catenulatus, L.; *Silurian*, Northern Regions.
12. Favosites Gothlandica, Lam.; *Silurian*, North.

only in the *Cyathophyllidæ*, but also in the *Milleporidæ*, *Favositidæ*, and other palæozoic families. Of the 129 Silurian corals, 121 belong to the tabulated divisions.

The Devonian system contains about 150 described corals, the carboniferous limestone 76, and the magnesian limestone only 5 or 6. The commonest forms of simple, turbinated corals, are *Cyathophyllum* (fig. 4, 2 and 3), which exhibits four slight *fossulæ* in its cup, and is often supported by root-like processes. In *Zaphrentis* (fig. 4, 5) there is but one deep fossula. *Amplexus* (fig. 4, 1) is a characteristic carboniferous fossil, nearly cylindrical, and often so straight and regular in its growth as to have been originally described as a chambered shell. The radiating septa are very slight, and the horizontal partitions simple, flat, and almost as regular as the septa of the *Orthoceras*. In the Silurian *Cystiphyllum* (fig. 4, 4) the lamellæ are also evanescent ; but the *tabulæ* are represented by numerous vesicular plates. The corals of these genera are not always solitary, or merely in groups ; some species of *Cyathophyllum* constantly form compound masses, with cups rendered polygonal by contact, like *C. regium* of the Bristol limestone. The allied genus *Acervularia* (fig. 4, 8) resembles an *Astræa*, and exhibits, in a remarkable manner, the multiplication of its corallites by calicular gemmation. The genus *Lithostrotion* (fig. 4, 7) of the carboniferous limestone is also compact and astræiform, but the new corallites are produced by lateral gemmation. Corals, with the same structure, but not compact, are known by the name *Lithodendron* (fig. 4, 6). The "chain-coral" *Halysites* (fig. 4, 11) and *Syringopora* (fig. 4, 10) resemble, at first sight, the recent asteroid *Tubiporidæ* : in *Halysites* the radiating septa are quite rudimentary ; and in *Syringopora* the tabulæ are funnel-shaped, forming a central axis to each tube. The *Favositidæ* (fig. 4, 12) are mostly very regular both as to their polygonal shape and transverse tabulæ ; the cells of adjacent corallites are connected by pores, either in the sides or angles of the walls ; the septa are rudimentary. In the genus *Chætetes* the tubes are always slender, and much elongated, and their walls imperforate. *Michelinia* resembles the fruit of the *Nelumbium;* it has vesicular tabulæ and root-like processes to its basal plate. *Heliolites* (fig. 4, 9), of which many species are found in the Silurian and Devonian limestones, is related to the recent *Milleporæ*. The radiating septa are distinct, and the tabulæ regular ; the interspaces between the stars are filled up with fine and regular tubes. One genus of *Fungidæ* (*Palæocyclus*) occurs in the Upper Silurian.

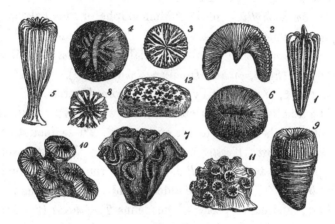

Fig. 5.

Secondary and Tertiary Corals (*Anthozoa*).

1. Turbinolia sulcata, Lam.; *M. Eocene*, Europe.
2. Diploctenium lunatum, Brug.; *Chalk*, France.
3. Micrabacia coronula, Goldf.; *U. Greensand*, Europe.
4. Aspidiscus cristatus, Lam.; *Cretaceous* (?), Algeria.
5. Cyclolites elliptica, Lam.; *L. Chalk*, France.
6. Parasmilia centralis, Mant.; *U. Chalk*, England.
7. Pachygyra labyrinthica, Mich.; *L. Chalk*, France.
8. Holocystis elegans, Lonsd. ; *L. Greensand*, Isle of Wight.
9. Montlivaltia caryophyllata, Lam. ; *Great Oolite*, France.
10. Stylina De la Bechei, M. Edw.; *Corallian*, Wilts.
11. Thecosmilia annularis, Flem.; *Corallian*, Wilts.

The British secondary corals are not very numerous ; for although specimens abound in the coral-rag districts, only fourteen species are found in that formation. Altogether, sixty-five species are found in the English oolites, and twenty-two in the chalk and greensands. These are mostly *Astrœidœ*, or related to *Fungia*. Three common forms in the oolites are *Montlivaltia* (fig. 5, 9), *Stylina* (fig. 5, 10) and *Thecosmilia* (fig. 5, 11). The English cretaceous strata afford the *Holocystis* (fig. 5, 8), which is the most recent coral with quadripartite septa ; *Trochocyathus* and *Parasmilia* (fig. 5, 6), resembling the recent *Cyathina*; and the little "Fungia" *coronula* (fig. 5, 3), described in two genera of distinct orders (*Micrabacia* and *Stephanophyllia*) in the "Monographs of the Palæontographical Society." The lower chalk of France and Germany contains many other corals, especially *Cyclolites* (fig. 5, 5), *Pachygyra* (fig. 5, 7), and *Diploctenium* (fig. 5, 2). The *Aspidiscus* (fig. 5, 4) was sent by Dr. Shaw from Algeria.

The English eocene strata contain twenty-five corals, all extinct, and belonging to fifteen genera. These include an *Astrœa* (*Litharœa Websteri*), which grows on the water-worn flint pebbles ; a *Balanophyllia*, similar to the existing coral ; a *Dendrophyllia*, which is the oldest member of the genus ; an *Oculina*; and eight species of the genus *Turbinolia* (fig. 5, 1). The corals of the English pliocene are mostly *Bryozoa*; only four true corals have been found in the coralline crag belonging to the genera *Sphenotrochus*, *Flabellum*, *Cryptangia*, and *Balanophyllia*, all reputed extinct, although the first is very closely related to the living *Sphenotrochus Macandrewi*.

The total number of fossil corals enumerated by M. d'Orbigny in the "Prodrome de Paléontologie," amounts to 1135, grouped under 216 genera. But notwithstanding all the labour which has been bestowed on this branch of palæontology by Goldfuss, Michelin, Lonsdale, and Milne Edwards, species are continually discovered or brought home from abroad which are altogether new, and cannot be placed in any of the constituted genera.

Class III.—BRYOZOA.

Char.—Tentacles of the polype hollow, with ciliated margins ; alimentary canal with stomach, intestine, and anus ; polypary, when present, external, horny, and calcareous.

The metamorphoses which the *Bryozoa* undergo are like those of the lower *Polypi ;* the embryo developed from the ovum is an oval, discoid, or subdepressed body, with a general or partial ciliated surface, by which it enjoys a brief locomotive life after its liberation from the parent. The *Bryozoa* are allied to the compound *Ascidia ;* but not one of the ascidian Molluscoids quits the ovum as a gemmule swimming by means of cilia ; and no Bryozoon quits the ovum in the guise of a Cercarian or tadpole, to swim abroad by the alternate inflexions of a caudal appendage. In a progressive and continuous series of teachings, by pen or word of mouth, the place of an osculent or transitional group is governed by convenience, by considerations of how best to teach by comparison and easy gradation. The real merits of the man who would make scientific capital by changing the position of such group, and by imputing error or ignorance to the author from whom he may differ in this respect, are easily weighed and soon understood.

The *Bryozoa,* whether regarded as the highest organized Polypes, or as the lowest organized *Mollusca,* or as an intermediate type, are treated of in systematic palæontology in the position here assigned to them. The practical palæontologist finds himself compelled to arrange and study the fossil *Bryozoa* along with the corals, if only on account of the difficulty he, in many cases, experiences of determining to which class of Polypi his specimens belong. M. d'Orbigny, who has devoted much attention to this class, enumerates 544 fossil species, distributed in 73 genera. This number must be very far below the real one, since the *Bryozoa* of the chalk, which alone have been carefully examined, amount to 213 species; while only two species are known from the trias, none at all from the lias, and only five from the upper oolites, so rich in corals and sponges. In the "Cours Elémentaire" of d'Orbigny the fossil Bryozoa are stated to amount to 1676.

Of the 19 or 20 palæozoic genera, none extend into the secondary strata; but of the 18 oolitic genera, *Entalophora* and *Defrancia* range onwards to the tertiaries; and *Alecto*, *Idmenea*, and *Eschara* still survive. The oldest known fossil, *Oldhamia* (fig. 3, 2), is supposed to be a Bryozoon. The most common palæozoic form is *Fenestrella* (fig. 3, 11), resembling the recent "lace-coral"; there are 35 species, ranging from the Lower Silurian to the Permian. One of its modifications resembles a feather *(Ptilopora*, fig. 3, 10), and is found in the carboniferous limestone. Another, more remarkable, has a spiral axis (*Archimedipora*, fig. 3, 9), and occurs in the same formation in Kentucky. One of the oldest genera is *Ptilodictya* (fig. 3, 8), of which seven species are found in the Lower Silurian formations. The slabs of Silurian limestone obtained at Dudley are covered with myriads of small and delicate fossils, including many *Bryozoa*. Some of these are spread like a film over other fossils, and have been doubtfully referred to the modern genera *Discopora* and *Berenicea*; others, with slender branches, and erect or creeping, are called Milleporas, Heteroporas, and Escharinas. The genus *Cœnites* (fig. 3, 7), perhaps belongs here. The magnesian limestone contains several large "lace-corals" of the genera *Fenestrella*, *Synocladia*, and *Phyllophora*; and two branching species of *Thamniscus* and *Acanthocladia*. The oolites afford many small incrusting species related to *Diastopora*, and branching forms like *Terebellaria* and *Chrysaora*. In the chalk, the *Escharas* are most numerous, and *Lunulites* and *Cupularia* first appear. Some thin beds of the lower chalk are almost composed of

Bryozoa, mingled with *Foraminifera.* The coralline crag of Suffolk takes its name from the great abundance of *Bryozoa* it contains, among which *Eschara, Cellepora, Fascicularia, Theonoa, Hornera, Idmonea, Flustra,* and *Tubulipora* are the most important.

Class IV.—ECHINODERMATA.

(*Star-Fishes, Sea-Urchins.*)

Char.—Marine ; commonly free, repent animals, with the integument in most perforated by erectile tubular tentacles, hardened by a reticulate deposit of calcareous salts, and in many armed with spines.

The fossil Radiata present a mine of comparatively unexhausted riches to the palæontologist. More difficult of study than shells, and less uniformly present in all strata, the enduring remains of echinoderms and corals are unsurpassed in beauty of form and structure, and in the value of the evidence they afford.

The present summary of the extinct forms of *Echinodermata* will commence with

Order 1.—Crinoidea.

Char.—Body with ramified rays, supported temporarily or permanently on a jointed calcareous stem ; alimentary canal, with mouth and vent, both, as in Bryozoa, approximated.

The " stone-lilies," or crinoid star-fishes, formed a numerous and important group in the palæozoic seas, where they obtained their maximum number and variety. M. d'Orbigny describes thirty-one palæozoic genera, two triassic, ten oolitic, and four cretaceous—of which latter three (*Pentacrinus, Bourgueticrinus* and *Comatula*) are found in the tertiaries and modern seas. The *Crinoidea* differ from the other echinoderms in having the generative organs combined with the arms, and opening into special orifices near their base. Nearly all the

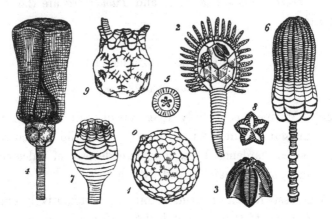

Fig. 6.

Crinoidea ; Blastoidea ; Cystoidea.

1. Sphæronites aurantium, Wahl. ; *L. Silurian*, Sweden.
2. Pseudocrinus bifasciatus, Pearce ; *U. Silurian*, Dudley.
3. Pentremites florealis, Say ; *Carboniferous*, Ohio.
4. Crotalocrinus rugosus, Mill. ; *U. Silurian*, Dudley.
5. Poteriocrinus (joint of column) ; *Carboniferous*, Yorkshire.
6. Encrinus entrocha ; *L. Muschelkalk*, Germany.
7. Apiocrinus Parkinsoni, Mill. ; *Bradford Clay*.
8. Pentacrinus basaltiformis, Mill. ; *Lias*, Lyme.
9. Marsupites ornatus, Mill. ; *Chalk*, Sussex.

genera, except *Comatula* and *Marsupites* (fig. 6, 9), appear to have been attached either by the expanded base of the column, as in *Apiocrinus*, or by jointed processes, as in *Bourgueticrinus*. In many instances the lower part of the column throws out innumerable root-like side-arms, which strengthen and support it. The column is comparatively short in *Apiocrinus Parkinsoni*, and extremely elongated in *Pentacrinus Hiemeri*. It is round in nearly all the palæozoic Crinoids ; and when five-sided, the articular surfaces of the joints are simply radiated, as in the rest. These joints are perforated in the centre, and when detached, are the "St. Cuthbert's beads" of story (fig. 6, 5).* In *Platycrinus* the stem is compressed, and the articular surfaces are elliptical. In the genus *Pentacrinus*, which commences in the lias, the sculpturing of the articulations is more complex (fig. 6, 8), but it is quite simple in the other modern genera. The body of the Crinoid is composed of polygonal plates forming a cup, which is covered by a canopy of smaller plates. The mouth is often proboscidiform ; the anal orifice is near it. The five arms which crown the cup are sometimes nearly simple, but feathered with slender, jointed fingers ; in other genera they divide again and again, dichotomously ; and in two remarkable Silurian forms, *Anthocrinus* and *Crotalocrinus* (fig. 6, 4), these subdivisions are extremely numerous, and the successive ossicles are articulated to each other laterally, forming web-like expansions, similar in appearance to the coral *Fenestrella* (fig. 3, 11). Other remarkable Silurian Crinoids belong to the genera *Glyptocrinus*, *Eucalyptocrinus*, *Geocrinus* (the "Dudley Encrinite") and *Caryocrinus*. Several are common to the Silurian and Devonian, as *Melocrinus*, *Cyathocrinus*, and *Rhodocrinus ;* the two last, and *Poteriocrinus*, extend into the carboniferous formations. *Cupressocrinus* and some others are peculiarly Devonian ; *Platycrinus*, common to Devonian and coal formations ; and many genera (including the "nave Encrinite"—*Actinocrinus*, *Gilbertsocrinus*, and *Woodocrinus*), are proper to the carboniferous limestone. The famous "lily Encrinite" (*Encrinus entrocha*, fig. 6, 6) is characteristic of the middle trias, or "muschelkalk ;" the "clove Encrinite" (*Eugeniacrinus*, fig. 7, 9) abounds in the Oxfordian oolites of Germany ; *Apiocrinus*, *Millericrinus*, and several forms related to *Comatula*—*e. g.*, *Pterocoma* and *Saccosoma*—are also peculiarly oolitic. The "tortoise

* Casts, in chert, of the canal which passes down the crinoidal column are called " screw stones;" and those limestones which abound in columns and detached joints are called " entrochal marbles."

Encrinite" (*Marsupites*, fig. 6, 9), is found only in the chalk, along with *Bourgueticrinus* (fig. 7, 10); and the bodies of *Co-matulæ*, which, when they have lost their arms and claspers, are called "Glenotremites." (Fig. 7, 7,—upper surface with sockets of the five arms; 8,—under surface, showing articulations of claspers, and the scar of the larval stem.)

Order 2.—CYSTOIDEA.

This order was established by Von Buch for a small group of palæozoic echinoderms formerly included with the *Crinoideæ*. They have a globular body covered with close-fitting polygonal plates attached by a simple jointed stem. The mouth is minute, and opposite to the stalk; close to it is the small anal opening; and a little more distant the generative orifice, covered by a pyramid of five or six little valves. Some of the genera, like *Pseudocrinus* (fig. 6, 2), have two or four tentaculiferous arms, bent down over the body and lodged in grooves, to which they are anchylosed. Others, like the *Sphæronites* (fig. 6, 1), have only obscure indications of tentacles situated close to the mouth. In *Pseudocrinus* and some other genera two or three pairs of lamellated organs, called *pectinated rhombs,* are placed on the contiguous margins of certain body-plates. They are supposed not to penetrate the interior, and no office has been conjecturally assigned to them; but Professor Forbes suggested that they might represent the "epaulettes" of the larval *Echinidæ*, to which group he supposed the Cystidean bore the same relation as the Crinoids hold to the star-fishes. There are nine genera, of which eight are found in the British strata — four in the upper and four in the lower Silurian.

Order 3.—BLASTOIDEA.

A separate order has been proposed for another small group of palæozoic fossils typified by *Pentremites* (fig. 6, 3). The body is globular or elliptical, and supported on a small, jointed stalk, with radiated articular surfaces and irregular side-arms. It is composed of solid polygonal plates, with a minute oral orifice at the summit surrounded by five other openings, four of which are double and ovarian, the fifth rather larger and anal. There are five petaloid ambulacra of variable length, converging to the mouth, furrowed down the centre, and striated across. According to the observations of Dr. Ferdinand Rœmer, these supported numerous slender, jointed tentacula, indicated by the rows of marginal pores. One species is found in the upper Silurian, six in the Devonian, and twenty-four in the Carboniferous, which has received the name of "pentremite limestone" in the United States, on account of the abundance of these fossils it contains.

Order 4.—ASTEROIDEA.

(Sea-Stars, Brittle Stars.)

Char.—Body radiate; integument hardened by calcareous pieces, and more or less armed with spines; no dental apparatus.

Asteriadæ and Ophiuridæ.—Fossil star-fishes, though less common, have a wider range than their allies the fossil urchins, being found amongst the earliest organic forms. *Palæaster*, *Protaster* (fig. 7, 6), and *Lepidaster* (fig. 7, 5), are Silurian star-fishes, presenting many anomalies, and scarcely referable to any existing families. *Tropidaster, Pleuraster, Aspidura, Ophi-urella*, and *Amphiura* are oolitic genera; *Ophioderma, Luidia, Astropecten* range from the lias to the present seas; *Stellaster* and *Arthraster* are peculiar to the cretaceous ; and *Ophiura,*

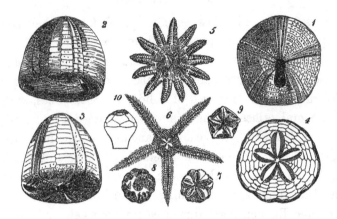

Fig. 7.

Galeritidæ; Asteriadæ; Crinoidea.

1. Pygaster semisulcatus, Ph.; *Inf. Oolite,* Cheltenham.
2. Ananchytes ovatus, Lam.; *U. Chalk,* Europe.
3. Galerites albogalerus, Lam.; *U. Chalk,* Kent.
4. Scutella subrotunda; *Miocene,* Malta.
5. Lepidaster Grayi, Forbes; *U. Silurian,* Dudley.
6. Protaster Miltoni, Salter; *L. Ludlow rock,* Salop.
7. Comatula (Glenotremites), upper surface of body.
8. Comatula (lower surface); *Chalk,* Sussex.
9. Eugeniacrinus quinquedactylus, Schl.; *Oxfordian,* Wurtemberg.
10. Bourgueticrinus ellipticus, Mill.; *Chalk,* Kent.

Astrogonium, Oreaster, and *Goniodiscus* are both cretaceous and living.

Order 5.—ECHINOIDEA.

(Sea-Urchins.)

Char.—Body spheroid or discoid, incased in a crust of in-
flexibly-joined calcareous plates, and armed with spines ;
dental system complex, arranged so as· to resemble a
" lantern."

The *Echinoidea* appear first in the carboniferous limestone
and attain their maximum in the cretaceous strata. In all

secondary and more modern *Echinidæ,* the shell is composed
of five double rows of ambulacral plates, and five inter-ambulacral ; but in the *Palæchinus* (fig. 8, 1) of the carboniferous
limestone there are six rows of inter-ambulacral plates, and in
Perischodomus five. Only detached plates of *Archæocidaris*

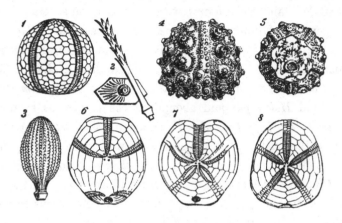

Fig. 8.

Echinidæ; Spatangidæ.

1. Palæchinus sphæricus, Scouler ; *Carboniferous,* Ireland.
2. Archæocidaris Urii, Flem. ; *Carboniferous,* Ireland.
3. Cidaris glandifera, Goldf. (spine); *Jura,* Mount Carmel.
4. Hemicidaris intermedia, Flem. ; *Corallian,* Calne.
5. Salenia petalifera, Desm. ; *U. Greensand,* Wilts.
6. Disaster ringens, Ag. ; *Inferior Oolite,* Dorset.
7. Hemipneustes Greenovii, Forbes; *U. Greensand,* Blackdown.
8. Catopygus carinatus, Goldf. ; *U. Greensand,* Wilts.

(fig. 8, 2), have been seen, and these, by their six-sided form,
seem also to have been arranged in more than double series.
Normal Echinidæ, of the existing genus *Cidaris,* abound in
the upper *trias.* Some of the secondary species of Cidaris
have the ambulacral pores widely separated (=*Rhabdoci-
daris*) ; in others the rows of pores are doubled (=*Diploci-
daris*). The genus *Hemicidaris* (fig. 8, 4), distinguished by
the large spine-bearing tubercles on the lower part of the
ambulacra, ranges from the trias to the chalk-marl. *Diademæ,*

with smooth, solid spines (=*Hemidiadema*), appear in the lias, and continue to the chalk, where the modern type, with annulated, hollow spines, appears. *Echinopsis* also occurs in the lias ; and *Acrosalenia*, a genus characteristic of the oolites, and distinguished from *Salenia* by its perforated tubercles. *Acrocidaris* and *Heliocidaris*, with *Glypticus*, and several other sub-genera of Echinus, are also peculiar to the oolites. *Salenia* (fig. 8, 5) with its ornamental disk, is characteristically cretaceous. *Arbacia* and *Temnopleurus* appear first in the eocene. The *Cassidulidæ* commence in the oolites, with *Pygaster* (fig. 7, 1) and *Holectypus*, and abound in the cretaceous system. *Galerites* (fig. 7, 3), *Discoidea*, *Pyrina*, and *Cassidulus* are peculiar to the chalk. The *Clypeastridæ* are represented in the oolites by numerous species of *Echinolampas* and *Nucleolites* (or *Clypeus*) ; the latter genus attains a large size. The sub-genus *Catopygus* (fig. 8, 8) is peculiar to the cretaceous series. *Conoclypeus* occurs in the chalk and tertiaries. *Clypeaster* flourished most in the miocene age ; many large species are found in the south of Europe, Madeira, and the West Indies. Numerous genera, remarkable for their flattened form, and popularly known as " cake-urchins," are peculiar to the tertiaries and existing seas. *Lenita* and *Scutellina* are eocene ; *Scutella* (fig. 7, 4) is miocene. *Mellita* and *Echinarachnius* are both fossil and recent. The heart-shaped urchins (*Spatangidæ*), are only remotely represented in the oolites by *Disaster* (fig. 8, 6) ; they are numerous in the chalk, to which *Micraster*, *Epiaster*, *Hemipneustes* (fig. 8, 7), *Archiacia*, *Holaster*, and *Ananchytes* (fig. 7, 2), are peculiar. *Toxaster* is characteristic of the lower neocomian. *Hemiaster* is cretaceous and tertiary. *Spatangus*, *Eupatagus*, *Brissus*, *Amphidotus*, and *Schizaster* are tertiary and recent forms.

The shell of the *Echinodermata* has the same intimate structure in all the orders and families, and in every part of the skeleton, whether " test," or " spine," or " tooth." The smallest plates resemble bits of perforated card-board, and the largest and most solid are formed of a repetition of similar laminæ. In a few membranous structures, minute spicula, curved, bi-hamate, or anchor-shaped, are met with. They are always composed of carbonate of lime ; but owing to their porosity, fossil examples are commonly impregnated with earth, or pyrites, or silica, and form bad subjects for microscopic investigation. Without, however, losing their organic structure, the fossil Echinoderms exhibit a cleavage like that of calcareous spar, by which the smallest ossicle of star-fish or Crinoid may be recognised : this peculiarity is most strikingly obvious in the great spines of the *Cidaris* (fig. 8, 3), or the enlarged column of the " pear Encrinite" (fig. 6, 7). Examples of the latter may be seen which had been crushed when recent, and before the sparry structure was superinduced.

Order 6.—Holothurioidea.

(Sea-Cucumbers, Trepang.)

Char.—Body vermiform ; integument flexible, with scattered reticulate calcareous corpuscles, or beset with small anchor-shaped spicula.

The Holothurioid order presents scarcely any examples likely to be met with in a fossil state, except the genus *Psolus*, of whose imbricated shield a fragment has been found by Mr. Richmond in the northern drift of Bute. Count Münster has figured the microscopic plates, apparently of a *Holothuria*, from the chalk of Warminster ; and the anchor of a *Synapta* from a still older formation,—the upper oolite of Bavaria.[*] Microscopic observers will doubtless meet with many such detached plates and spines when searching for Polycystineæ and other Rhizopods in the oolitic and cretaceous strata ; but it is scarcely probable that the order has dated far back in time.

[*] Beitrage, heft 6, 1843.

Province II.—ARTICULATA.

In the great division of invertebrate animals called *Articulata* the brain is in the form of a ring encircling the gullet. A double ganglion above the tube supplies the chief organs of sense. The ganglions below the tube are connected with two chords which extend along the ventral surface of the abdomen, and are in most species united at certain distances by double ganglions, which are connected with the nerves supplying the body segments and their appendages. The body presents a corresponding symmetrical form. The skeleton is external, and consists of articulated segments of a more or less annular form. The articulated limbs, in the species possessing them, have a like condition of the hard parts, in the form of a sheath which incloses the muscles. The jaws, when present, are lateral, and move from side to side.

The worm, the lobster, the scorpion, and the beetle, exemplify this province.

The articulate division of the animal kingdom, most universally distributed and numerically abundant at the present day, is least perfectly represented amongst the relics of the former world. Their chitinous integuments, often hardened with earthy salts, are quite as capable of preservation as the shells of the *Mollusca*, and remains of them are met with in all aqueous deposits ; but that manifold, complex organization, which in the recent state fits them so admirably for generic and specific comparisons, is fatal to their entire preservation, and the fossil examples are often so fragmentary as to admit of little more than the determination of their class and family.

The most ancient fossiliferous rocks bear imprints which have been regarded as the tracks and burrows of marine worms. With these are found Crustacea of the lowest division, and of a group which is wholly extinct. A little later appear the

Phyllopods, Copepods, and other existing orders of *Entomostraca*. Only a few obscure forms, doubtfully referred to the higher division, *Malacostraca*, have been found in the carboniferous and Permian systems. The secondary strata contain abundant remains of Isopods, and of lobsters and hermit-crabs. True crabs (*Brachyura*) abound in the oldest tertiaries. Air-breathing insects and Arachnida existed even in the palæozoic age ; the " sombre shades" of the carboniferous forests were not " uncheered by the hum of insects ;" nor were the insects blind, like those which now inhabit the vast caverns of Kentucky and Carniola. The *Articulata* which come latest are the Cirripedes, whose lowest family appears in the lias ; while the *Balanidæ* are only found in the tertiaries.

The number of fossil Articulata catalogued and described forms but a very small proportion of those which have probably existed. Bronn enumerates 1551 fossil insects, 131 arachnids, 894 crustaceans, and 292 anellides. Darwin describes 69 fossil cirripedes, 12 of which are living species.

CLASS I.—ANNULATA.

(*Worms, Tube-Worms, Nereids.*)

Char.—Body soft, symmetrical, vermiform, annulated, with suckers, or setæ, or setigerous tube-feet ; blood of a red colour in most.

The peculiar markings on the surface of the old Cambrian slate rocks, conjectured to afford the earliest indications of the existence of marine worms, are not without suspicion as to their origin. The so-called " Nereites" bear considerable resemblance to other equally ancient impressions which have been described as Zoophytes, under the name of *Protovirgularia* (fig. 3, 1). No such doubt attaches to the worm-tracks which abound in the thin-bedded sandy strata of the forest-marble ; and the " Cololites" of the lithographic limestone are

most probably the castings of worms. Long calcareous tubes
occur in the upper Silurian and carboniferous strata, which
have received the name of *Serpulites*. The *Microconchus* of
the carboniferous period is now regarded as an Anellide ; and
in all the later formations, tubicolar Anellides, especially of
the genera *Serpula*, *Spirorbis*, and *Vermilia* abound. Some of
these, although attached and gregarious, are so regular in their
growth as to have been usually called *Vermeti*, but are now
placed in the genus *Vermicularia*. *Spiroglyphus*, and some
other shell-excavators, are indicated in the tertiaries. Amongst
the problematic fossils of the palæozoic strata, two are supposed
to be anellidous, viz., the *Tentaculites* (fig. 10, 7), which was
apparently free, and almost always regular in its growth, so
as more to resemble one of the gregarious Pteropods ; and the
Cornulite (fig. 10, 8), which is attached when young, singly or
in groups, to Silurian shells and corals : the structure of its
shell is vesicular, and the cavity resembles a series of inverted
cones. The unattached and gregarious *Ditrupa* appears in
the upper chalk, and abounds in the London clay and crag.

Class II.—CIRRIPEDIA.

(*Barnacles, Acorn-Shells.*)

Char.—Body chitinous or chitino-testaceous, subarticulated,
    mostly symmetrical, with aborted antennæ and eyes ;
    thorax attached to the sternal surface of the carapace,
    with six pairs of multiarticulate, biramous, setigerous
    limbs ; metamorphosis resulting in a permanent para-
    sitic attachment of the fully-developed female to some
    foreign body.

    The fossil Cirripedes belong chiefly to the sessile division,
and consist of the ordinary forms of the still-existing *Balanidæ*.
They are rare in the eocene tertiary, but more abundant afterwards. The *Balanus porcatus* attains a great size in the shelly

beds of northern drift; its large basal plate, when detached, is a puzzling fossil, and has caused some mistakes. A *Coronula* has been found in the middle division of the crag which has afforded so many cetaceous bones. Remains of pedunculated Cirripedes occur in older deposits, but are mostly scarce and fragmentary. A species of *Pollicipes* is found adhering to drift-wood, perforated by bivalves, in the lias; another occurs in the Oxford clay, attached in groups to drift-wood, and the shells of Ammonites, which probably floated in the sea after death. The chalk affords many species of *Pollicipes* and *Scalpellum*, a species of the anomalous genus *Verruca*, and the only extinct genus of Cirripedes—*Loricula* (fig. 10, 6). This remarkable fossil is found attached to Ammonites, and exhibits only one side in any of the examples hitherto found. In this unsymmetrical development and the imbrication of its valves it more resembles *Verruca* than any other Cirriped. " During the deposition of the great cretaceous system, the *Lepadidæ* arrived at their culminant point : there were then three genera, and at least thirty-two species ;" whereas at the present day the Philippine Archipelago, which is the richest marine province, affords but five species.

Class III. — CRUSTACEA.

Char.—Body articulated, with articulated limbs ; head with antennæ ; branchial respiratory organs ; sexes distinct ; metamorphosis in most, in none resulting in fixed individuals.

Sub-*Class* 1.—ENTOMOSTRACA.

Char.—Body with more or fewer segments than fourteen ; integument chitinous, forming in some a bivalve shell, eyes sessile.

Small bivalve entomostracous Crustacea are found in all

strata, and attain their maximum size in the older rocks. Minute *Ostracoda*, related to the recent *Cypris* (fig. 10, 5), swarm in the laminated fresh-water clays of the Wealden; whilst the marine *Cytheridæ* assist with their multitudinous atoms in building up the chalk. Amongst the Phyllopods, the gregarious *Estheria* covers the slabs of Wealden and of Keuper with crowds of bivalve shells which have been commonly mistaken for *Cyclades* and *Posidonomyæ*. *Estheria* abounds in the

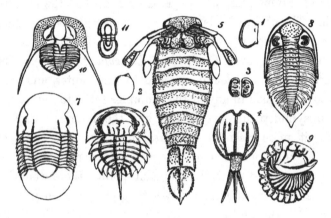

Fig. 9.

Palæozoic Entomostraca.

1. Leperditia Baltica, Wahl.; *U. Silurian*, Gothland.
2. Entomoconchus Scouleri, M'C.; *Carboniferous*, Ireland.
3. Beyrichia complicata, Salter; *L. Silurian*, Wales.
4. Dithyrocaris Scouleri, M'C.; *Carboniferous*, Ireland.
5. Pterygotus Anglicus, Ag.; *Old Red Sandstone*, Ludlow.
6. Bellinurus bellulus, König.; *Carboniferous*, Coalbrookdale.
7. Illænus Davisii, Salter; *L. Silurian*, Bala.
8. Phacops caudatus, Brun.; *U. Silurian*, Dudley.
9. Calymene Blumenbachii, Br.; *U Silurian*, Dudley.
10. Trinucleus ornatus, Sternb.; *L. Silurian*, Britain.
11. Agnostus trinodus, Salter; *L. Silurian*, Britain.

Caithness flags of the middle Devonian series. The globose *Entomoconchus* (fig. 9, 2) is found in the carboniferous limestone; *Leperditia* (fig. 9, 1) in the Silurian rocks of the north; and *Beyrichia* (fig. 9, 3), which is characteristically Silurian, may be distinguished from the young forms of Trilobites by the unsymmetrical shape of its separated valves. Other palæozoic Phyllopods (*Ceratiocaris* and *Hymenocaris*) related to the recent *Nebalia*, and having a conspicuous tail, occur in the upper and lower Silurian strata ; the genus *Leptocheles* (M‘C.) was founded on the tail-spines of these *Crustacea*. *Dithyrocaris* (fig. 9, 4), which resembles the recent *Apus* in the horizontal compression of its carapace, is found in the carboniferous limestone. The lower coal measures also contain, in their nodules of clay-ironstone, frequent examples of *Bellinurus* (fig. 9, 6), a small Pœcilopod, differing from the recent king-crab (*Limulus*) in the moveable condition of the body-segments. But the most extraordinary of the palæozoic Crustacea are the *Eurypterus, Himantopterus,* and *Pterygotus* (fig. 9, 5*), from the Upper Silurian and Old Red Sandstone, of which some far surpassed the largest living lobster or king-crab in size. They have been considered an extinct family, related to the *Limuli ;* or as the representatives of the larval condition of the stalk-eyed *Malacostraca*. But the following structures show an affinity to the *Ostracoda*. Their carapace is comparatively small, with compound eyes on the antero-lateral margins ; the body segments are eleven or twelve in number, without appendages, and terminated by a pointed or bilobed tail. *Eurypterus* has eight feet ; the others have three pairs of limbs—viz., the chelate antennæ, the foot-jaws, and the natatory feet, with their fin-like palettes, which spring from the under side of their cephalo-thorax. The surface of the body and limbs often presents a peculiar imbricated sculpture, which caused them at one time to be regarded as fishes by Agassiz. The *Pterygotus problematicus* is supposed to have attained a length of seven feet, and some of the others were a yard long. Crustacea of this magnitude' may have formed tracks on the

---

* This figure (by Mr. Salter) as well as several others, are taken from the *Siluria* of Sir R. Murchison, P.G.S.

sea-bed, like those on the Potsdam sandstone of America, called " Protichnites," subsequently to be described.

Order TRILOBITES.

Char.—Trunk segments trilobed ; sessile compound eyes in
most ; limbs aborted.

The great family of Trilobites is entirely confined to the
palæozoic age ; none are found even in the upper coal measures
or Permian system.　Above 400 species have been described,
and grouped in 50 genera.　Of these 46 are Silurian, 22 Devonian, and 4 carboniferous.　According to Bronn, 13 genera
are peculiarly Lower Silurian, 3 Upper Silurian, 1 Devonian,
and 3 carboniferous.

The skeleton of the Trilobite consists of the cephalic
shield, a variable number of trunk-rings or segments, and the
pygidium or tail composed of a number of joints more or less
anchylosed.　In some species a *labrum* (or " hypostome") has
been discovered, but no indications of antennæ or limbs have
ever been detected ; still there can be no doubt they enjoyed
such locomotive power as even the limpet and chiton exhibit
when requisite.　Variations in the length of the cephalic and
caudal spines (*e.g.* in *Asaphus caudatus* and *longi-caudatus*),
and in the prominence of the head-lobes, have been considered
indications of difference of sex.　One of the oldest and simplest
forms is the minute *Agnostus* (fig 9, 11) ; it is usually found
in little shoals, with only the cephalic shield preserved, as if it
were the larval form of some large Trilobite.　According to the
observations of M. Barrande, the *Sao* passes through twenty
stages of growth, being first a simple disc, and ultimately
having seventeen free thoracic segments and two caudal joints ;
the additional segments are developed between the thorax and
abdomen.　The *Trinucleus* (fig. 9, 10) with its ornamental border, and *Illænus* (fig. 9, 7), in which the trilobation is less con-

spicuous than in most genera, are characteristic of the Lower Silurian strata. Two others from the Wenlock limestone have long been celebrated, viz., *Calymene* (fig. 9, 9), or the "Dudley Trilobite," so compactly rolled up ; and *Asaphus* (or Phacops) *caudatus* (fig. 9, 8), in which the lenses of the large eyes are frequently well preserved, and visible without a glass. Each eye has at least 400 facets, and in the great *Asaphus tyrannus* each is computed to have 6000. In one species (*Asaphus Kowalewskii*) the eyes are supported on peduncles. The largest Trilobite is *Asaphus gigas ;* some of the fragments indicate a creature eighteen inches long.

Sub-Class 2.—MALACOSTRACA.

Char.—Body divided into thorax and abdomen, with seven segments in each.

The *Isopods* are represented in the upper oolite by *Archæoniscus Brodiæi*, which is gregarious, in large numbers in the slabs of Purbeck limestone ; and in the Permian system by the *Prosoponiscus* (or *Palæocrangon*). The problematic *Pygocephalus*, and the "*Apus dubius*," both from the carboniferous strata, are doubtfully referred to the *Stomapoda*, and, with the exception of the *Gitocrangon* of Richter, are the oldest of the known stalk-eyed Decapods.

Macrourous Crustacea are of constant occurrence throughout the oolites and cretaceous strata. One of the most remarkable forms, *Eryon* (fig. 10, 3), is found in the lias (with the closely-allied *Tropifer* and *Coleia*) and in the Oxford clay. The small lobsters of the genus *Glyphea*, in the oolites, and *Meyeria*, in the Speeton clay and greensand, are commonly the nucleus of hard nodules of phosphate of lime. The larger species of the chalk form the genus *Enoploclytia*. The Oxfordian oolite of Solenhofen, with its finely-laminated lithographic slates, opens like a book filled with compressed and wonderfully preserved shrimps and lobsters. One of them, remarkable for its long and slender arms (*Megachirus*, fig. 10, 4), is also found in the Oxford clay of Wiltshire. One of the most remarkable repositories of fossil Crustacea is the Isle of Sheppy, where the " London clay" has afforded countless examples of

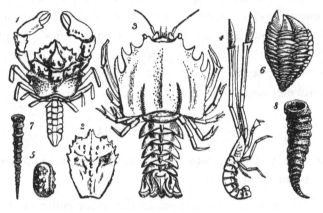

Fig. 10.

Crustacea; Anellida.

1. Dromilites Lamarckii, Desm.; *London Clay*, Sheppy.
2. Notopocorystes Stokesii, Mant.; *Gault*, Folkestone.
3. Eryon arctiformis, Schl.; *Oxfordian*, Solenhofen.
4. Megachirus locusta, Germar.; *Oxfordian*, Solenhofen.
5. Cypridea tuberculata, Sby.; *Weald*, Sussex.
6. Loricula pulchella, G. B. Sby.; *L. Chalk*, Sussex.
7. Tentaculites ornatus, J. Sby.; *U. Silurian*, Dudley.
8. Cornulites serpularius, Schl.; *U. Silurian*, Dudley.

the higher organized division, including nine Brachyura, three Anomura, and five macrourous species. The island of Hainan, on the coast of China, abounds with fossil crabs of the genus *Macropthalma*, which are sold in the drug-market of Shanghae. Others are found in the miocene of Malta, and of Perim Island in the Red Sea. The reputed instances of secondary *Brachyura* are open to doubt; in England we have only the little *Etyus Martini* (or *Reussia*) from the gault, for the *Podopilumnus* (M'C.) is probably from some foreign tertiary deposit. Pairs of chelate claws occur in the upper chalk, which are referred to a hermit-crab (*Mesostylus Faujasii*). Small Crustaceans, resembling in form the living *Corystes*, abound in the gault (fig. 10, 2), but they are known to be anomourous by the small size and dorsal position of the posterior legs, and by the little plates intercalated between the last joints of the tail, as seen also in the *Dromilites* (fig. 10, 1) from the London clay.

Class IV.—INSECTA.

Char.—Body chitinous, articulated, with articulated and uncinated limbs ; head provided with jointed antennæ ; respiratory system tracheal.

The fossil insects hitherto examined have afforded no new types or forms of unusual interest. , The oldest known, those from the lower coal measures, resemble the *Curculionidæ* and *Blattidæ* or *Locustidæ* of the present day. The lias limestones have afforded a greater variety to the persevering skill of Mr. Brodie : species of the genera *Berosus*, *Elater*, *Gyrinus*, *Laccophilus*, and *Melolontha*, and undetermined genera of the families *Carabidæ*, *Buprestidæ*, *Chrysomelidæ*, and *Telephoridæ ;* Panorpa-like insects of the genus *Orthophlebia ;* dragon-flies, *Nepadæ* and *Cimicidæ*, *Cicada*, and the dipterous genus *Asilus*. Next in age is the insect depository of the Stonesfield slate, which affords the large wing-covers of *Buprestis Bucklandi*, species of *Prionus* and *Coccinella*, and the great neuropteran *Hemerobioides*. The Purbeck limestone has supplied, in addition, species of *Cerylon* and *Colymbetes*, *Cyphon*, *Helophorus*, and *Limnius ;* and examples of *Staphylinidæ*, *Cantharidæ*, *Harpalidæ*, *Hydrophilidæ*, and *Tenebrionidæ*, *Libellula* and *Phryganea*, *Acheta* and *Blatta*, *Aphis*, *Cercopis*, and other Homoptera, and ten dipterous genera. In the newer pliocene fresh-water formations the recent *Copris lunaris* has been detected, and the elytra of *Donacia* and *Harpalus*. The principal foreign sources of fossil insects have been the lithographic slates of Solenhofen, and the tertiary deposits of Aix in Provence, and Œningen, near Constance, on the Rhine. Remains of species of *Tinea* and *Sphinx* are said to have been found in the lower Jura, and of a diurnal Lepidopteran in the Molasse. Numerous examples of insects in true amber have been obtained, and much more abundantly in " gum animi," a more modern fossil resin. These are all unknown to entomologists, and are probably extinct, since no department of recent natural history has been so closely worked, although the fossil insects have been comparatively neglected. It has been suggested by Mr. Westwood that the lias insects have a subalpine character, and may have been brought down by torrents from some higher region. But no attempt has been made to show whether these or any other group of fossil insects most nearly resemble those of any particular zoological province of the present day.

Much has been said of the " indusial limestone " of Auvergne, supposed to be built up of the fossilized cases of caddisworms (*Phryganeidæ*) ; but Mr. Waterhouse, the only entomologist who has visited the country and examined the formation, entertains doubts of the correctness of this interpretation.

Of the *Myriapoda*, 17 fossil species have been found, commencing in the oolitic system. And of the *Arachnida*, 131 species are catalogued ; the earliest and most interesting of these is the fossil scorpion (*Cyclopthalmus senior*) of the Bohemian coal measures (figured in Buckland's *Bridgewater Treatise*). Fossil spiders are found in the Solenhofen slates and in the tertiary marls of Aix.

PROVINCE III.—MOLLUSCA.

Remains of the *Testacea*, or shell-bearing molluscous animals, are the most common of all fossils, and afford the most complete series of " medals," or characteristic signs for the. identification of strata. The duration of types and species, as a general rule, is inversely proportional to rank and intelligence. The most highly organized fossils have the smallest range, and mark with greatest exactitude the age of the deposit from whence they have been derived. But the evidence afforded by shells, if less precise, is more easily and constantly obtained, and holds good over larger tracts of country.

Class I.—BRACHIOPODA.

The lamp-shells (*Brachiopoda*), more than any other group, have suffered with the lapse of time. Of 1300 known species, only 75 are living ; and of the 34 genera, the larger part (21) are extinct. The number of generic forms is greatest in the Devonian period and least in the upper oolites, after which a second set of new types gradually appears. The preponderance of fossil *Brachiopoda* is contrasted with the scarcity of the recent shells even more strongly by the abundance of individuals than by the number of species ; for the living shells mostly inhabit deep water and rocky situations inaccessible to the dredger, and are seldom obtained in large numbers.

The genus *Terebratula*, as now restricted to shells with a short internal loop, musters above 100 fossil species, of which only one survives (*T. vitrea*), an inhabitant of the Lusitanian province. The Waldheimias, or *Terebratulæ* with long loops, are widely distributed in our present seas, although only nine living species are known ; individuals of one or more of these are found on the coast of Spitzbergen and Labrador, at Cape Horn, and most abundantly in New South Wales and New Zealand : there are sixty fossil species dating from the trias. The *Terebratellæ*, having the loop fixed to a mid ridge, commenced in the lias, and occur in small numbers throughout the cretaceous and tertiary periods, and are the only lamp-shells which attain their climax in recent seas. Five species of *Argiope* occur in the greensand, chalk, and tertiaries. The allied genus *Thecidium* is represented by one species in the carboniferous and one in the triassic system, becomes comparatively common in the secondary period, and dwindles again to a single species in the newer tertiary ; this species survives within still narrower limits in the Mediterranean Sea. The sub-genus *Terebratulina* is represented by twenty species in the secondary and

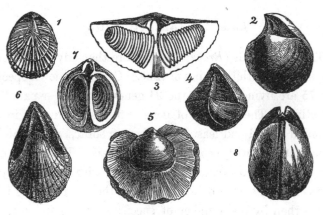

Fig. 11.

Brachiopoda.

1. Trigonosemus Palissyi, Woodw.; *U. Cretaceous*, Ciply.
2. Stringocephalus Burtini, Defr.; *Devonian*, Eifel.
3. Spirifera striata; *Carboniferous*, Britain.
4. Cyrtia trapezoidalis; *U. Silurian*, Dudley.
5. Athyris Roissyi, Ler.; *Carboniferous*, Ireland.
6. Uncites gryphus, Schl.; *Devonian*, Belgium.
7. Atrypa reticularis, L.; *U. Silurian*, Malvern.
8. Pentamerus lævis; *Caradoc S.*, Salop.

tertiary formations. *T. striata* of the chalk is so like the recent *T. caput serpentis* as to be with difficulty distinguished from it. Several extinct sub-genera occur in the cretaceous strata, of which the most remarkable are *Trigonosemus* (fig. 11, 1) and *Lyra*, shaped like a violin. The genus *Stringocephalus* (fig. 11, 2) is peculiar to the Devonian strata, and has a large internal loop, and a very prominent cardinal process, forked at the end, and fitting over the central plate of the opposite valve.

The shell of *Terebratula* and some of its allies (*Argiope, Thecidium, Cyrtia,* and *Spiriferina*) is dotted with minute quincuncial perforations, sometimes visible to the naked eye, as in *T. lima,* but usually requiring a lens of low power. They are smallest in *T. carnea.*

The lamp-shells with sharp beaks and plaited valves have been separated from the *Terebratulæ* under the name *Rhynchonella* (Fisch.) Their shells do not exhibit the punctate structure under a magnifying-glass, and they have no internal skeleton to support their arms, which in the recent species are coiled up spirally, and directed towards the concavity of the smaller valve, like the spires of the extinct *Atrypa* (fig. 11, 7). Of the three living species of *Rhynchonella,* one is found throughout the Arctic Seas, a second in New Zealand, and the third at the Feejees (?). The fossil species exceed 250, and are found in all parts of the world ; those from the palæozoic strata may prove distinct from the rest, since the permian species are known to be provided with large internal processes (*Camarophoria,* King). Casts of these shells are frequently impressed with the narrow and angular pallio-vascular impressions. The extinct genus *Atrypa* differs from *Rhynchonella* solely in having calcareous spires, which are preserved in many instances, and may be cleared to some extent by the application of acid. The foramen is separated from the hinge-line by a *deltidium* ; and the interior of the valve is marked by ovarian and vascular spaces exactly as in *Rhynchonella.* The lower Silurian rock contains another genus, *Porambonites* (Pander), as yet imperfectly understood, but having the valves marked externally by impressed dots, which are not perforations. The genus *Pentamerus* occurs in all the strata below the carboniferous limestone, and is remarkable for its great internal partitions, causing the shell to split readily across the middle ; and giving rise to deep incisions in those casts of the interior which are so common in the Caradoc sandstone (fig. 11, 8).

The extinct *Spiriferidæ* are a family characterized by the possession of internal calcareous spires extending from the centre of the shell outwards (fig. 11, 3). These spires, like the shell itself, are frequently silicified, and may be disengaged from the matrix by the action of acid. At other times the shell is imbedded in soft marl, removeable by careful washing, so as to show the calcareous lamina of the spire fringed with hair-like processes, formerly the support of cirri. In the genus *Spirifera* the shell has a long straight hinge-line, and the flattened *area* of the larger valve has a deltoid byssal notch.* The typical species are characteristic of the palæozoic strata, and have a shell-structure like *Rhynchonella*. The liassic species (*Spiriferina*, d'Orb.), have punctate shells, and the byssal opening is closed (at least in the adult) by a thin arched plate or " pseudo-deltidium." In the sub-genus *Cyrtia* (fig. 11, 4), the hinge-area is ultimately as long as it is wide, and the deltidium is perforated in the centre by a byssal tube ; some of the species have a punctate shell. The genus *Athyris* (Dalman), not always easily distinguished from *Terebratula*, has usually a smooth and rounded shell, ornamented with concentric lamellæ or wing-like expansions (fig. 11, 5) ; the beak is truncated by a round foramen ; the hinge-area is obsolete ; and the spires are as in *Spirifera*, with the addition of some further complications near the hinge. There are twenty-five species, mostly from the Devonian and carboniferous rocks. The species of *Retzia* (King) are still more like plaited *Terebratulæ*, but have lateral spires ; they range from the Silurian

---

* The term *deltidium*, applied by Von Buch to this foramen, has, by misconception of his meaning, become constantly used for the plates which partially close it.

strata to the trias. *Uncites gryphus* (fig. 11, 6), a peculiar Devonian fossil, has a prominent beak, perforated in the young shell by a minute apical foramen ; the hinge-area is filled up by a deeply concave deltidium, on each side of which (but only in some specimens) there is a lateral pouch formed by an inflection of the margin of both valves.

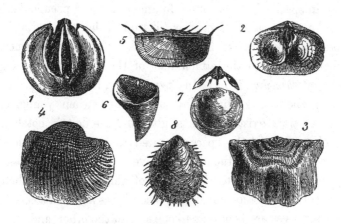

Fig. 12.

Brachiopoda.

1. Orthis hysterita, L. (cast); *Devonian*, Rhine.
2. Davidsonia Verneuili, Bouch.; *Devonian*, Eifel.
3. Strophomena rhomboidalis, Wahl.; *U. Silurian*, Dudley.
4. Producta semireticulata, Martin; *Carboniferous*, Derbyshire.
5. Chonetes striatella, Dalm.; *U. Ludlow rock*, Herefordshire.
6. Calceola sandalina, Lam.; *Devonian*, Eifel.
7. Obolus Apollinis, Eichw.; *L. Silurian*, Northern Europe.
8. Siphonotreta unguiculata, Echw.; *U. Silurian*, Britain.

The family *Orthidæ* consists of shells with a straight hinge-line, bordered by a flat, narrow area, with a central notch in each valve ; the ventral valve is furnished with articulating hinge-teeth, and the dorsal valve has short processes for the support of the oral arms, which appear to have been horizontally spiral (as in *Atrypa*). Between the oral processes there is a central projection for the attachment of the cardinal muscles. Internal moulds of the *Orthis* (fig. 12, 1) exhibit on the ventral side the single attachment of the adductor muscles in the centre, and on each side of it the cardinal muscles ; these are surrounded by the punctate ovarian spaces and impressions of the large pallial sinuses. The genus *Orthis* includes 100 species, ranging upwards to the Permian, but it is most abundant in the Silurian rocks. Some of the lower Silurian species have a round foramen in the " pseudo-delti-dium," and are called *Orthisinœ* (d'Orb.) Other species in the upper palæozoic rocks have the beak twisted or deformed, probably owing to the attachment of the shell when young (= *Streptorhynchus*, King). In *Strophomena*, Rafin (= *Leptœna*, Dalm.), there is a minute byssal foramen when young, of which no trace exists in the adult ; and the deltoid notch is also closed, except the space required to receive the divided cardinal process of the dorsal valve. The oral processes appear to be shifted to the centre of the valve. The shell, when young, is plano-convex, but when it has attained a certain size the valves are bent over to one side or the other, and more or less suddenly. The pallial impressions are the same as in *Orthis*.

The genus *Davidsonia*, peculiar to the Devonian limestones, resembles an *Orthis* attached, like *Thecidium*, by the ventral valve to corals, and sometimes taking the markings of the body on which it grows, like the oyster and *Anomiœ*. The pallial impressions are like those of *Orthis*, and the form of the spiral arms is indicated by prominences which almost fill up the interior of the shell in aged examples. Some indications have been obtained of slender calcareous spires for the support of the arms in this genus ; and also in *Koninckia*, a small shell from the trias of St. Cassian, in which there are always spiral grooves in the interior of the valves crossed by the impressions of the pallial sinuses.

The anomalous fossil called *Calceola sandalina* (Lam.) is also peculiar to the Devonian limestones. In shape it resembles

Cyrtia, but has no hinge, and neither foramen nor internal processes, except a row of small projections along the hinge-line, and two small lateral groups of ridges in the smaller valve. The interior is punctato-striate, but has no recognisable muscular markings.

The *Productidæ* are altogether palæozoic fossils, and most abundant in the carboniferous limestones. Their valves are concavo-convex, the hinge-line straight, and the interior marked with distinct impressions of the muscles for opening and closing the valves, and simple vascular spaces. There are 60 species of *Producta* found in the upper palæozoic rocks, and having a very wide range in North and South America, and from Spitzbergen to Thibet and Tasmania. Some of them are extremely variable in form ; many are armed with long tubular spines, and others completely clothed with short, hair-like processes ; they have no hinge-teeth, and the hinge-area is extremely narrow, except in the sub-genus *Aulosteges* of the Russian zechstein. *Producta proboscidea* has its convex-valve prolonged into a tube, as if for the constant supply of respiratory currents. The Permian genus *Strophalosia* has its valves articulated by hinge-teeth, and covered with long and slender hollow spines ; the shell is attached when young by the umbo of the large valve. *Chonetes* is distinguished from *Producta* by a row of spines along the hinge-margin of the convex-valve ; it also has a narrow hinge-area with a covered notch, and small hinge-teeth. There are 25 species in the Silurian and carboniferous, usually of small size, and finely striated.

Crania is one of the oldest living types, ranging upwards from the lower Silurian. One of the earliest species appears to have been unattached, and another to have had hinge-teeth. *Crania Ignabergensis*, of the chalk of Sweden, has the valves externally alike, being attached only when very young. The internal markings of *C. antiqua*, and other fossil species, are remarkably grotesque. Lower valves of this genus and *The-cidium* are not uncommon, attached to the tests of sea-urchins, in the chalk ; but upper valves are scarce, either detached or *in situ.*

The *Discinidæ* are also ancient fossils, few in number, but appearing in every period. Some of the palæozoic *Discinæ* (= *Orbiculoidea,* d'Orb.) cannot be generically distinguished from the recent species by any characters with which we are as yet acquainted ; but others (= *Trematis,* Sharpe) are orna-mented with quincuncial punctures, and the casts exhibit indications of diverging internal plates, which imply very considerable difference in the organization of the animal. The genus *Siphonotreta* (Verneuil), peculiar to the Silurian forma-tions, is covered with moniliform tubular spines.

Lingula, which has given its name to one of the oldest fos-siliferous rocks, is another form occurring unchanged in strata of every period. Only 34 species are known, and none of them are very common. The latest British *Lingula* is found in the coralline crag (older *pliocene*) of Suffolk : the nearest living species is as far off as the Philippines. *L. Davisii,* of the "lingula flags" in North Wales, has a pedicle groove in the ventral valve, by which the posterior adductor (or cardinal muscle) must have been divided into two elements, as in the genus *Obolus ;* externally it has all the appearance of an ordi-nary existing shell. From the fragments of *Lingula* in the lower Silurian stiper stones of Shropshire, they appear to belong to a species distinct from *L. Davisii. Obolus,* Eichw. (= *Ungula,* Pander) is so abundant in the lower Silurian sandstones of Sweden and Russia as to have given its name to the "obolite grit." In England it occurs only in the upper Silurian of Dudley. The shell is horny in texture, and often stained blue, like the *Lingula,* by the presence of phosphate of iron. In shape it is regularly oval, and differs from *Lingula* in the character of the internal muscular impressions.

Class II.—LAMELLIBRANCHIATA.

(*Bivalve Shells.*)

More than a third part of the known fossil shells are ordinary bivalves (*Conchifera*, Dh.) They amount to nearly 6000, while the recent species scarcely exceed half that number. Nevertheless, it is a group which attains its maximum in the present seas. The genera are seven times more numerous in the newer tertiary than in the oldest geological system; and the number of species found in the entire Silurian series is less than 100, while the chalk contains 500, and the miocene 800. Out of 150 genera, 35 have become extinct,

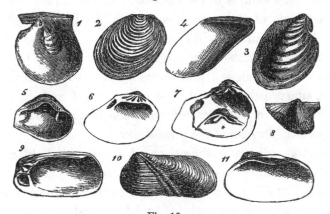

Fig. 13.

Palæozoic Bivalves.

1. Aviculopecten, sp.; *Carboniferous*, Belgium.
2. Posidonomya Becheri; *Carboniferous*, Hesse.
3. Ambonychia vetusta, Sby.; *Carboniferous*, Belgium.
4. Myalina Goldfussi, Dkr.; *Carboniferous*, Vise.
5. Ctenodohta cuneata, Hall; *L. Silurian*, Canada.
6. Lyrodesma plana, Conrad; *L. Silurian*, Hudson River.
7. Axinus obscurus, Sby.; *Magnesian limestone*, Durham.
8. Conocardium armatum, Ph.; *Carboniferous*, Tournay.
9. Pleurophorus costatus, T. Br.; *Magnesian limestone*, Durham.
10. Grammysia cingulata, His.; *Ludlow rocks*, Kendal.
11. Edmondia, sp.; *Carboniferous*, Belgium.

besides numerous sub-genera. The families *Cyprinidæ, Astartidæ,* and *Anatinidæ,* have passed their maximum; the

Trigoniadæ are nearly extinct ; and the *Hippuritidæ* have no living representatives.

The monomyary bivalves, and others with an open mantle, attain a degree of importance at an early period ; and with them some of the burrowing families (*Myacidæ* and *Anatinidæ*) ; while the highest organized siphonated shells (*e. g.,* *Veneridæ* and *Tellinidæ*), unknown in the older rocks, are most abundant now.

The family *Ostreidæ,* distinguished from the Pectens and *Anomiæ* by resting on the *left* valve, contains two fossil forms. Of these, *Exogyra* resembles an oyster with spiral umbones, directed backward, or to the left hand ; it is an attached shell, characteristic of the cretaceous strata. The genus *Gryphœa* (fig. 14, 1) abounds in the oolites, and is gregarious, but unattached, the umbo of the larger valve being curved inward like a claw. A single *Ostrea* occurs in the carboniferous limestone, after which the species become abundant, and are with difficulty distinguishable from the smooth and plaited, or "cocks-comb," oyster of the present day.

Several curious modifications of *Anomia* and *Placuna* have been obtained in a fossil state. *Limanomia* (Bouchard) has ears like *Lima,* and is attached to shells and corals of the Devonian age. *Placunopsis* (M. and L.), found in the oolites, has a transverse ligamental groove, which, like the umbo of the upper valve, is some way within the margin of the shell. And *Carolia* (Cantr.), a tertiary form of *Placuna,* has a byssal plug passing through a foramen like that of *Anomia* when young, but closed in the adult.

Fossil *Pectinidæ* are very numerous. Some of them in the carboniferous limestone (*e. g., P. Sowerbyi*) cannot be distinguished generically from the living Pectens, and retain diverging bands of colour. But the greater part of these old species are somewhat aviculoid in form (fig. 13, 1), and their hinge-area is grooved with cartilage-furrows, like those of *Arca.*

The most beautiful forms occur in the chalk and greensand, and resemble the recent scallop (*Janira*, Schum.) in the

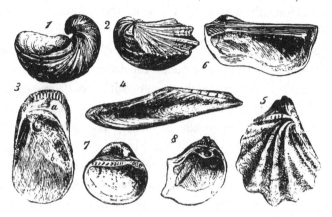

Fig. 14.

Secondary Bivalves.

1. Gryphæa arcuata, Lam.; *Lias*, Charmouth.
2. Pecten (Neithea) quinquecostata, Sby.; *Chalk*, Sussex.
3. Pulvinites Adansoni, Defr. (internal mould); *Corallian*, Rochelle.
4. Gervillia anceps, Dh.; *L. Greensand*, Isle of Wight.
5. Inoceramus sulcatus, Park.; *Gault*, Folkestone.
6. Cucullæa (Macrodon) Hirsonensis, D'Arch.; *Great Oolite*, Minchinhampton.
7. Isoarca cordiformis, Schloth.; *Corallian*, Nattheim.
8. Myophoria decussata, Münt.; *Trias*, S. Cassian.

inequality of their valves, but are further characterized by the possession of articulating hinge-teeth like *Spondylus*. These constitute the genus *Neithea* (fig. 14, 2). *Plicatula* exist in the trias and oolites, along with shells referred dubiously to *Hinnites* and *Spondylus*. True *Hinnites* (a sub-genus of *Pecten*) are characteristic of the *miocene*. *Spondyli* appear in the greensand and chalk. Some of them (like the so-called "*Plagiostoma spinosum*") are unattached ; others resemble the recent deep-water *S. Gussonii*, and have been called "Dianchora." The inner layer, including the hinge of these shells, is seldom preserved. *Lima proboscidea* first appears in the lower oolite, and reappears in the great oolite, and in the Kelloway rock. *Lima duplicata*,

and some other oolitic species, have two ranges of little hinge-
teeth, but not like those of the recent species of *Limœa*. The
large and smooth or striated Limas of the oolites have been
called *Plagiostoma*, a name originally given by Llhwyd.

The pearl-oysters (*Aviculidœ*) are also very abundant
fossils : but owing to the frequent repetition of similar forms,
it is difficult to determine the genera with any degree of certainty by the aid of external characters alone. The Silurian
species mostly belong to the genus *Pterinea* (Goldfuss), and
are broadly winged, and have the hinge-area striated lengthwise, and a few diverging hinge-teeth. *Amboncychia* (Hall)
resembles *Inoceramus*, and ranges from the Silurian to the
carboniferous strata (fig. 13, 3). The Silurian genus *Cardiola*
is ridged like a cockle ; and *Posidonomya*, which is found in
all the palæozoic rocks, is very thin and concentrically furrowed (fig.13, 2). Many other genera have been proposed whose
characters are even more imperfectly understood. *Monotis*
(*Salinarius*), one of the common shells of the trias, has no
anterior ear. *Pteroperna* (Lycett), an oolitic form, has a
winged shell, with numerous small anterior teeth and long
posterior laminæ. The genus *Gervillia* (fig. 14, 4), ranging
from the carboniferous limestone to the chalk, consists of elongated shells, with several cartilage-pits in the ligamental area.
Bakewellia, found in the Permian, has an anterior muscular
impression like *Arca*. The recent genus *Perna* commenced
in the lias or preceding formation, and exhibits great variety
of shape. *Pulvinites Adansonii* (fig. 14, 3) appears to have
been a Perna with a byssal foramen like *Anomia ;* and *Inoceramus* (fig. 14, 5), characteristic of the cretaceous strata and
oolites, differs from Perna chiefly in form, the larger valve
being sometimes completely involute, and resembling a
Nautilus. The genus *Pinna*, which appears to belong to this
family, although provided with two adductor muscles, occurs
fossil in the Devonian and all subsequent strata. Some of the

oolitic species, distinguished by the name *Trichites*, are inequi-valve and irregular, and attain a thickness of more than an inch, resembling mineral masses of fibrous carbonate of lime.

Amongst the *Mytilidæ* are many Silurian species distin-guished by their large, round, anterior muscular scar (*Modio-lopsis,* Hall), and others which have a straight hinge-line and plaited valves (*Orthonotus,* Conrad). *Myalina* has the cartilage-groove repeated (fig. 13, 4), and is found in the upper palæozoic rocks. Sometimes the anterior adductor is supported on a shelf, as in the recent *Septifera* and *Dreissenia.* True *Mytili* and *Modiolæ* abound in the oolitic strata. *Dreissenia,* now confined to the rivers of the Aralo-Caspian region, or only naturalized in Western Europe, was represented by many species, and some of large size, in the *eocene* of Hampshire and *miocene* of Vienna.

Fossil *Arcadæ* are far more numerous than the recent shells, and mostly belong to the division *Cucullæa,* of which a single species survives in the Coral Sea. The palæozoic Arks have anterior teeth like *Arca,* and posterior teeth like *Cucullæa,* and differ from both in the reduction of the hinge-area to a narrow tract corresponding with the posterior half only in the recent shells. The casts of Ark-like shells in the Silurian rocks are further distinguished by a deep furrow behind the front muscular impression. These constitute the genus *Ctenodonta* (Salter), which has hinge-teeth like *Nucula,* and a prominent external ligament (fig. 13, 5). Some of the oolitic Arks, with a byssal sinus, and the posterior teeth very long and parallel, form a sub-genus called *Macrodon* (fig. 14, 6). Others, with prominent umbones, teeth like *Nucula,* and a striated ligamental area, form the genus *Isoarca* of Münster (fig. 14, 7). Above 200 species of *Nucula* and *Leda* are known only as fossils, and range through all the rock systems. The palæozoic species are anomalous in form, and when better understood, will certainly be considered distinct as genera.

Yoldia is a newer tertiary form characteristic of high northern latitudes ; and *Solenella* occurs fossil in Patagonia and New Zealand. The problematic genus *Solemya* is supposed to have existed in the carboniferous period. *Pectunculi* appear first in the cretaceous strata, being less ancient than *Limopsis*, which occurs in the Bath oolite. A member of the latter genus found in the Belgian eocene has the ligamental area entirely *behind* the cartilage-pit, and is called *Nucunella* by d'Orbigny. The "Stalagmium" of Conrad (= *Myopara*, Lea) is identical with *Crenella* (T. Br.), a sub-genus of *Modiola*, found in the cretaceous and tertiary strata.

The *Trigoniadæ* are represented in the lower Silurian strata by *Lyrodesma* (fig. 13, 6), a shell with several radiating hinge-teeth, striated transversely ; and in the upper palæozoics by *Axinus* (fig. 13, 7) and several other imperfectly-known genera. *Schizodus* occurs in the permian at Garford (with *Turbo* and *Rissoa*), near Manchester. The trias contains true *Trigoniæ* associated with the genus *Myophoria* (fig. 14, 8), which has the umbones turned forwards, and a posterior hinge-tooth. The only member of this family which has yet been found in tertiary strata is the little genus *Verticordia* (Wood) of the crag. No *Trigoniæ* have been met with, although 100 species are known in the secondary rocks, and two are still living on the coasts of South Australia.

Fresh-water mussels (*Unionidæ*), of large size and various form, occur in the Wealden formation, and are not generically distinguishable from recent shells ; but those of the coal measures and older rocks are extremely problematic, and may even belong to marine genera.

Of the genus *Chama* there is one species in the upper greensand and chalk of England, and another in the London clay. Elsewhere they are more abundant, amounting to thirty species. Closely allied to *Chama* is the *Diceras* (Lam.), of which the remarkable casts attracted attention at an early period (fig. 14, 1). They are found in the coral rag of France and Germany, and resemble the horns of some animal. The shell is attached by the *umbo* of either valve, indifferently, like some of the recent Chamas. The posterior adductor muscle is supported on a prominent ridge (as in *Pachydesma*, *Megalodon*, and the recent *Cardilia*), which causes a spiral furrow in each horn of the cast. The shells which succeed *Diceras*, in the lower cretaceous strata, have the right valve usually much smaller than the left, and in one instance (fig. 14, 2) it is like the operculum of a spiral univalve. The only British species of this group is *Requienia Lonsdalii*, found in the ironsand of Bowood. In France, and also in Texas, another form occurs, with the attached valve simple and conical, like a Hippurite. The ligamental groove is straight, and the umbo of the free valve marginal.

These shells are so intimately allied to the *Hippuritidæ*, that *Requienia* has been frequently included with them in the apocryphal order " Rudista." The members of the Hippurite group are attached and gregarious, like oysters, often occurring in great numbers, and filling large tracts of rock. Their valves are different in structure and sculpturing, and are articulated by two prominent teeth above and one below ; the cartilage is internal, but there is a conspicuous ligamental furrow outside. There are nearly 100 species characteristic of the cretaceous strata, and especially of the lower chalk, or "hippurite limestone." Only one species (*Radiolites Mortoni*) is found in England ; the rest are from the West Indies, Southern Europe, Algeria, and the East. The form which approaches nearest to *Chama* is the little genus *Caprotina* (fig. 15, 7), whose upper valve has a marginal umbo, but is in other respects like a miniature *Radiolite*. *Caprina* (d'Orb.) has the free valve perforated by canals which open in the inner margin, and in *Caprinella* the outer lamina of both valves possesses this structure. One valve is sometimes spiral (fig.

15, 6), and partitioned off internally by numerous *septa*, like
the water-Spondylus, but so regularly as to resemble the
chambered shell of a *Nautilus*. In the *Radiolite* (fig. 15, 5),

Fig. 15.
Secondary Bivalves.

1. Diceras arietinum, Lam.; *Corallian*, France.
2. Requienia ammonia; *Neocomian*, S. France.
3. Monopleura trilobata, d'Orb.; *Neocomian*, Orgon.
4. Hippurites Toucasiana, d'Orb.; *L. Chalk*, France.
5. Radiolites angeiodes, Lam.; *L. Chalk*, Gosau.
6. Caprinella Boissyi, d'Orb; *L. Chalk*, Valley of Alcantara.
7. Caprotina semistriata, d'Orb.; *U. Greensand*, Le Mans.

both valves are conical, and the *umbo* of the free valve (marginal in the very young shell) becomes central in the adult.
The structure of the hinge is modified by the absence of any
spirality in the valves, but is essentially the same as in *Caprotina* and *Diceras;* the prominent teeth of the upper valve
support curved plates for the attachment of the adductor
muscles, which become continually more undercut in the
course of their growth. In *Hippurites* the anterior muscular
plate projects horizontally, the posterior vertically, like a third
tooth, for which it has been mistaken. In this genus there
are two longitudinal inflexions of the outer shell-wall beside
the ligamental furrow, one corresponding to the posterior

muscular plate, the other (or third) apparently a siphonal inflexion like that in *Trigonia* and *Leda* (fig. 15, 4).

The cockle-shells (*Cardiadœ*), as they have a world-wide distribution now, had a corresponding range in time, and are found in all strata from the Silurian upwards. The commonest fossil type of *Cardium* is ribbed concentrically on the sides, and radiately on the posterior slope, a style of ornament almost unique amongst the 200 recent species. The Caspian cockles, distinguished by a *sinus* in the pallial line, appear to have inhabited the Aralo-Caspian region almost from the middle tertiary period; the hinge-teeth are reduced to one (*Monodacna*) or two (*Didacna*) in each valve, and are sometimes quite wanting even in the young shell (*Adacna*, Eichw.) *Lithocardium aviculare* (fig. 16, 7) is a characteristic shell of the Paris basin, and appears to have spun a byssus, like the fry of some recent cockles; it also resembles the oriental *Tridacna*, of which a species is found in the miocene of Poland. The genus *Conocardium* (fig. 13, 8) of the upper Silurian and carboniferous systems is remarkable for the prismatic cellular structure of its shell, and the truncation of the posterior (?) side of the valves, which are furnished in some species with a slender siphonal process.

The *Lucinidœ*, allied to the cockles in their hinge-structure, are also plentiful in the fossil state, and have as wide a range. They are usually recognisable, even when in the condition of internal casts, by their circular form and the oblique ridge on their disk. Casts of *Lucina* also exhibit the peculiar narrow outline of the anterior adductor detached from the pallial line. *Cryptodon, Diplodonta, Kellia,* and *Pythina* are found in the eocene tertiary. *Corbis,* under the sub-generic form of *Sphœra,* commences in the trias; another modification, found in the oolites and chalk (*Unicardium,* d'Orb.), is edentulous; and *Tancredia* (Lycett), a compressed triangular shell, with a dentition like *Corbis,* is frequent in the lias and oolite.

The fresh-water *Cycladidæ* are represented in the *Wealden* and *eocene* by many species of *Cyrena,* mostly of small size. The recent *Corbicula fluminalis* of eastern rivers is a common fossil of the pliocene tertiary in England and Sicily.

The *Cyprinidæ* and *Astartidæ* are more abundant as fossil shells, and had a wider range of old than at the present day. Nearly 100 species of *Cyprina* have, been catalogued, commencing in the trias ; the dentition of the older species is, however, somewhat peculiar. The *Isocardiæ* are almost as numerous, and have the same range, but many of the fossil Isocardia-looking shells are really related to the *Anatinidæ.* A yet higher antiquity has been assigned to *Cypricardia,* a genus now very scarce and difficult to obtain, on account of its habit. The palæozoic *Pleurophorus* (fig. 13, 9) is distinguished by the prominent ridge behind the anterior muscular impression ; and *Megalodon* (J. Sby.), by the plate supporting the posterior adductor. This genus is represented in the oolites by *Pachyrisma* (fig. 16, 1), and in the tertiaries and modern seas by *Cardilia.*

The genus *Astarte,* now limited to a dozen species in the North Atlantic and Arctic seas, has an almost world-wide geological distribution, and counts 200 species in d'Orbigny's catalogue, commencing with the lias period. *Crassatella,* now almost a southern form, is common in the cretaceous and tertiary strata of Europe. Closely allied to *Astarte* is the extinct genus *Opis* (fig. 16, 3), of which there are 42 species in the secondary series ; and *Cardinia* (fig. 16, 2), characteristic of the lias and oolites. The so-called Unios of the coal measures (*Anthracosia,* King) are probably members of this group.[*] One hundred species of *Cardita* (including *Venericardia*) are found in the secondary and tertiary strata ; of the

---

[*] " They occur in the valuable layers of clay-ironstone called ' mussel-bands,' associated with *Nautili, Discinæ,* etc. In Derbyshire the mussel-band is wrought, like marble, into vases."— *Woodward.*

50 recent forms, one only is Arctic, and this occurs in the glacial deposits of England. The allied genus *Myoconcha* is characteristic of the older secondary rocks, and *Hippopodium* of the lias.

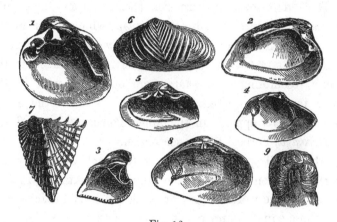

Fig. 16.

Secondary and Tertiary Bivalves.

1. Pachyrisma septiferum, Bur.; *Corallian*, Meuse.
2. Cardinia hybrida, Sby.; *Lias*, Gloucester.
3. Opis tumulatus, Mill.; *Inf. Oolite*, Bayeux.
4. Tancredia secariformis, Dkr.; *Lias*, Saxony.
5. Sowerbya crassa, d'Orb.; *Oxfordian*, Ardennes.
6. Goniomya scripta, Sby.; *Kelloway rock*, Wilts.
7. Lithocardium aviculare, Lam.; *Eocene*, Paris.
8. Grateloupia irregularis, Bart.; *Miocene*, Bordeaux.
9. Teredina personata, Lam.; *Eocene*, Bognor.

The *Veneridæ* are pre-eminently characteristic of the tertiary and present period. Some obscure species of Venus are found in the oolites : *Cytherea* occur in the greensands ; *Artemis*, *Trigona*, *Lucinopsis*, *Venerupis*, and *Tapes* appear in the middle tertiary ; *Petricola* in the eocene. The only extinct form is *Grateloupia* (fig. 16, 8), which differs but little from *Trigona*.

The Mactras and Tellens are also comparatively modern groups ; most of the supposed oolitic species belong to *Lucinidæ*, except *Sowerbya* (fig. 16, 5), which has a pallial sinus, and is found in the oolites of Malton and Portland. *Psam-*

mobiæ and *Mesodesmæ* occur in the greensand ; *Donax* and *Syndosmya* in the eocene ; *Gastrana,* (= *Venerupis,* Lam.) and *Lutraria* in the *miocene.* *Lutraria rugosa,* still living on the coast of Portugal, is fossil in the raised beaches of Sussex.

The oldest forms of razor-fish (*Solenidæ*) are those with the transverse internal rib (*Solecurtus*), which occur in the neocomian, whilst true Solens and *Glycimeris* appear first in the eocene strata. The genus *Mya,* as now restricted to the species resembling *M. arenaria,* are only met with in the newer tertiary. *Corbula* ranges upwards from the lower oolites ; *Neæra* appears in the upper greensand ; and *Thetis* (= *Poromya,* Forbes) in the neocomian.

Above 100 species of *Panopæa* (a genus essentially like *Mya*) have been obtained from oolitic and tertiary strata in all parts of the world. They are with difficulty distinguished from those equally numerous forms of *Anatinidæ* which have been associated with *Pholadomya* on account of the tenuity of their finely-granulated valves ; they constitute the genus *Myacites* (Bronn), and occur in all the palæozoic and secondary rocks ; some of the oolitic and cretaceous species are distinguished by V-shaped furrows (fig. 16, 6). Still more numerous are the fossil forms of *Pholadomya,* which range upwards from the lias, but are reduced to a single species now living in the Caribbean seas. Shells with the umbones fissured like *Anatina* also occur in the oolites. *Pandora* first appears in the older tertiary. Amongst the extinct genera referred to this family are the Silurian *Grammysia* (fig. 13, 10), with valves folded transversely ; the carboniferous *Edmondia* (fig. 13, 11), with large oblique cartilage plates ; and *Cardiomorpha,* shaped like *Isocardia ;* and the oolitic *Ceromya* (Ag.), which also resembles the heart-cockle in form. *Cercomya* is an oolitic *Anatina,* with the posterior end of the valves much attenuated.

The genus *Gastrochæna* appears in the lower oolites ; and casts of its burrows are frequently preserved after the decomposition of the coral in which they were made. *Clavagella* dates from the upper greensand, and *Aspergillum* from the miocene. *Saxicava* is found in the newer tertiary and raised beaches of Northern Europe ; and the great species commonly called "*Panopœa*" *Norwegica* is a characteristic fossil of the newer pliocene of Britain and Greenland.

The Pholades and ship-worms appear first in the oolitic strata. Forms resembling the recent *Martesia striata* have been discovered in fossil wood of the lias and Speeton clay. *Jouannetia* (Desm.) was first known as a miocene fossil ; and *Pholas* occurs in the older tertiary. Extinct species of *Teredo* are found in the silicified wood of the greensand of Blackdown and in the fossil palm-fruits of Brabant and Sheppy. The drift-wood of the London clay is usually perforated by the ship-worm, and also by an extinct form (*Teredina*, fig. 16, 9), which resembles *Martesia* in possessing an umbonal shield : when adult it not only closes the anterior pedal opening, but also cements its valves to the shelly lining of its burrow, like an *Aspergillum*. Specimens have been obtained in which the whole interior of the valves and tube had been excessively thickened towards the close of life by successive layers of shell.

Class III.—GASTEROPODA.

Fossil univalves—the remains of spiral and limpet-like shells—are not wanting in any but the very oldest fossiliferous rocks ("lingula flags"). From the lower Silurian, where less than 100 species, referable to scarcely more than ten genera, are found, they increase in number and variety slowly and regularly up to the newer tertiaries, which have afforded ten times as many genera and twenty times as many species. The total number of fossil marine univalves is less than 6000 ; the recent exceed 8000 ; and although we may expect to discover more new fossil species than recent, yet it is evident that, in comparison with past conditions, the group of univalves has only now attained its maximum development,

Between the extinct and living air-breathers the numerical discrepancy is still greater. About 300 land-snails, and half as many fresh-water *Pulmonifera*, are enumerated in the fossil catalogues; but the greater part of these are recent species, and the whole bears no proportion to the number of living land-snails, which exceeds 4000. That many more have formerly existed is indicated by the fact, that the fossil land-snails of the older tertiaries of Europe are entirely different from their living successors, and most nearly represented at the present time in the West Indies and Brazil. The generic forms peculiar to oceanic islands (remains of old continents) are more numerous than those of the mainlands, as if this order had once been more important. But the circumstances favourable to their petrifaction must have been of such rare occurrence as to preclude the probability of attaining more than the scantiest information concerning them.

From the large and proportional number of living Gasteropods, and the great amount of information which has been obtained of late years respecting their structure and habits, it might be expected that the affinities of the fossil univalves would be easily worked out, and their indications fully interpreted. Such, however, is not the case. Univalve shells present no internal markings, easily accessible like those of bivalves, and exhibiting the essential characters of the soft parts; and their external forms are often so overlaid with ornament, and disguised by mimetic characters, as to mislead upon a first examination. Shells of any family may be limpet-shaped, or turreted, or discoidal, plain or ornamented. It is more desirable to ascertain whether they have been nacreous or porcellanous; whether the apex (or *nucleus*) presents any peculiarities; and if operculated, whether the operculum was pauci- or multi-spiral.

The earlier describers of fossil univalves unhesitatingly recognised many familiar recent genera, even in the older rocks. But their *Melaniæ* were marine shells ; the sup-

Fig. 17.

Palæozoic Univalves.

1. Loxonema Lefeburei, Lév.; *Carboniferous*, Tournay.
2. Macrochilus Schlotheimi, d'Arch.; *Devonian*, Eifel.
3. Scoliostoma expansilabrum, Sdgr.; *Devonian*, Nassau.
4. Euomphalus sculptus, Sby.; *Wenlock Limestone*, May Hill.
5. Murchisonia angulati, Ph.; *Devonian*, Eifel.
6. Porcellia Puzosi, Lév.; *Carboniferous*, Tournay.
7. Bellerophon bi-carinatus, Lév.; *Carboniferous*, Tournay.
8. Tubina armata, Barr.; *U. Silurian*, Bohemia.
9. Maclurea Peachii, Salter; *L. Silurian*, Sutherland.
10. Conularia quadrisulcata, Sby.; *Carboniferous*, Lanark.

posed *Buccinum* had no notch ; the *Solariæ* were pearly ; the *Neritæ* assumed, when adult, the irregular aperture of *Pileopsis ;* the *Naticæ* had non-spiral opercula ; and the *Maclurea* was figured upside down.

The more closely palæozoic univalves are examined, so much the more do they appear to differ from ordinary recent types ; and the search for allied forms has to be conducted amongst the rare and minute and least understood of recent shells.

Pteropoda.—The fragile shells of *Hyalea* and *Cleodora* are found in the newer tertiary of Italy, with *Vaginella* (fig. 19, 12), a form allied to *Cuvieria*. But the occurrence of *Pteropoda* in the older rocks is attended with considerable obscurity. *Ecculiomphalus* is like an incompletely convoluted *Euomphalus*; *Maclurea* is like *Euomphalus* with a depressed spire; the shells called *Theca* are slender and conical; *Pterotheca* has a wing-like expansion; and *Conularia* (fig. 17, 10) is a four-sided sheath, with the apex partitioned off, as in the recent *Cuvieria*. If really pteropodous, these shells are the giants of the order.

Nucleobranchiata.—Those fossil univalves, which in their symmetry resemble the *Nautilus*, but are unfurnished with air-chambers, have been compared to the recent *Heteropoda* (or *Nucleobranchiata*, Bl.), and especially to that division typified by the tiny *Atlanta*, in which the animal can withdraw itself completely into its shell, and close the aperture with an operculum. The genus *Porcellia*, characteristic of the carboniferous age, has a discoidal shell, with a spiral nucleus projecting, as in *Atlanta*, from the right side; the whirls are exposed, and marked with a narrow band along the back, ending in a deep slit (fig. 17, 6). Another genus (*Bellerophon*) resembles the recent *Oxygyrus* in its more globose form, with a similar narrow umbilicus on either side (fig. 17, 7); sometimes the shell is thin and the aperture expanded, like a trumpet, whilst other species are globular and solid; the former may have been tenanted by large animals living at the surface of the open sea, the latter seem to have been more adapted to protect their owners crawling over the bottom, for it can scarcely be insisted that all were necessarily floaters on account of their organization. The species of *Bellerophon* are numerous in all the palæozoic rocks, and some of the smaller kinds appear to have been gregarious : those with disconnected whirls have been called *Cyrtolites* (Conrad.) The *Bellerophina* of d'Orbigny (fig. 18, 11), is a minute shell found in the gault. The other division (*Firolidæ*) consists of Mollusks in which

the shell is wanting or rudimentary, and small compared with
the bulk of the animal.   A single species of the genus *Cari-
naria* has been found in the middle tertiary of Turin.

Strombidæ.—The Strombs with their massive shells, nevertheless, resemble the fragile Heteropods in some respects.
They have the same lingual dentition, and the same carnivorous habits ; and though living on the sea-bed, they rather leap
than glide, having a narrow sole and a deeply-divided opercu-
ligerous lobe.   Characteristic of the warmer zones of existing
seas, they are only found fossilized in the newer tertiary strata
of countries south of Britain; but there is a group of little shells
related to the recent *Strombus fissurellus* in the older tertiaries
of London, Paris, and America, to which Agassiz has given the
name *Rimella.*   The allied genus of scorpion-shells (*Pterocera*),
now peculiar to eastern seas, has been described as occurring
fossil in the secondary strata of Europe ; but the extinct
species appear to be more nearly related to *Aporrhais.*   This
genus, now confined to the western shores of Europe, occurs in
all the tertiaries, and is represented in the secondary rocks by
many remarkable forms.   Some have been separated under
the name *Alaria ;* and to this group the so-called *Pterocera
Bentleyi* may perhaps be referred (fig. 18, 2).   *Rostellaria* and
Serapis (or *Terebellum*), now peculiar to the Red or eastern
seas, are conspicuous fossils of the European eocene, at which
time their range extended to America.   Some of the ancient
Rostellarias have the outer lip enormously expanded, as in
the *R. ampla* (*Hippocrena*) of the London clay.   In the oolites
and chalk there are slender fusiform shells (*Spinigera,* d'Orb.,
fig. 18, 1) with spines on the sides of the whirls, as in some
recent *Ranellæ.*

Muricidæ.—The great family of whelks, by far the most
important group of living sea-shells, is scarcely of higher antiquity than the eocene tertiary.   The *Purpurina* of the oolites
(fig. 18, 3), and *Columbellina* of the chalk, are extinct genera,

somewhat resembling *Purpura* and *Columbella*. But since the so-called "cones" of the oolites have proved to be *Tornatellæ*,

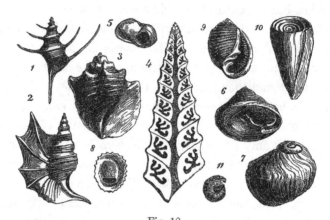

Fig. 18.

Secondary Univalves.

1. Spinigera, sp.; *Oxford Clay*, Chippenham.
2. Alaria Bentleyi, M. and L.; *Great Oolite*, Collyweston.
3. Purpurina Morrisii, Buv., *Great Oolite*, Minchinhampton.
4. Nerinæa Bruntrutana, Thurm.; *Corallian*, Poland.
5. Crossostoma Pratti, M. and L.; *Great Oolite*, Minchinhampton.
6. Trochotoma conuloides, Desl.; *Great Oolite*, Minchinhampton.
7. Neritoma bisinnata, Buv.; *Oxfordian*, Ardennes.
8. Pileolus plicatus, Sby.; *Great Oolite*, Ancliff.
9. Cinulia incrassata, J. Sby.; *U. Greensand*, Blackdown.
10. Acteonina concava, Desl.; *Lias*, Normandy.
11. Bellerophina minuta, Sby.; *Gault*, Folkestone.

it may not be unreasonable to distrust these other presumed affinities. The huge univalve of the chalk, which Sowerby called a *Dolium*, has been described as a *Pterocera* by d'Orbigny. In the tertiaries siphonated univalves abound, and are mostly referable with certainty to recent genera. The only marked change consists in the comparative abundance of some scarce existing forms, and the absence or rarity of many now most conspicuous. Moreover, the geographical distribution of the genera has undergone a great change since the close of the eocene period. This change is most noticeable in the cold-temperate zone, and is evidently the result of altered climate.

The northern seas must ever have been inclement, and the tropical seas always tropical ; but the latitude of England being most liable to vicissitudes of climate, might be expected to show the greatest variety, and the most complete and rapid alterations of organic life. In the London clay are found many

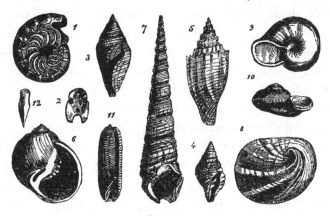

Fig. 19.

Tertiary Univalves.

1. Nautilus (Aturia) zic-zac, Sby.; *Eocene*, Britain.
2. Nautilus zic-zac, front view of a *septum.*
3. Conorbis dormitor, Sol.; *Eocene*, Britain.
4. Borsonia lineata, T. Edw.; *M. Eocene*, Hants.
5. Volutilithes luctator, Sol.; *Eocene*, Britain.
6. Natica (Deshayesia) cochlearia, Brongn.; *Eocene*, N. Italy.
7. Turritella (Proto) cathedralis, Brongn.; *Miocene*, Bordeaux.
8. Nerita (Velates) perversa, Gm.; *Eocene*, France.
9. Helix (Lychnus) Matheroni, Req.; *Eocene*, S. France.
10. Ferussina tricarinata, M. Br.; *Miocene*, Hockheim.
11. Volvaria bulloides, Lam.; *Eocene*, Grignon.
12. Vaginella depressa, Bast.; *Miocene*, Bordeaux.

species of *Clavella, Typhis, Mitra, Pseudoliva, Oliva,* and *Ancillaria ;* and some extinct forms (*Leiostoma* and *Strepsidura*) related to *Fusus.* The middle tertiary, wanting in England, but largely developed in Central and Southern Europe, also contains many genera belonging now to warmer latitudes, and many species still living in the south. In the newer tertiaries of Europe these southern forms disappear, and are gradually replaced by others of an opposite character (*Trophon, Neptunia,* and *Trichotropis*), now inhabiting the Arctic and boreal coasts. The entire number of fossil *Muricidæ* amounts to 1000, or about half as many as the recent. The older tertiaries of England also contain species of *Triton, Cassidaria, Cancellaria,* and *Pyrula,* shells (now foreign to our seas), which have formerly been included in this family. As regards bulk, there are no fossil species of *Fusus, Triton,* and *Cassis* (or *Strombus* and *Voluta*) to compare with those of the present day.

Conidæ.—The Cones and Pleurotomas appear first in the chalk, and are abundant in the eocene, accompanied by an intermediate form (*Conorbis,* fig. 19, 3), and another extinct sub-genus (*Borsonia,* fig. 19, 4), in which the column is plaited, as in *Mitra.* The genus *Terebra* is more common in the miocene.

Volutidæ.—The Volutes also appear as cretaceous fossils in Europe and Southern India ; they are very abundant in the London clay, and one occurs in the English crag. The ancient species are mostly distinguished by their spires being acute, as in *Mitra* (fig. 19, 5), a peculiarity only found in one very rare living (?) species, dredged from a bed of dead shells in 132 fathoms water (792 feet) off the Cape. The crag Volute resembles the Magellanic form. *Cymba olla,* the only living European Volute, is a fossil in the *pliocene* of Majorca.

Cypræidæ.—The Cowries form another group of subtropical shells once common in the temperate zone. Several large species are found in the London clay, most nearly related to the southern *Cyprovula;* whilst the crag contains only members of the sub-genus *Trivia,* one of which still lives on our coast.

The round-mouthed shells (*Holostomata*), whether animal-feeders or vegetarians, make a conspicuous figure amongst the fossils of an earlier period than that in which the last group began to flourish. The carnivorous *Naticidæ* and *Pyramidellidæ* are represented in the palæozoic strata by *Naticopsis,*

Loxonema (fig. 17, 1), and *Macrochilus* (fig. 17, 2). The violet-snail (*Ianthina*), so unlike any other existing shell-fish, seems related to the Silurian *Scalites, Raphistoma,* and *Holopea.* Shells like *Scalaria* and *Solarium* occur in the trias and oolites associated with *Chemnitziæ* (?) of extraordinary size, and species of *Eulima* and *Niso.* These families of shells and the *Cerithiadæ* are more abundant fossil than recent, the known numbers being 1500 extinct and 900 living forms. *Solaria,* with disconnected whirls and pyramidal opercula (*Bifrontia,* Dh.), are common in the eocene tertiary, and a single living species (*B. zanclæa*) has been discovered by M'Andrew.

Amongst the tertiary Naticas are many with an oblique aperture and peculiar perforation (*Globulus,* J. Sby., = *Ampullina,* Bl.), and others with prominences on the pillar (*Deshayesia,* fig. 19, 6.) The *Nerinæas* of the oolites are remarkable for the spiral ridges (like the "worm" of a screw) winding round their interior, and giving rise to the variety of singular patterns seen in sections (fig. 18, 4). A similar structure exists in the recent "telescope-shell" (*Terebralia*). The fresh-water univalves of the Wealden and older tertiaries differ but little from their recent congeners of the genera *Paludina, Potamides, Melania,* and *Melanopsis.* Fossil *Turritellæ* are of doubtful occurrence before the tertiary; the Silurian species have the peristome complete (*Holopella,* M'C.); another form (*Proto,* fig. 19, 7) is characteristic of the miocene.

The bonnet-limpets (*Calyptræidæ*) are common in the old rocks, which also contain a few species of *Chiton* and shells like *Dentalium.* Fossil *Trochidæ* are very numerous, but hitherto many *Litorinidæ* have doubtless been included with them. Perhaps no true *Turbo* is known from strata before the cretaceous. The *Euomphali* (fig. 17, 4), which characterize the older rocks, have multispiral calcareous opercula, like the recent *Cyclostrema* (= *Adeorbis*). The genus *Maclurea* (fig. 17, 9), which has been regarded as a "left-handed" *Euompha-*

lus, is probably very different ; it has a thick shelly operculum, sinistrally spiral, and furnished with an internal process, as the Nerites are ; the spire is sunk and concealed, whilst the whirls are exposed on the flattened under-side ; it occurs in the older Silurian rocks of Scotland and North America. One common feature of the palæozoic spiral shells is their tendency to become irregular towards the conclusion of their growth : in *Serpularia* (= *Phanerotinus*, Sby.), the whirls are all disunited ; in *Scoliostoma* (fig. 17, ₃) and *Catantostoma* the aperture is expanded. Some small oolitic shells have a thickened peristome (*Crossostoma*, fig. 18, ₅), like the recent *Lietia*, which commences in the older tertiary. A large proportion of the trochiform fossil shells have their whirls, whether round or angular, marked by a peculiar band, terminating in a deep slit at the aperture ; most of these were solid nacreous shells belonging to the genus *Pleurotomaria*, of which but a single species survives ; others in their slenderness resemble *Turritellæ*, and have been named *Murchisonia* (fig. 17, ₅). The carboniferous shell called *Polytremaria* has a row of holes in place of a slit ; and the Silurian *Tubina* (fig. 17, ₈) has three rows of tubular spines. The *Cirrus* of the inferior oolite is a reversed shell with one row of similar ornaments ; and *Trochotoma* (fig. 18, ₆) has a perforation near the margin of the aperture, which is carried onward as the shell grows. *Scissurella*, which is always diminutive and not pearly, makes its first appearance only in the newer tertiary. *Haliotis* occurs in the *miocene* of Malta. The *Neritidæ* appear in the oolites ; besides true Nerites, there are *Neritomæ* (fig. 18, ₇), with a channeled outer lip ; *Pileolus*, which is perfectly limpet-like above (fig. 18, ₈) ; and *Neritopsis*, with its angular columellar notch most distinctly marked. Key-hole limpets (*Fissurellidæ*) occur as early as the carboniferous period, but are very scarce at first, and never become numerous. The oolitic *Rimula* is a minute shell supposed to be related to a very rare living species. Ordinary limpets (*Patellidæ*) of unequivocal form are found in the Bath oolite, but are afterwards less plentiful, and almost disappear from the tertiaries ; M. d'Orbigny regarded them as generically distinct, but employed for them a name (*Helcion*, Mont.) synonymous with *Patella*.

Pulmonifera.—The existence of air-breathing snails in the palæozoic rocks is shown by a small "chrysalis-shell, with a round, not toothed, aperture, (*Dendropupa*), discovered by Dr. Dawson and Sir C. Lyell in·a hollow coal tree of Nova Scotia." Upwards of 40 species of *Pupa* have been found fossil in eocene strata. The Purbeck limestone contains a modern-looking *Physa ;* and other species of extraordinary size are found in the older tertiary of France, and also in Central India, where the genus does not exist at the present day. The fresh-water eocene of the Isle of Wight and Paris has afforded many species of *Limnæa* and *Planorbis ;* a *Glandina* rivalling in size the *G. truncata* of South Carolina ; a *Cyclostoma*, with a sculptured operculum like the *Cyclotus Jamaicensis ;* and an elongated species of the section *Megalomastoma*, which is now living in both East and West Indies. At Hordle has been found the little *Helix labyrinthicus*, still living in Texas ; and in the south of France occur representatives of the Brazilian genera *Megaspira* and *Anastoma* (fig. 19, 9). In the *miocene* is found another genus (*Ferussina*, fig. 19, 10) resembling the lamp-snail, but supposed to be operculated. The *Pulmonifera* of the English *pliocene* are in a few instances extinct, at least in England; nearly all are still living here, but more or less abundant now than they were in the times of the mastodon and elephant. The extinct land-snails of the Atlantic islands Madeira and Porto Santo are associated with remains of many recent species occurring in numbers which have relatively altered, telling the same tale of gradual changes, affecting some species prejudicially, but favourable to the increase of others. The fossil land-snails of St. Helena were supposed by Mr. Darwin to have become finally extinct only in the last century, owing to the destruction of the native woods by the instrumentality of goats and swine.

Tectibranchiata.—The families typified by *Tornatella, Ringicula*, and *Bulla* played a more important part in the secondary and tertiary periods, but their affinities have been seldom understood. The cone-like *Acteonina* appeared in the carboniferous rocks, and attained a remarkable development in the lias (fig. 18, 10). They were succeeded by the *Acteonellæ*, with a plaited columella, in the cretaceous strata ; and by *Volvaria* (fig. 19, 11) in the eocene. The diminutive *Ringiculæ* of our seas were preceded by large species of the same genus in the tertiaries, and by *Cinulia* (fig. 18, 9), *Globiconcha*, and *Tylostoma*, in the cretaceous strata. The genus *Varigera* has varices recurring twice in each whirl, like *Eulima ;* and *Pterodonta* is winged like *Strombus.*

CLASS IV.—CEPHALOPODA.

Order 1.—TETRABRANCHIATA.

(Nautiloid Cephalapoda.)

Of the lower group of Cephalopods, possessing chambered shells similar to the pearly *Nautili*, there are 1400 extinct species, belonging to above 30 genera, while 3 or 4 species alone exist in modern seas. These fossils resemble the *Nautilus*, and differ from the dibranchiate *Spirula* in the structure of their shell, which is composed of two layers, the outer porcellanous, the inner pearly ; whereas the *Spirula*—an internal shell—is entirely nacreous. They also agree with the *Nautilus* in the relative capacity of their last chamber, which seems obviously large enough to contain the whole animal. Moreover, it appears, from the position of the siphuncle and the form of the aperture, that these shells were revolutely spiral, or coiled over the back of the animal, and not involute like the *Spirula*. No traces of fossil ink (*sepia*) or horny claws have been found associated with them, nor any indications of

dense muscular tissue, even in the same matrix which has preserved so completely the mummy cuttle-fish. By their form
and size they were ill adapted for rapid locomotion, and must
have depended for safety on the shelter afforded by their solid
shell. The discoidal *Ammonites* attained a diameter approaching 3 feet, and the straight-shelled *Orthocerata* were sometimes
not less than 6 feet in length. These latter must have lived
habitually in a position nearly vertical ; whilst the discoidal
genera would creep over the sea-bed with their air-chambers
above them, like a snail-shell reversed. The *Ammonites* appear to have been provided with an operculum, more secure
than the "hood" of the Nautilus, composed, like it, of two
elements, not, however, fibrous and confluent, but calcified
and united by a straight suture. These opercula, which have
been mistaken for bivalve shells, have a porous structure
altogether peculiar, and are frequently sculptured on their
outer convex surface ; whilst their concavity exhibits only
lines of growth (fig. 21, 7). Special forms are associated, in all
localities, with particular species of ammonite ; and their size
is adapted exactly to the specimens in which they are found.
Calcareous mandibles occur in all the secondary strata, but
not, hitherto, in such numbers or circumstances as to imply
that they belonged to any other genus beside the true *Nautilus*.
They are of two forms : those corresponding to the upper
mandible (fig. 21, 8) have been called "Rhyncholites" (*Palæo-
teuthis* and *Rynchoteuthis* of d'Orbigny) ; whilst the lower mandibles constitute the genus *Conchorhynchus* of De Blainville
(fig. 21, 9). The arms of the extinct Tetrabranchs may have
been organized like those of the Nautilus, but were probably
less numerous in the genera with slender shells, and in those
early forms with a small many-lobed aperture. The length of
the body-chamber is greatest when its diameter is least ; and
the prominent spines which ornament the exterior are partitioned off internally by a nacreous lamina, indicating con-

siderable motion of the animal in its shell. When the outer
shell of the fossil is removed by decomposition or the hammer, the margins of the internal septa (or partitions of the
air-chambers) are exposed : these marginal lines are called
" sutures."

The chambered shells may be divided into two principal
groups, viz., those with simple sutures, like the recent Nautilus ; and those in which the margins of the septa are lobed
and foliaceous, the Ammonites. In the former the siphuncle
is central or internal (*i.e.*, at the margin next the spire) ; in
the latter it is external (*i.e.*, at the back of the shell, but ventral
as regards the animal). There are, however, *Nautili* with lobed
sutures (*Aturia*, Bronn, fig. 19, 1) ; and some with an external siphuncle (*Cryptoceras*, d'Orb.) And on the other hand,
the sutures of the Ammonite are at first very slightly lobed,
and become progressively more complex ; so that specimens
of the same species have been referred to three genera—*Goniatites*, *Ceratites*, and *Ammonites*—according to their age.

With the exception of *Goniatites*, the *Ammonitidæ* are
peculiar to, and co-extensive with, the secondary strata ; while
the *Nautilidæ*, with the exception of *Nautilus* and *Aturia*, are
confined to the palæozoic rocks. But the palæozoic so-called
Nautilidæ exhibit peculiarities suggesting very wide differences
from the modern pearly Nautilus. It has been proposed to
associate the greater part of them with the *Orthocerata* as a
distinct family, but at present the data are defective. Like
the *Ammonitidæ*, their shells assume almost every conceivable
form and curvature, and the genera founded on these characters are very ill defined.

The simplest form of *Orthoceras* is like a Nautilus unrolled;
and *Lituites* (fig. 20, 2) is the same with the apex spiral. Some
of the carboniferous *Nautili* have a square back, and the whirls
either compact or open in the centre (fig. 20, 1) ; whilst the
last chamber is more or less disunited. The species with the

whirls quite disunited constitute the genus *Trigonoceras*, M'C. (= *Nautiloceras*, d'Orb.) The Silurian genus *Trochoceras*, Barr, is a spiral Nautilus. *Clymenia*, a characteristic Devonian fossil, has angular sutures and an internal siphuncle ; it may perhaps be coiled up ventrally like the *Spirula*. The tertiary shell called *Nautilus zic-zac* (*Aturia*, Br., fig. 19, 1, 2), which is so widely distributed in Europe, America, and India, has a siphuncle nearly marginal when young, but gradually becoming more central in the adult : it has no special relation to *Clymenia*.

Those species of *Orthocerata* in which the aperture is contracted form the genus *Apioceras*, Fischer (= *Poterioceras*, M'C.), or when also curved, the *Oncoceras* of Hall. In Barrande's genus *Ascoceras* (fig. 20, 9), the shell is flask-shaped, the chambered and siphunculated apex being apparently deciduous ; the aperture is contracted, and the air-chambers occupy only the dorsal half of the shell. In *Phragmoceras* (fig. 20, 7), the shell is slightly curved to the ventral side, and the aperture is remarkably contracted, the opening for the respiratory funnel being nearly distinct from the cephalic aperture. In *Cyrtoceras* the curvature is dorsal.

In some other members of this family the siphuncle attains a remarkable size or extraordinary complexity.

In *Camaroceras* (fig. 20, 4), the siphuncle is lateral, quite simple, and equal to half the diameter of the shell. Casts of these great siphuncles were called "Hyolites" by Eichwald ; they frequently contain small shells of *Orthoceras*, *Bellerophon*, and other genera. In some species the siphuncle is strengthened internally by repeated layers of shell, or partitioned off by a succession of funnel-shaped diaphragms ; these constitute the genus *Endoceras* of Hall. The same author has given the name *Discosorus* to a fossil which is evidently the siphuncle of some very delicate and perishable chambered shell (fig. 20, 6). In those *Orthocerata* with siphuncles most nearly resembling

the *Discosorus* they diminish rapidly towards the last chamber.
Perhaps the most remarkable fossil of this group is the *Huronia*
(fig. 20, 5), found in the upper Silurian limestone of Drum-
mond Island. Siphuncles 6 feet in length and 1½ inch in
diameter, were seen by Dr. Bigsby in the cliffs ; they are silicified, and stand out in bold relief from the matrix, but are un-
accompanied by any vestige of the shell, except in one or two
instances, where the septa are faintly indicated by coloured
lines. They are sometimes overgrown with coral, and were

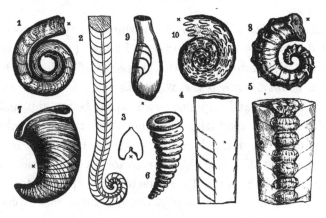

Fig. 20.

1. Nautiloceras Omalii, de Kon.; *Carboniferous*, Belgium.
2. Lituites (Breynius), *U. Silurian*, Sweden.
3. Section of Clymenia, showing internal siphuncle ; *Devonian*, Petherwin.
4. Section of Camaroceras duplex, Wahl. ; *L. Silurian*, Russia.
5. Siphuncle of Huronia Bigsbyi, Stokes ; with outline of shell, and septa.
6. Siphuncle of Discosorus, Hall ; *U. Silurian*, Lake Huron.
7. Phragmoceras ventricosum, Sby.; *L. Ludlow rock*, Herefordshire.
8. Gyroceras Eifeliense, d'Arch. ; *Devonian*, Prussia.
9. Ascoceras Bohemicum, Barr.; *U. Silurian*, Prague.
10. Goniatites, Henslowi, Sby. ; *Carboniferous*, Asturias.

evidently so durable as to remain on the sea-bed long after the
shell itself had decayed. The joints of the siphuncle are
swollen at the upper part, and the interior is filled with an
irregularly-radiated structure, apparently produced by the
plaiting and calcification of the lining membrane. This struc-

ture also exists and is very regular in the siphuncle of the Devonian *Orthoceras trigonale*, and in the shells referred to *Gyroceras* by d'Orbigny (fig. 20, 8) ; also in *Actinoceras*, a subgenus of *Orthoceras*, discovered by Dr. Bigsby, and described by Stokes (*Geol. Trans.*, vol. i., 1825). The plication of this interior structure takes place in segments corresponding to the septa, and meeting in the centres of the siphuncular beads, leaving spaces or foramina for the passage of blood-vessels to the lining membrane of the air-chambers.* The vascularity of the latter is well shown in the impression of septa on the fine mudstones of the Ludlow rock, often mistaken for *Spongaria*, which they somewhat resemble.

Towards the conclusion of its growth the air-chambers of the *Orthoceras* frequently become shallower, and the siphuncle diminishes in size. These indications of changed or diminished energies are accompanied by a diminution or disappearance of the internal radiated structure in the last part of the siphuncle.

In *Orthoceras bisiphonatum* (*Tretoceras*, Salter) the body-chamber is prolonged in the form of a marginal lobe, simulating a second siphuncle.

The genus *Bactrites* of Sandberger also resembles an *Orthoceras* with single-lobed sutures.

Ammonitidæ.—In the second division or family of chambered shells—those with lobed sutures and a marginal siphuncle—we find a similar series of forms, straight, spiral, and discoidal, but more varied and more highly ornamented.

One large genus (*Goniatites*, fig. 20, 10) is found in the Devonian, carboniferous, and triassic strata, and permanently resembles the youngest form of the *Ammonites*, having the sutures lobed but not foliated. They seldom exceed 6 inches in diameter and are usually very much smaller. The whirls are

---

* In the carboniferous species of *Actinoceras* (*e.g.*, *A. giganteum*), these foramina form a cross on the ventral side of the siphuncle.

most frequently concealed to some extent, and often marked by cross furrows or "periodic mouths."

The *Ceratites* are distinguished by having the lobes of the sutures serrated, while the intervening "saddles" (or curves directed towards the aperture) are simple. They are found in the trias of Europe, Thibet, and South America ; and again, though rarely, in the cretaceous strata of France and Syria—a circumstance quite anomalous in the history of the geological

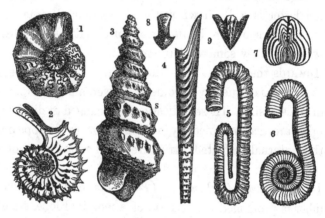

Fig. 21.

1. Ceratites nodosus ; *Muschelkalk*, Bavaria.
2. Ammonites Duncani (spinosus, Sby.) ; *Oxford Clay*, Wilts.
   3. Turrilites.   5. Hamites.
   4. Baculites.   6. Ancyloceras.
7. (Trigonellites or Aptychus), operculum of Ammonites.
8. (Rhyncholites hirundo), upper mandible of Nautilus arietis, Rein. *Muschelkalk.*
   Lower mandible (Conchorynchus avirostris).

distribution of life. Many Ammonites, perhaps all, are like Ceratites when young.

A bisected specimen of the *Ammonites obtusus*, in the Hunterian collection (No. 188), shows well the extent of the last, or inhabited, chamber of the shell, and the effects of the influence of the animal matter of the decaying cephalopod upon the petrifactive processes after death. The liassic clay has penetrated as far as the retracted soft parts of the ammonite permitted : the decomposing mollusk had been partially replaced by crystals of silex, discoloured by the pigmental or carbonized parts of the animal. The silex, which has more slowly infiltrated through the pores of the shell into the air-chambers, is of a much lighter colour. In the same collection may be seen exemplifications of injury and repair of the shell. In No. 195, *Ammonites Goliathus*, from " Oxford clay," a portion of the shell, at the period when it formed the dwelling-chamber, "had been broken away during the life-time of the animal, and repaired by fresh nacreous material, wanting the ribbed structure of the originally formed shell."[*]

The species of Ammonite exceed 500 ; and their range is co-extensive with that of the secondary rocks. They are found throughout Europe, and at the Cape, in Kamtschatka, Thibet, and S. India. They are absent from a large area of the United States, but are found in the cretaceous strata of New Jersey, Missouri, and the West Indian Islands ; also in Chili and Bogota.

The sections into which, for the sake of convenience, this extremely natural group has been broken up, are very ill-defined, and have no pretension to be considered sub-generic. The group (called *Cassiani*) characterising the triassic period, is remarkable for many-lobed and elaborately-foliated sutures—a circumstance more important because it is the oldest group, and associated with *Ceratites* and the last-surviving *Goniatites* and *Orthocerata*. They abound in the "alpine limestone" of St. Cassian, and Hallstatt in Austria. A second group (*Arietes*), having the back keeled,

---

[*] Catalogue of Fossil Invertebrata, Mus. College of Surgeons, London, 4to, p. 43, in which work the writer has described upwards of 350 specimens, illustrative of the different sections of Ammonitidæ, collected by John Hunter in the last century.

with a furrow on each side of the keel, as in the great Ammonites called *Bucklandi* and *Coneybearei*, mark the lias period;
they are less plentiful in the oolites, and are represented in
the greensands by the *Cristati*, which are keeled, but not
furrowed, and develope a "beak," or process, from the keel
when adult. The *Arietes* pass by many intermediate forms
into the *Falciferi* (*e. g.*, *A. serpentinus*), also characteristic of
the upper lias, and these are represented by a few quoit-shaped
species (*Disci*), with sharp backs, in the oolites.

Ammonites with serrated keels (*Amalthei*), exemplified by
A. spinatus and *margaritatus*, abound in the middle and upper
lias, and again in the oolites (*e. g.*, *A. cordatus* and *excavatus*).
They are succeeded by the *Rothomagenses* in the chalk—
thick Ammonites with a line of tubercles in the place of the
keel.

Ammonites with channelled backs (*Colliciati*) are represented in the lias (*A. anguliferus*), inferior oolites (*A. Parkinsoni*), and middle oolite (*A. anceps*), and in the cretaceous
strata by numerous species (*e. g.*, *A. serratus, lautus*, and *falcatus*), remarkable for their elegance.

Of the species with backs more or less squared, *armatus*
and *capricornus* occur in the lias, *athleta* and *perarmatus* in
the Oxfordian. But the oolitic forms which have the back
square, and ornamented with two rows of spines when young,
like *Goweri, Duncani* (fig. 21, 2), and *Jason*, become rounded
and unarmed in their old age.

Round-backed Ammonites abound in the lias and oolites.
The snake-like *annulatus*, the spine-bearing *coronatus*, and
fimbriatus with its ornamented fringes, have been regarded
as types of small groups. A more important division (*Ligati*)
is distinguished by nearly smooth whirls, constrictions recurring at regular intervals. These are seen in *A. tatricus*, and
others related to *Heterophyllus;* in many neocomian Ammonites, and in *A. planulatus* of the lower chalk.

These constrictions, often accompanied by a prominent rib, undoubtedly indicate periods of rest, when the Ammonite ceased for a while to grow. They may be traced in species belonging to other groups, as well, *e.g.*, in *biplex* and *triplicatus*, as in the *Ligati;* but most frequently all indications are obliterated by subsequent growth. It has been a question whether the lateral processes of *Ammonites Duncani* (fig. 21, 2), are formed and removed periodically, or whether they are peculiar to the adults, and mark the close of their outward growth. The first conclusion is more probable from analogy ; and they are commonly found with small and apparently young shells, but not (any more than the lateral spines of the living Argonaut) in those of adult size and condition.

It was remarked by the elder Sowerby that Ammonites were most beautiful when of middle growth, the ornamental characters being less developed in the young, and lost in the adult. The ribs and spines, and even the keel or furrow of the back disappear, in many instances, from the body-whirl of the full-grown shell.

Varieties of form, such as marked the palæozoic *Nautilidæ*, are met with in the *Ammonitidæ*, chiefly towards the close of their reign. The *Baculite* (fig. 21, 4), with its straight shell, is characteristic of the upper chalk ; and the *Turrilite*, which is spiral, and usually a left-handed spiral, abounds in the lowest beds of the same formation. In *Hamites* the shell is straight, returning upon itself after a certain space, and forming a simple or complex hook. In *Ptychoceras* these limbs of the hook-like shell are in close contact. The *Toxoceras* is curved like a bow ; in *Crioceras* the discoidal whirls are separate ; and in *Scaphites* (including *Ancyloceras*) the shell, at first compact like an Ammonite, or open-whirled like *Crioceras*, lengthens out finally, and returns upon itself like the crozier of the *Hamite*. *Helicoceras*, again, connects the last with the *Turrilite* by its elevated spire terminating in a prolonged crozier.

Of these forms, *Ancyloceras* alone is found in the oolites ; all the rest are cretaceous ; and most abound in the alpine districts of the south of France.

Order 2.—DIBRANCHIATA.

(*Cuttle-fishes.*)

Of the two great divisions of cephalopodous Mollusca, that which is represented at the present day by the pearly *Nautilus* was developed in the greatest profusion and variety in the palæozoic and secondary periods ; whilst the more active and intelligent cuttle-fishes and squids have not been (certainly) found in rocks older than the lias, and scarcely above 100 are found in the whole secondary and tertiary series, while twice as many have been obtained in existing seas.

The *Sepiadæ* are represented in the middle and upper oolites by the genus *Coccoteuthis* (fig. 22, 6), whose strong and granulated bone is furnished with broader lateral expansions than the recent cuttle-fishes. In the older tertiaries of London and Paris, many species of *Sepia* appear to have existed, but only the solid *mucro* (fig. 22, 5) of the shell is usually preserved. In the miocene tertiary of Malta, a diminutive cuttle-bone is not rare ; and at Turin a remarkable form (*Spirulirostra,* fig. 22, 7) has been discovered, in which the apex is provided with a chambered and siphonated cavity like the shell of the *Spirula.* Two other genera, *Beloptera* (fig. 22, 8) and *Belemnoris,* very imperfectly known by rare and fragmentary examples, occur in the eocene tertiary.

Remains of the Calamaries (*Teuthidæ*) are often found in the fine-grained and laminated argillaceous limestones of the lias and Oxford clays, as at Lyme Regis, and Böll of Solenhofen. Some of these are slender, like the pens of the recent *Ommastrephes,* and furnished with a small conical appendix, as in that genus ; whilst others are broad, and pointed at each end (*Beloteuthis*). The most common form has the shaft wide and longer than the wings, and is truncated posteriorly. It has a nacreous lining, and is usually accompanied by a large and well-preserved ink-bag (fig. 22, 4). These were called *Belemnosepia* by Agassiz and Buckland, who supposed them to belong to the same animal with the Belemnite. They have also been called *Loligo-sepiæ* and *Loliginites;* but the name *Geoteuthis,* given by Count Münster, appears less objectionable. One species (*Mastigophora brevipinnis**) is of frequent occurrence in the Oxford clay near Chippenham, which retains not only the horny (chitinous) pen and ink-bag, but also the muscular mantle, the rhombic terminal fins, and at least the bases of the arms, with the minute hooks, and traces of the mandibles. Horny claws, like those of the uncinated Calamary (*Onychoteuthis*), have been observed arranged in double series in the lias of Watchett, and they sometimes occur in great numbers in the coprolitic remains of the *Enaliosauri.* The most remarkable examples of this kind are preserved in the lithographic limestones of Solenhofen, and show that the extinct Calamary had ten nearly equal arms, the tentacles, in their retracted condition, being undistinguishable from the rest—each furnished with 20 to 30 pairs of formidable hooks. What further evidence was needed respecting the nature of this creature has been supplied by the Chippenham fossils, which in all probability are identical in genus, if not in species, with the *Acanthoteuthis* described by Münster. One of these extraordinary fossils— the mummy of a cuttle-fish more ancient than the chalk formation and the upper oolites—is represented in (fig. 22, 2,) reduced to one-sixth from the original in the British Museum. Nine of the arms are preserved, the sclerotic plates of the eyes, the bases of the large lateral fins, the small ink-bag, and the conical shell. This shell, which is chambered internally,

---

* Catalogue of Fossil Invertebrata, in the Museum of the College of Surgeons, London. 4to, 1856, p. 1.

like the *phragmocone* of the Belemnite (fig. 22, 1), has an outer
sheath of fibrous structure, one-fourth of an inch thick at the

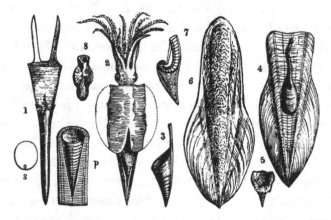

Fig. 22.

1. Belemnites Oweni ; *Oxford Clay*, Chippenham. *p. Phragmocone*
exposed by the removal of the fibrous guard from one side; *s, septum,*
showing the marginal *siphuncle.*
2. Acanthoteuthis antiquus (Cunnington) *Oxford Clay*, Chippenham ;
dorsal aspect.
3. Conoteuthis Dupinii ; *Gault*, Folkestone.
4. Geoteuthis.
5. Sepia.
6. Coccoteuthis.
7. Spirulirostra.
8. Beloptera anomala.

apex, and furnished with two converging ridges on its dorsal
side ; the external surface, however, is horny (or chitinous),
like the pen of the Calamary. These chambered shells occur
in great numbers, and are so like the phragmocones of the
associated Belemnites, both in structure and proportions, that
they were originally described by me as such,[*] and I still view
them as evidences of the close affinity of the cephalopod
possessing them to the true Belemnite : hitherto they have
only been noticed in the laminated Oxford clay of Wilts, and
the equivalent lithographic shales of Solenhofen.

[*] Philosophical Transactions, 1844; and Cat. Fossil Invert., Mus. Coll. of
Surgeons. 4to, p. 5.

Species of Belemnite are found in all the oolitic and cretaceous strata, from the lowest lias to the upper chalk. In its ordinary imperfect state, it is a cylinder pointed at one end (fig. 22, 1), and truncated or excavated by a funnel-shaped cavity (*alveolus*, ib. *p*) at the other, and has a radiating fibrous structure, with less distinct concentric laminæ of growth. But even this " guard," which corresponds simply to the "mucro" of the cuttle-bone (ib. 5), exhibits such remarkable modifications of form, that nearly 100 species have been founded upon no higher evidence. In some Belemnites of half an inch diameter, the guard is scarcely an inch longer than the phragmocone, whilst in others it attains a length of ten inches, and is tubular, as in *B. acuarius*. Some are fusiform, others laterally compressed ; some have a longitudinal groove extending from the apex along the upper or under side, and in others the apex is furrowed laterally as well. The Belemnites of the chalk have been called *Belemnitellæ* (d'Orb.), because they have a slit in the ventral side of the alveolar border of the guard ; their external surface also exhibits more distinct traces of vascular impressions.

Specimens of Belemnite have been discovered in which the guard had been broken during the lifetime of the animal ; but the broken portions, being held together by the investing organised integuments, had been re-united by the deposition of new layers of the fibrous structure peculiar to the guard. Several examples of Belemnites, with the apex injured and healed during life, are preserved in the British Museum. In all perfect Belemnites, the *alveolus* is occupied by a *phragmocone* (fig. 22, *p*), with tender nacreous walls and septa, terminating in a minute globular apex, and perforated by a ventral *siphuncle* (fig. 22, 1). The last chamber is rarely preserved, and appears to have thinned off into a mere horny sheath, with sometimes two pearly bands like knife-blades on the dorsal side. It must have been sufficiently capacious to contain all the viscera. The ink-bag has been very rarely found, and is even smaller than in the last genus, as if in relation to the more greatly developed shell.

The *Conoteuthis* (fig. 22, 3) of the Gault has an oblique phragmocone, with a very thin shell, and seems to have been attached to a slender style, like the funnel-shaped appendix of the gladius in the recent sagittated Calamary.

Mr. Dana has described, under the name *Helicerus Fugiensis*, a belemnitoid fossil from the "slate" rock of Cape Horn. It is half an inch in diameter, has a thick fibrous guard, and the slender phragmacone terminates in a fusiform spiral nucleus.*

Subjoined is a table of the extinct genera of the molluscous province :—

BRACHIOPODA.—Trigonosemus, Lyra, Magas, Rhynchora, Zellania, Stringocephalus, Meganteris ; Spirifera, Cyrtia, Suessia, Athyris, Merista, Retzia, Uncites ; Camarophoria, Porambonites, Pentamerus, Atrypa, Anoplotheca; Orthis, Orthisina, Strophomena, Koninckia, Davidsonia, Calceola ; Producta, Chonetes, Aulosteges, Strophalosia ; Trematis, Siphonotreta ; Obolus.

CONCHIFERA.—Gryphæa, Exogyra, Limanomia, Carolia, Placunopsis, Neithea, Eligmus ; Pteroperna, Aucella, Ambonychia, Cardiola, Eurydesma, Pterinea, Monotis, Posidonomya, Aviculopecten, Gervillia, Streblopteria, Pulvinites, Inoceramus, Trichites ; Megalina, Orthonotus, Modiolopsis, Hoplomytilus ; Macrodon, Isoarca, Bakewellia, Nuculina, Nucinella, Cucullella, Ctenodonta ; Myophoria, Axinus, Lyrodesma ; Diceras, Monopleura, Requienia ; Hippurites, Radiolites, Caprinella, Caprina,

* For the drawings and most of the facts, or their verification, relating to invertebrate fossils, the writer is indebted to his experienced colleague in charge of that department of the British Museum, Mr. S. P. Woodward, F.G.S.

Caprotina ; Lithocardium, Conocardium, Corbicella, Sphæra, Unicardium, Tancredia, Volupia ; Pleurophorus, Myoconcha, Anthracosia, Megalodon, Pachydomus, Pachyrisma, Cleobis, Mæonia, Opis, Cardinia, Hippopodium, Megaloma ; Grateloupia, Sowerbya, Quenstedtia, Goniophora, Redonia ; Cercomya, Myacites, Goniomya, Grammyria, Ceromya, Cardiomorpha, Edomondia, Ribeiria.

GASTEROPODA.—Bellerophon, Porcellia, Cyrtolites, Ecculiomphalus ; Rimella, Hippocrena, Alaria, Spinigera, Amberlya ; Leiostomus, Strepsidura, Purpurina, Columbellina, Borsonia, Conorbis ; Euspira, Naticopsis, Globulus, Deshayesia, Loxonema, Macrochilus ; Diastoma, Nerinæa, Brachytrema, Ceritella, Vicarya, Scoliostoma, Proto, Holopella, Catantostoma, Naticella ; Platyceras, Metoptoma, Hypodema, Deslonchampsia ; Euomphalus, Ophileta, Phanerotinus, Serpularia, Discohelix, Platystoma, Crossostoma, Pleurotomaria, Murchisonia, Polytremaria, Cirrus, Trochotoma, Platyschisma, Scalites, Rhaphistoma, Holopea, Maclurea ; Neritoma, Velates, Pileolus ; Helminthochiton ; Lychnus, Dendropupa, Ferussina ; Cylindrites, Acteonina, Acteonella, Cinulia, Globiconcha, Varigera, Tylostoma, Pterodonta, Volvaria, Chilostoma ; Vaginella, Theca, Pterotheca, Conularia.

CEPHALOPODA.—Aturia, Discites, Nautiloceras, Trigonoceras, Temnochilus, Lituites, Trocholites, Trochoceras, Clymenia ; Orthoceras, Camaroceras, Huronia, Actinoceras, Discosorus, Gonioceras, Tretoceras, Apioceras, Gomphoceras, Phragmoceras, Cyrtoceras, Gyroceras, Ascoceras ; Goniatites, Bactrites, Ceratites, Ammonites, Crioceras, Toxiceras, Ancyloceras, Scaphites, Helicoceras, Turrilites, Hamites, Ptychoceras, Baculites ; Mastigophora, Teudopsis, Beloteuthis, Geoteuthis, Leptoteuthis, Bel-

emnites, Acanthoteuthis, Helicerus, Conoteuthis, Coccoteuthis, Belosepia, Spirliurostra, Beloptera, Belemnosis.

Province IV.—VERTEBRATA.

There is an enormous series of subaqueous sediment,
originally composed of mud, sand, or pebbles, the successive
bottoms of a former sea, derived from pre-existing rocks,
which has not undergone any change from heat, and in which
no trace of organic life has yet been detected. These non-
fossiliferous, non-crystalline, sedimentary beds form, in all
countries where they have yet been examined, the base-rocks
on which the Cambrian or oldest Silurian strata rest.

Whether they be significative of ocean abysses never
reached by the remains of coeval living beings, or whether
they truly indicate the period antecedent to the beginning of
life on this planet, are questions of the deepest significance,
and demanding much farther observation before they can be
authoritatively answered.

It has been shown that every type of invertebrate animal
is represented in the superimposed stratified deposits called
Cambrian and lower Silurian.

An important work,[*] embodying the labours of the
accomplished naturalist and acute observer, Dr. Christian H.
Pander, has recently been published by the Russian govern-
ment, descriptive of the fossil fishes of the Silurian formations
of that empire. Of some hundred fossils described and
beautifully figured in this work, and referred to different
genera and species of fishes, from lower Silurian rocks, the
writer, after the closest comparison and consideration of the
evidence, is disposed to regard only those referred by Pander
to the genera *Ctenognathus*, *Cordylodus*, and *Gnathodus*, as

---

[*] *Monographie der Fossilen Fische (Untersilurische Fische Conodonten,*
etc.), 4to, Petersburgh, 1856.

having any probable claims to vertebrate rank ; and to this admission must be appended the remark, that the parts referred to jaws and teeth may be but remains of the dentated claws of *Crustacea*. With regard to the fossils called " Conodonts," on which the main part of M. Pander's evidence of lower Silurian fishes rests, the following remarks, penned after microscopic examination of original specimens, are applicable to them.

Minute, glistening, slender, conical bodies, hollow at the base, pointed at the end, more or less bent, with sharp opposite margins, might well be lingual teeth of Gastropods, acetabular hooklets of Cephalopods, or teeth of cartilaginous fishes. Against the latter determination is the minute size of the "Conodont" bodies. Their basal cavity doubtless contained a formative pulp, but the proof that the product of such pulp was "dentine" is wanting : the observed structure of the hooklet presents concentric conical lamellæ of a dense structureless substance, containing minute nuclei or cells.

In some specimens the base is abruptly produced and divided from the body of the hooklet by a constriction—a form unknown in the teeth of any fishes, but presented by certain lingual teeth of Gastropods—*e. g.*, the lateral teeth of *Sparella*. In other Conodonts the elongated base is denticulate or serrate, as in the lateral teeth of *Buccinum* and *Chrysodomus*. It is improbable, however, that they belong to any conchiferous toothed Mollusk, the shells of such being wanting in the deposit where the Conodonts are most abundant.

The more minute hooklets have a yellowish, transparent, horny appearance ; the larger, perhaps older ones, present a harder whitish appearance. Their analysis by Pander yielded " carbonate of lime," carbonic acid being evolved by application of dilute nitric acid, and oxalic acid producing an obvious precipitate. Some English analysts have believed that the Conodonts yielded a trace of phosphate of lime.

H

The detached condition of the hooklets, and the integrity of the thin border of the basal pulp-cavity, indicate that they have not been broken away from any of those kinds of attachment to a bone which the minute villiform teeth of osseous fishes would show signs of. The Conodonts have been supported upon a soft substance, such as the skin of a mollusk or worm, the mucous membrane of a mouth or throat, or the covering of a proboscis; but to select the teeth of cyclostomous or plagiostomous fishes as the exclusive illustration of the above condition, would be to take a partial and limited view of the subject.

In comparing the Conodonts with the teeth of fishes, they present most resemblance to the minute conical recurved teeth of the genus *Rhinodon* of Smith : they more remotely resemble the conical, pointed, horny teeth of Myxinoids and Lampreys in that class : and the absence of any other hard part in the strata containing the Conodonts tallies with the condition of the cartilaginous skeleton ; but not more than it does with the like perishable soft condition of annelidous worms and naked mollusks. *Rhinodon* has very small teeth, " en brosse," of a simple conical recurved form : there are 12 or 13 teeth in each vertical row, and about 250 such rows in each jaw : so that each fish may have from 6000 to 7000 teeth. But the teeth of *Rhinodon* have not the basal extensions and processes of many of the Conodonts ; and the teeth of all known Cyclostomes are much less slender and are less varied in form than in the Conodonts. Certain lingual plates of Myxinoids are serrate, but not with a main denticle of much greater length—such as shown in the form of the Conodont called *Machairodus* by Pander. Most cyclostomous teeth are simple, thick cones, with a subcircular base ; and every known tooth of a cyclostomous fish is much larger than any of the forms of *Conodon*, which rarely equal half a line in length. This minuteness of size, with the peculiarities of form, supports a reference of the Conodonts rather to some soft invertebrate genus. Certain parts of small Crustacea— *e. g.*, the pygidium or tail of some minute *Entomostraca*— resemble in shape the more simple Conodonts ; but when we perceive that these bodies occur in thousands, detached, with entire bases, and that any part of the carapace, or shell of an Entomostracan or other Crustacean, has been rarely detected in the lower Silurian Conodont beds, it is highly improbable that they can have belonged to an organism protected by a substance as susceptible of preservation as their own substance. Much more likely is it that the body to which the minute hooklets were attached was as soluble and perishable as the soft pulp upon which the Conodont was sheathed. The writer finds no form of spine, denticle, or hooklet in any Echinoderm, and especially in any soft-bodied one, to match the Conodonts ; and concludes that they have most analogy with the spines, or hooklets, or denticles of naked Mollusks or Annelides. The formal publication of these minute ambiguous bodies of the oldest fossiliferous rocks, as proved evidences of fishes, is much to be deprecated.

CLASS I.—PISCES.

Order 1.—PLAGIOSTOMI.

(*Sharks, Rays.*)

Char.—Endo-skeleton cartilaginous or partially ossified ; exo-skeleton placoid ; gills fixed with five or more gill-apertures ; no swim-bladder ; scapular arch detached from the head ; ventrals abdominal ; intestine with spiral valve.*

The earliest good evidence which has been obtained of a vertebrate animal in the earth's crust is a spine, of the nature

---

* For an explanation of the technical terms in these characters, see the article ICHTHYOLOGY, especially of the scales, Enc. Brit. vol. xii., p. 216.

of the dorsal spine of the dog-fish (*Acanthias*), and a buckler
like that of Cephalaspis. Both have been found in the most
recent deposits of the Silurian period, in the formation called
"upper Ludlow rock." The discovery of the first is due to
Murchison ;* its determination to Agassiz, who assigns it to
a genus of plagiostomous cartilaginous fishes called *Onchus*.
The buckler was discovered by Mr. Banks, in the "passage-
beds" of Kington, Herefordshire, and is referred to the genus
Pteraspis, Knerr, with microscopic evidence of its piscine
nature, by Huxley.

The *Onchus* spines from the upper Ludlow bone-beds are
compressed, slightly curved, less than two inches in length, with
no trace at their base of the joint characteristic of the dorsal
spines of the "sheat-fishes" (Ganoids of the family *Siluridæ*),
or "file-fishes" (*Balistidæ*). The sides of the spine are finely
grooved lengthwise, with rounded ribs between the grooves.
They are referred to two species—*Onchus Murchisoni*, and *O.
semistriatus*. Sir P. Egerton has lately figured another species
from the argillaceous beds near Ludlow, which is more curved,
and is armed along the posterior edge ; the longitudinal ribs
are fine and numerous, but are constricted at intervals, as in
the genus *Ctenacanthus*, and become subtuberculate at the
base. He deems them significant of a distinct genus of shark-
like fishes.†

With the dorsal spines of *Onchus* are found petrified portions of skin, tubercular and prickly, like the shagreen of
shark's skin, and referred to a genus called *Sphagodus ;* also
coprolitic bodies of phosphate and carbonate of lime, including
recognisable parts of the small Mollusks and Crinoids which

---

* Silurian System, ch. xlv., p. 606.

† In a formation in Indiana, United States of America, referred by Messrs.
Norwood and Dale Owen to the Silurian formation, a badly-preserved fossil,
considered as an Ichthyolite, and referred to a genus allied to *Pterichthys*, has
been discovered, and called *Macropetalichthys raphiidolabis*. (Silliman's
Journal, 1846, p. 367.)

inhabited the sea-bottom in company with the Onchus-fish. No vertebræ, or other parts of the endo-skeleton of a fish, have been discovered, unless the fragments of a calcified bar, with tooth-like processes, called *Plectrodus*, be truly jaws with teeth. They resemble, however, parts of the pincer claws of Crustaceans, as well as of the jaws and teeth of fishes, and do not indicate that class so satisfactorily as the *Onchus* spines and *Sphagodus* shagreen. Yet the denticles are confluent with an outer ridge of the bone, according to the "pleurodont" type, and consist of separated large teeth, with minute serial teeth in the interspaces; and the large teeth are grooved longitudinally.[*]

If the Plectrodonts be jaws with anchylosed teeth, they belong to an order distinct from the *Plagiostomi*. If they should belong to any of the fishes indicated by the dorsal spines and shagreen skin, a combination of characters would be exemplified not known in other formations or in any existing fishes.

No detached teeth unequivocally referable to a plagiostomous genus, nor any true ganoid scale of a fish, have yet been found in the formations that have revealed these earliest known evidences of vertebrate animals. What, then, it may be asked, were the conditions under which so immense an extent, as well as amount, of sediment was deposited—including chambered Cephalopods, Gastropods, Lamellibranchs, Brachiopods, various and large trilobitic and entomostracous Crustaceans, with Crinoids, Polypes, and Protozoa—that precluded the preservation of the fossilizable parts of fishes, if that class of vertebrate animals had existed in numbers, and under the variety of forms, comparable to those that people the ocean at the present day? Bonitos now pursue flying-fishes through the upper regions of an ocean as deep as any known part of the Silurian seas of which the deposits afford an idea of greatest

* Egerton, Proc. Geol. Soc., March 1857, p. 288, pl. x., figs. 2-4.

depth. If fishes of cognate habits with the present deep-sea
fishes, under whatever difference of form such Silurian fishes
may have been manifested, had really existed, we might
reasonably expect to find the remains of some of the countless
generations that succeeded each other during that vast and
indefinite period, sufficing for the gradual deposition of sedimentary beds of thousands of feet in depth or vertical thickness.

The evidences of plagiostomous fishes afforded by fossil
spines will be here pursued. In most of the existing cartilaginous fishes of this order the defensive spine which stands
erect in front of the dorsal fin is smooth ; such is the case in
the dog-fishes (*Spinacidæ*) in which each dorsal fin is fronted
with a spine. In the Port-Jackson sharks (*Cestraciontidæ*)
the spine in front of each dorsal is bony, and is armed along
its hinder or concave border with bent spines. The fin is
connected with this border, and its movements are regulated
by the elevation or depression of the spine during the peculiar
rotatory action of the body of the shark. This action of the
spine in raising and depressing the fin, resembles, Dr. Buckland
has remarked, that of the moveable or jointed mast, raising
and lowering backwards the sail of a barge. But their more
obvious use, in the small Plagiostomes possessing such spines,
is as defensive weapons against the larger and stronger voracious
fishes. The spine of the *Onchus* indicates its danger from some
larger, and as yet unknown, predatory fish.

Certain bony fishes are similarly armed—*e. g.*, sticklebacks
(*Gasterostei*), sheat-fishes (*Siluridæ*), trigger-fishes (*Balistes*),
and some species of snipe-fishes (*Fistularidæ*). In the latter
family the *Centriscus humerosus* (fig. 23) shows a dorsal spine,
denticulated behind, as in the Cestracionts ; but the base of
the spine in bony fishes is peculiarly modified for articulation
with another bone. In the Plagiostomes the base of the spine
is hollow, becomes thin and smooth when the body of the spine
is sculptured, and is in the recent fish implanted in the flesh.

The following genera of plagiostomous fishes have been
founded on the fossil spines, or "ichthyodorulites," which have
been discovered in the "Devonian," or "Old Red Sandstone

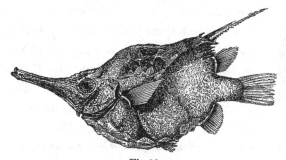

Fig 23.

Centriscus humerosus.

series." *Onchus* (represented by *O. semistriatus, O. heterogyrus*),
*Dimeracanthus, Haplacanthus, Narcodes, Naulas, Byssacanthus,
Cosmacanthus, Homacanthus* (fig. 24), *Ctenacanthus, Ptyacan-
thus, Climatius, Parexus, Odontacanthus,* and *Pleuracanthus.*

The genus *Homacanthus* is founded on small compressed
spines, with fine recurved teeth on the back edge,
and longitudinal striæ on the sides. Specimens
of *Homacanthus arcuatus* (fig. 24) have been found
in Devonian formations near St. Petersburg.

The carboniferous series of formations includes
the mountain limestone, millstone grit, and the
coal measures (see fig. 1). In this series the
genus *Onchus* is still represented by the *O. sulcatus, O. rectus,* and *O. subulatus;* and the genus
Homacanthus, by *H. macrodus* and *H. microdus,*
from the carboniferous limestone of Armagh.
Ptyacanthus, Ctenacanthus, and *Pleuracanthus*
are also forms common to the Devonian and

Fig. 24.

*Homacanthus
arcuatus.*
(Devonian,
Russia.)

carboniferous periods. The spine of the latter genus is denticulated along both margins, a structure which is presented,
in existing Plagiostomes, only by species of the ray family;

Pleuracanthus, therefore, as Agassiz concludes, may offer the
earliest example of the flat form of cartilaginous fish which is
represented by the sting-rays (*Trigon, Myliobates*) in the

present seas. The ichthyodorulite (*ichthys*, a
fish ; *dora*, a spear ; *lithos*, a stone) here selected
to illustrate this fossil, is a portion of the spine
of the *Pleuracanthus levissimus* (fig. 25), from the
carboniferous beds near Dudley. The other
plagiostomous genera based upon fossil spines
from the coal formations are,—*Oracanthus, Gyra-*

Fig. 25.   *canthus, Nemacanthus, Cosmacanthus, Leptacan-*

Pleuracanthus   *thus, Homacanthus, Trystichius, Asteropterychius,*
levissimus.
(Coal, Dudley.)   *Physonemus, Sphenacanthus, Platyacanthus, Dip-*
riacanthus, Erismacanthus, Orthacanthus, Cladacanthus,
Lepracanthus.

Immediately above the coal measures lie a variable series
of sands and clays of different colours, including the coal
plants ; above this, a marl slate in thin layers, containing
scanty evidences of fishes ; but these are more abundant and
instructive in the superincumbent magnesian limestone, in
which formation, near Belfast, ichthyodorulites of the genus
Gyropristis (Ag.) have been found. Above this are the penean
red sandstones, in which, at Westoe, have been found fossil
spines closely allied to, if not identical with, the *Gyracanthus
formosus* (Ag.) The foregoing formations constitute the uppermost of the palæozoic series called " Permian," from the
Russian province in which these strata are most extensively
developed. Their relative position is known by the term
" magnesian limestone " in the " Table of Strata," fig. 1.

The superimposed strata, marked " new red sandstone,"
includes also a varied series of red and white sands, marls,
and conglomerates, forming collectively the system called
" triassic." The ichthyodorulites of this system are referable
to the genera *Nemacanthus, Leiacanthus,* and *Hybodus.* In

the "lias," which is the oldest or lowest of the great "oolitic" system, the dorsal spines of the genus *Hybodus* (fig. 26), are the largest and most abundant; this genus, however, is represented by detached teeth in the keuper and muschelkalk members of the "trias." The lias formations give evidence that the dorsal spines and fins of *Hybodus* were two in number ; and the genus is shown, both by the structure of the spine and the form of the teeth, to have had its nearest affinities with the *Cestracion* amongst existing Plagiostomes. *Hybodus* continued to be represented by successive and varying specific forms up to, and including, the cretaceous period. *Hybodus* is therefore a genus of cartilaginous fishes eminently characteristic of the secondary or mezozoic period in palæontology, and ranges through every formation of that period. The specimen selected for the illustration of the dorsal spine of *Hybodus* is that of the *H. subcarinatus,* from the Wealden of Tilgate Forest.

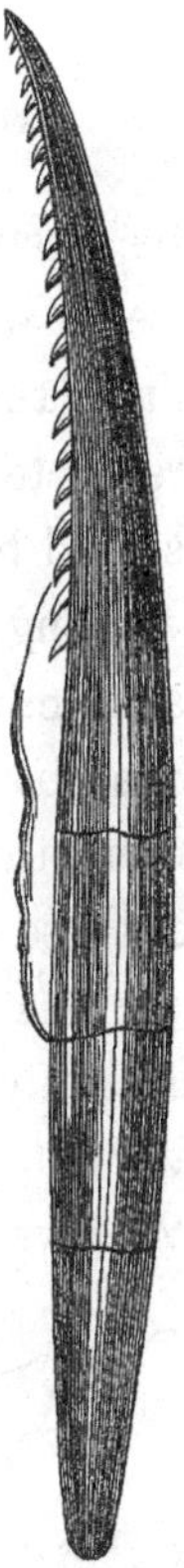

Large fossil spines, longitudinally grooved, have been found associated with the teeth of the extinct cestraciont genus (*Ptychodus*) of the chalk formations.

In the tertiary formations, the fossil spines present for the most part the generic characters of those of existing Plagiostomes—*e. g., Spinax, Trigon,* and *Myliobates ;* but one form, found in the eocene beds near Paris, is the type of the extinct genus *Aulacanthus* of Agassiz.

Fig. 26.
Hybodus subcarinatus.
(Wealden.)

The teeth of the plagiostomous fishes—viz., sharks (*Squalidæ*), rays (*Raiidæ*), and Cestracionts, are very numerous, and, being attached only by ligament to the membrane of the mouth, they must soon fall off in the decomposition of the dead fish, become scattered abroad by the movements of the body through the action of the waters, and sink into the sediment.

FAMILY I.—CESTRACIONTIDÆ.

(*Port-Jackson Shark.*)

The existing genus which has thrown most light upon the fossil teeth which have thus become imbedded in the oceanic deposits of the palæozoic and mezozoic periods, is the *Cestracion*, now restricted to the Australian and Chinese seas, where it is represented by two or three species, and suggests the idea of a form verging towards extinction. It formerly flourished under a great number of varied generic or family modifications, represented by species, some of which attained dimensions far exceeding the largest known living Cestracions. The dentition of these fishes is adapted to the prehension and mastication of crustaceous and testaceous animals ; they are of a harmless, timid character ; and have the before-described denticulate dorsal spines given to them as defensive weapons. Fig. 27 gives a side view of the upper and lower jaws of

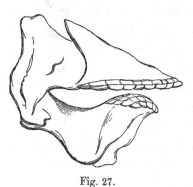

Fig. 27.

Cestracion Philippi (recent).

the "Port Jackson shark," showing the oblique disposition of the large crushing teeth, which cover like a pavement the working borders of the mouth. The anterior teeth were small and pointed. Behind the cuspidate teeth the five consecutive rows of teeth progressively increase in all their dimensions, but principally in their antero-posterior extent. The sharp point is converted into a longitudinal ridge traversing a convex crushing surface, and the ridge itself disappears in the largest teeth. As the teeth increase in size, they diminish in number in each row. The series of the largest teeth includes from six to seven in the upper, and from seven to eight in the lower jaw. Behind this row the teeth, although preserving

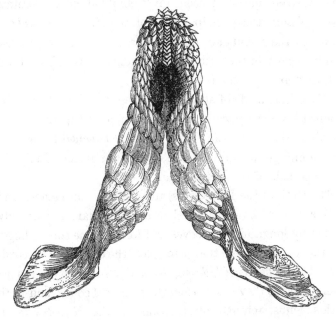

Fig. 27 a.

Upper jaw and teeth of Port-Jackson Shark (*Cestracion*), half nat. size.

their form as crushing instruments, progressively diminish in size, while at the same time the number composing each row decreases. From the oblique and apparently spiral disposition of the rows of teeth, their symmetrical arrangement on the opposite sides of the jaw, and their graduated diversity of form, they constitute the most elegant tesselated covering to the jaws which is to be met with in the whole class of fishes.

The modifications of the form of the teeth above described, by which the anterior ones are adapted for seizing and retaining, and the posterior for cracking and crushing alimentary substances, are frequently repeated, with various modifications and under different conditions, in the osseous fishes. They indicate, in the present cartilaginous species, a diet of a lower organized character than in the true sharks ; and a corresponding difference of habit and disposition is associated therewith. The testaceous and crustaceous invertebrate animals constitute most probably the principal food of the Cestracion, as they appear, by their abundant remains in secondary rocks, to have done in regard to the extinct Cestracionts, with whose fossil teeth they are associated.

From their mode of attachment, these teeth would become detached from the jaws of the dead fish, and dispersed in the way above described ; and it is by such detached fossil teeth that we first get dental evidence of the Cestraciont family in former periods of the earth's history.

The teeth of the Hybodonts are conical, but broader and less sharp than those of true sharks. The enamel is strongly marked by longitudinal grooves and folds. One cone is larger than the rest, and called the "principal ;" the others are " secondary." In one genus (*Cladodus*, Ag.), the secondary cones go on enlarging as they recede from the principal cone ; and teeth of this genus, referred by Eichwald to the *Hybodus longiconus*, have been discovered in the old red sandstone in the vicinity of Petersburg.

In the *Orodus*, the cones are more compressed, trenchant,

Fig. 28.
Orodus cinctus (tooth).
(Carboniferous.)

and distinct from the body of the tooth than in *Hybodus ;* but they present a principal and secondary cones. Fig. 28 is a tooth of the *Orodus cinctus* (Ag.), from the carboniferous beds near Bristol. The O. *porosus* and O. *compressus* are from deposits of similar age near Armagh.

If fig. 29 be compared with fig. 27 *a*, it would seem as if the several teeth of each oblique row in *Cestracion* had been

welded into a single dental mass in *Cochliodus*, the proportions
and direction of the rows being closely analogous. Whether
in *Cochliodus*, there were any small anterior prehensile teeth,
is hypothetical; the large crushing dental plates must have

been admirably adapted to crack and
bruise the shells of mollusks and crustaceans. The *Cochliodus contortus*
(Ag.) (fig. 29) has been found in the
carboniferous formations near Bristol
and Armagh, and the genus is peculiar
to that geological period.

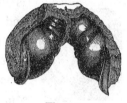

Fig 29.
Cochliodus contortus, Ag.
(Carboniferous.)

Teeth referable to the genus *Hy-*
bodus occur in all the secondary rocks from the trias to the
chalk inclusive.

A form of tooth which more closely resembles the crush-
ing-teeth of Cestracion, is that on which the genus *Acrodus* is

founded, and which also ranges from
triassic strata to the upper chalk of
Maestricht. The species here selected
(fig. 30) is the *Acrodus nobilis*, from
the lias of Lyme Regis. The upper
figure shows the grinding surface,
which, from its finely and transversely
striated character and dark colour, has
suggested to the quarrymen the name
of "fossil leeches." The older fos-

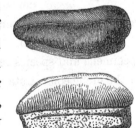

Fig. 30.
Acrodus nobilis (tooth).
(Lias.)

silists regarded these teeth as petrified Vermes; but the structure, as shown by the microscope, is closely similar to that of
the teeth of Cestracion.* Portions of the jaw of the *Acrodus*
have been discovered which show that these teeth were arranged, as in *Cestracion*, in oblique rows, with at least seven
teeth in each row. *Acrodus lateralis* is a muschelkalk fossil,
A. hirudo a Wealden fossil, and *A. transversus* a cretaceous

* See Owen's Odontography, vol. i., p. 54, pls. 14 and 15.

fossil. No tooth referable to the genus has been found in any tertiary stratum.

The genus *Ptychodus* is founded on teeth usually of large size, and of a more or less square form (fig. 31). The crown

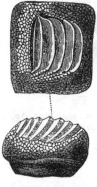

is deeper than the root, which is obtuse and truncate. The enamelled summit of the crown is granulate at the margin, and raised in the middle into an obtuse eminence, disposed in large transverse, parallel, sometimes wavy and rather sharp ridges. With teeth of this form are sometimes found others of smaller size, with more convex rounded crowns, doubtless forming the extremes of the multiserial pavement which, as in modern sharks and rays, covered the broad jaws of the Ptychodonts. Large dorsal spines have been found so associated with the above-described teeth as to indicate the affinity of the *Ptychodus* to the Cestraciont family of sharks. All the specimens and species referable to this genus have been found in the cretaceous strata.

Fig. 31.
Ptychodus latissimus.
(Chalk).

FAMILY II.—SQUALIDÆ.
(*Sharks.*)

The well-marked, saw-shaped tooth (fig. 32), so closely resembles the lower jaw-teeth of the sharks, called "grisets" by the French (*Notidanus*, Cuv.), as to be referred to that genus by Agassiz. Such teeth nevertheless occur in strata of oolitic age (*Notidanus Münsteri*, Ag., fig. 32). Other species—*e. g., N. pectinatus*—are found in the chalk of Kent; and *N. serratissimus*, in the eocene clay at Sheppy.

The tooth (fig. 33) on which Agassiz has founded the genus *Corax*, indicates by its close resemblance to those of *Carcharias*, its relationship with the true sharks (*Squalidæ*). Most of the species of *Corax*, including *C. falcatus*, are cretaceous ; a few are tertiary : all are extinct.

Fig. 32.

Notidanus Münsteri.
(Upper Oolite.)

Fig. 33.

Corax falcatus.
(Chalk.)

Fig. 34.

Galeocerdo aduncus.
(Miocene.)

Another form of shark's tooth, deeply notched at one margin, and with the rest of the border finely denticulate, resembles more that of the "topes" or gray sharks (*Galeus*, Cuv.), and is referred by Agassiz to the genus *Galeocerdo*. The species are found in both the cretaceous and tertiary formations ; *Galeocerdo aduncus* (fig. 34) is from the miocene of Europe and America. In the same tertiary series are found the teeth of the *Hemipristis serra*, Ag. (fig. 35).

Fig. 35.

Hemipristis serra.
(Miocene.)

Odontaspis (Ag.), presents a form of tooth most like that in the blue sharks (*Lamna*) of the present seas. Species of *Odontaspis* occur in the cretaceous and tertiary beds. The *O. Hopei* (fig. 36) is from the London clay of Sheppy. It indicates a very destructive and formidable species of shark.

Teeth shaped like those of the white sharks (*Carcharias*), but solid and usually of large size, are referred to the genus *Carcharodon*. One of these teeth, from miocene beds, Malta, in the Hunterian Museum, London, measures 5 inches 10 lines at its longest side, and 4 inches 8 lines across the base. By the side of it is placed a tooth of an existing *Carcharias*, 2 inches 3 lines at its "longest side," from a shark which measured 20 feet in length. If the tooth of the fossil *Carcharodon* bore the same

Fig. 36.

Odontaspis Hopei.
(Eocene.)

proportion to the body of the fish, this must have been about sixty feet in length.* Teeth of *Carcharodon* have been obtained from the Red Crag of Suffolk, measuring upwards of six inches in length. The microscopic structure of the teeth in sharks is illustrated by the longitudinal section of a fossil from Sheppy, showing the outer hard layer of "vitrodentine," and the "vaso-dentine" forming the body of the tooth. With these fossil teeth of sharks are found, though

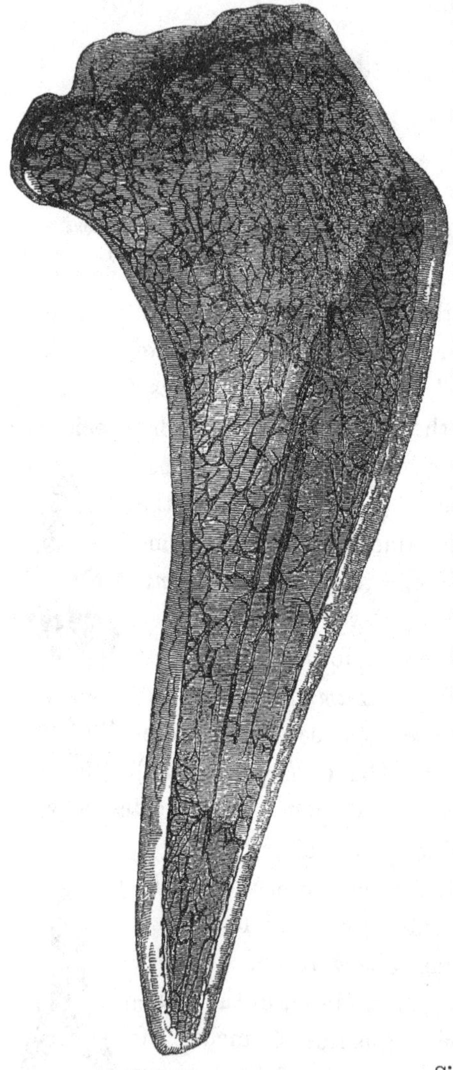

Fig. 36 a.

Magn. section of a tooth of a fossil Shark (*Lamna*).

Fig. 37.

Side view and back view of the body of a vertebra of a Shark, *Lamna* or *Odontaspis*. (London clay, Sheppy.)

sparingly, in both the cretaceous and tertiary beds, petrified

* See the Author's Catalogue of Fossil Reptilia and Pisces, in Mus. R. Coll. of Surgeons. Lond. 4to, 1854, p. 124.

bodies of vertebræ, showing by their extreme shortness in comparison with their breadth, by their bi-concavity, and the fissures on the external surface (as shown on the lower figure of cut 37), that they belonged to a shark closely allied to the Porbeagle (*Lamna,* Cuv.)

FAMILY III.—RAIIDÆ.
(*Rays.*)

Fossil evidences of this peculiar family of cartilaginous fishes have been discovered in oolitic (*Spathobatis Belemnobatis*), cretaceous, and tertiary formations, and consist of defensive spines, dermal tubercles, and teeth, but chiefly the latter. The most peculiar and distinctive modifications of the dental system, presented by the eagle-rays (*Myliobatidæ*) are unequivocally shown by fossils of the tertiary formations, and have not been found in earlier strata.

The teeth of the rays are in general more numerous than those of the sharks ; they have less mobility, are more closely impacted, and in some cases are laterally united together by fine sutures, so as to form a kind of mosaic pavement on both the upper and lower jaws. The *Myliobates,* or eagle-rays, which present the last-mentioned condition, unique in the vertebrate sub-kingdom, have large and massive teeth (fig. 38) ; but in the rest of the present family of cartilaginous fishes, the teeth (fig. 38) are remarkable for their small size as compared with those of the sharks. The teeth in some species of rays are adapted for crushing, but in others they have the middle or one of the angles of the crown produced into a sharp point. In all genera of the ray tribe, whatever the diversity of size and shape of the teeth, they are placed in several rows, and succeed each other uninterruptedly from behind.

The modification of the plagiostomous type of teeth, for the purpose of crushing alimentary substances, is most complete in the genus *Myliobates.* A view of this armature of the mouth,

I

as seen from behind in the *Myliobates aquila*, is given in fig. 38. Both jaws are covered with a pavement of broad teeth,

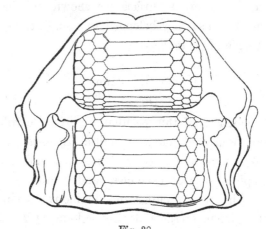

Fig. 38.

Jaws and teeth of an Eagle-Ray (*Myliobates aquila*).

having a flat grinding surface. To the genus *Myliobates*, as now restricted, certain fossils from the London clay of Sheppy (*Myliobates toliapicus*, Ag., fig. 39) belong.

In *Zygobates* (fig. 40), the middle series of teeth is less

Fig. 39.
Myliobates toliapicus
(Eocene, Sheppy).

Fig. 40.
Zygobates Woodwardi
(Miocene).

broad; and a narrower series is interposed between the middle and the small lateral teeth. Existing rays showing this modification are found in Brazilian seas; fossil teeth of this genus, *e. g.*, *Zygobates Woodwardi*, Ag. (fig. 40), occur in the

tertiary crag (probably miocene) of Norfolk, and in the miocene mollasse of Switzerland.

When the teeth form broad transverse undivided plates, as
in fig. 41, they characterize the genus *Ætobates*. Fossils of
this genus occur in the English eocenes and the Swis smol-
lasse.

In the "crag" of Norfolk and Suffolk, and in marine pliocene beds, fossils have been found which closely resemble

Fig. 41.   Fig. 41 a.
Ætobates subarcuatus   *Raia clavata*
(Eocene, Bracklesham).   (Dermal spines).

the osseous and spinigerous plates that beset the skin of
the kind of ray called "thornback" (fig. 41 a), and which
indicate the existence of a pliocene species allied to the *Raia
clavata*.

Thus we obtain evidence of fishes of the plagiostomous
order in the marine deposits of every formation from the
upper Silurian beds to the present period. But none of the
palæozoic fossils are referable to any existing genus. A
few only of the mezozoic Plagiostomes, and those chiefly
from the chalk, are so determinable : but most of them
belong or are allied to a family (*Cestraciontidæ*), now nearly
extinct. The evidence of the generic forms of Plagiostomes
characteristic of the present time become common only in
the tertiary periods. No fossil species is the same with
any existing one.

ORDER II.—HOLOCEPHALI.

(*Chimæroid Fishes.*)

Char.—Jaws bony, traversed and encased by dental plates ;
endo-skeleton cartilaginous ; exo-skeleton as placoid gran-
ules ; most of the fins with a strong spine for the first ray ;
ventrals abdominal ; gills laminated, attached by their
margins ; a single external gill aperture.

To judge from the paucity of existing representatives of this
order of cartilaginous fishes, it would seem, like the Cestracionts,
to be verging towards extinction.   One genus (*Chimæra*, Linn.)
is founded on a single known species of the northern seas
called " king of the herrings" (*Chimæra monstrosa*) ; another
genus (*Callorhynchus* of Gronovius) is represented by two
known species in the Australian and Chinese seas.   The only
parts of chimæroid fishes likely to be fossilized are the jaws
and spines.   The bony and dental substances are so combined
in the more or less beak-shaped jaws, that they characterize
the order, and are never found separate.   It is chiefly on such
fossil mandibles, and portions of them, that the evidence of the
Holocephali in former geological periods rests.   These singular
fishes ranged, under different generic and specific modifications,
from the bottom of the oolitic series to the present period.

Genus CHIMÆRA.—The premaxillary teeth, one in each
bone, are oblong, about twice as high as they are broad, and
terminate below in a transverse trenchant edge ; they present,
exteriorly, vertical columns of alternately harder and softer
substance, occasioning a notched margin when worn by use ;
interiorly, they have oblique laminæ which do not extend to
the margin.   The maxillary dental plates, one in each bone,
are triangular, and present a broad surface to the lower jaw.

Genus ISCHIODUS, Egerton.—Each upper maxillary has
four dental columns ; the lower jaw is less produced and

deeper than in *Edaphodus*. Of this genus, *I. Johnsoni* is from the lias of Dorsetshire ; *I. Egertoni* from the Kimmeridge of Shotover ; and *I. Townshendi*, a magnificent species, from the Portland stone. Two species (*I. Agassizii* and *I. brevirostris*) are from the cretaceous beds ; at which period the genus appears to have perished.

Genus GANODUS, Egerton (including *Ganodus* and *Psittacodus* of Agassiz).—This genus is exclusively represented by species from the oolitic slate of Stonesfield—*e. g.*, *G. Bucklandi, G. Colei, G. Owenii.*

Genus EDAPHODUS, Egerton (including *Edaphodon* and *Passalodon* of Buckland).—Each upper maxillary has three dental columns ; the lower jaw is more produced, but less deep, than in *Ischiodus :* the premaxillary dental mass consists of five vertical and slightly bent series of oblique and curved transverse plates ; the median and longest series being strengthened by a supplementary dental column behind : it represents the genus *Passalodon* of Buckland. The large *E. Sedgwickii* is from the greensand near Cambridge ; the still larger *E. gigas* from the chalk of Kent and Sussex. The ichthyodorulite called *Psittacodus Mantelli* by Agassiz may be the dorsal spine of this species. Three species, including the *E. Bucklandi,* are found in the eocene of Bagshot and Bracklesham ; and one species (*E. helveticus*) is from the mollasse of Switzerland.

Genus ELASMODUS, Egerton.—Each upper maxillary has three dental columns, but the dentine is confluent, "being rolled round like a scroll on the substance of the bone, one edge forming the margin of the tooth, the other buried deep in its centre.[*] The premaxillary has a thin incurved scalpriform tooth, rounded at the cutting edge, of a lamellate structure, with a columnar arrangement of the plates, which are juxtaposed. This genus is exclusively represented by species —*e. g.*, *E. Hunteri*—from the London clay of Sheppy.

[*] Egerton, Proc. Geol. Soc., May 12, 1847.

ORDER III.—GANOIDEI.

Char.—Endo-skeleton in some osseous, in some cartilaginous, in some partly osseous and partly cartilaginous ; exo-skeleton formed by enamelled bones ; fins usually with the first ray a strong spine.

Sub-Order 1.—PLACOGANOIDEI.

(*Ostracostei*, Ow. Cat. Mus. Coll. of Surgeons, 4to, 1854, p. 166.)

Char.—Endo-skeleton cartilaginous, or retaining the noto-chord ; head and more or less of the trunk protected by large ganoid, often reticulated, plates ; heterocercal.

The last term signifies a form and structure of tail illus-trated by fig. 42, and to be seen in the sharks, dog-fishes, and sturgeons of the present day : it results from a prolongation

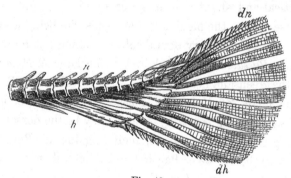

Fig. 42.
Heterocercal tail (*Lepidosteus osseus*).

of the vertebral column into the upper lobe *dn*, producing an unsymmetrical form of the caudal fin, which is contrasted with the symmetrical form of the same fin presented by most fishes of the present day, and illustrated by the *Leptolepis spratti-formis* (fig. 56, p. 144), in which the vertebral column termi-nates at the middle of the base of the caudal fin. There are a few exceptional intermediate forms and structures of this fin.

Fam. *Placodermata*.—The fossil remains of the singular

fishes of the extinct order *Placoganoidei* were first discovered about 1813, in formations of the " old red" or Devonian age in Russia, and are preserved in museums at St. Petersburg and Dorpat. The relation of these specimens to the class of fishes was first announced by Professor Asmuss,* and shortly after, the generic names *Asterolepis* and *Bothriolepis* were invented by Professor Eichwald,† to express certain modifications of the external surface of portions of the ganoid plates, subsequently recognised as constituting the buckler of the fore-part of the extinct fishes. In September 1840 Mr. Hugh Miller submitted to the geological section of the British Association at Glasgow the first discovered specimens, affording a recognizable idea of the form of one of these "old red" fishes, and for this form Professor Agassiz assigned the generic name *Pterichthys* (*pteron*, a wing, *ichthys*, a fish). Although, therefore, the term *Asterolepis* had been attached to a fragment of the cuirass of this fish a few months previously, yet, as no recognizable generic characters were associated with such name, and as *Asterolepis* has been applied also to other genera —*e. g.*, *Homosteus* and *Heterostius* of Asmuss—the example of British palæontologists will be here followed, in retaining the name *Pterichthys* for the present genus. " Of all the organisms of the system," wrote the lamented Hugh Miller in his work on the *Old Red Sandstone*, " one of the most extraordinary, and the one in which Lamarck would have most delighted, is the *Pterichthys*, or winged fish, an ichthyolite which the writer had the pleasure of introducing to the acquaintance of geologists nearly three years ago (1840), but which he first laid open to the light about seven years earlier" (1833).

Genus PTERICHTHYS (fig. 43).—The head and the anterior

* Bulletin Scient. par l'Acad. Imp. des Sciences de St. Petersburg, 1840, t. vi., p. 220.

† Ibid, t. vii., p. 78, communicated March 13, 1840. Dr. Fleming had recognized certain fossil scales as those of fishes in the " Old Red " of Fifeshire, in 1827.

half of the trunk are defended by ganoid plates—*i. e.*, plates composed of a hard bone coated with enamel ; those of the trunk forming a buckler composed of a back plate (fig. 43) and breast-plate (fig. 44), articulated together at the sides. The rest of the trunk was defended by small ganoid scales, flexible, like scale-armour, and bore a small dorsal fin (fig. 43, *d*), and a terminal heterocercal fin, very rarely displayed in fossil specimens. The pectoral spines, *c*, are formed of ganoid material, like the buckler. The armour of the head, or helmet, appears to have been articulated by a movable joint to the trunk-buckler. One of the few existing ganoid fishes (*Lepidosteus*) is remarkable for the degree in which the head moves upon the trunk. The component dermal plates of the helmet correspond in some measure with the position of the cranial bones in osseous fishes, but not sufficiently to sanction the application to them of corresponding names. They are indicated by figures in the cut 43 : $_2$ is the front terminal or *rostral* plate ; it is followed in the median line by four other plates in the following order :—$_4$, *premedian ;* $_6$, *median ;* $_8$, *postmedian ;* $_{10}$, *nuchal ;* $_3$ is the *marginal,* and $_7$ the *postmarginal ;* $_5$ is the *prelateral,* and $_9$ the *postlateral.* The dorsal shield of the trunk-cuirass is composed of two mid-plates and two on each side. $_{12}$ is the " *dorsomedian,*" $_{14}$ the *post-dorsomedian ;* $_{11}$ is the *dorsolateral,* $_{13}$ the *post-dorsolateral.* The ventral shield (fig. 44) consists of one mid-plate and two side-plates : $_{15}$ is probably a part of the cephalic shield or of the mandible : $_{19}$ is the *ventrolateral,* $_{21}$ the *postventrolateral ;* the small supplementary plate marked $_{17}$ is usually confluent with $_{19}$; $_{16}$ is the *ventromedian* plate ; its margins are bevelled off and overlapped by the lateral plates.

The pectoral spines (fig. 43, *c*) are long and slender, and consist of two principal segments, both defended by finely tuberculated ganoid plates, like those of the head and trunk. From their form, they would seem to have served to aid the fish in shuffling along the sandy bottom or bed, if left dry at low-water. The fins attached to the flexible part of the body

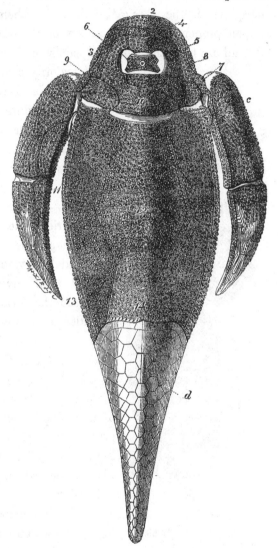

Fig. 43.
Pterichthys Milleri, dorsal surface (Devonian), after Pander.

indicate a certain power of swimming, though not with any great rapidity ; they include a small dorsal and a pair of

ventrals—these latter were first observed by Sir P. Egerton.
The jaws are small, and possess confluent denticles.

The type-species is the *Pterichthys Milleri;* others have
been based upon proportions of the cuirass, of the pectorals,
and the tail; all are from the "old red sandstone," and the
great majority have been
found in the Devonian
strata of Ross-shire, Caithness, and other Scotch
localities.

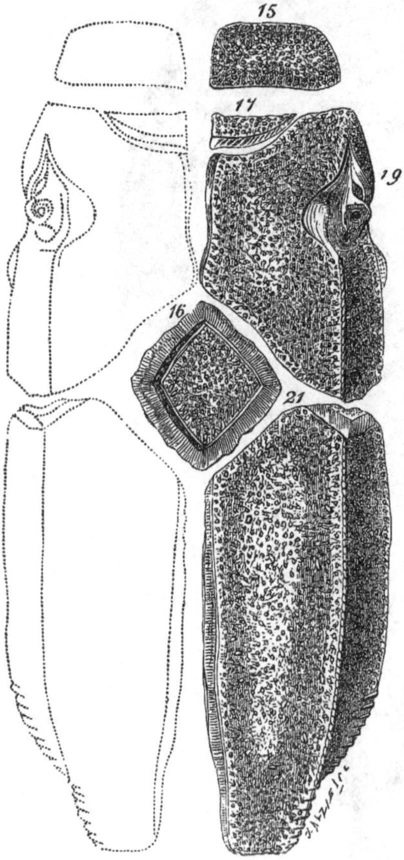

Genus CEPHALASPIS
(*kephale,* head ; *aspis,*
buckler).—In this genus
the posterior angles of the
shield-shaped helmet are
produced backward in a
pointed form, giving to
the head the form of a
" saddler's knife ;" in other
respects the genus closely
resembles *Pterichthys.*

Mr. D. Page has re-
cently acquired specimens
of *Cephalaspis* from Lan-
arkshire tile-stones, forming the base of the Devo-
nian system, which show
a dorsal fin, pectoral fins,
and a large heterocercal
fin, besides a well-marked
capsule of the eye-ball.

Fig. 44.

Pterichthys; Plastron or Ventral Shield.
(Devonian), after Pander.

Cephalaspis Murchisoni occurs in the passage beds from the
Silurian to the Devonian systems.

Genus PTERASPIS.—The buckler of *Pteraspis truncatus* has
been found in a Silurian stratum below the Ludlow bone bed ;

it is the earliest known indication of a vertebrate animal. *Pteraspis Lloydii* occurs in the lower "old red" of Britain.

Genus COCCOSTEUS (*kokkos*, berry ; *osteon*, bone).—If a heterocercal fin were added in outline to the restoration of the fish of this genus, given in fig. 45, a correct idea would be given of the "old red" fossil, which, in the progress of its reconstruction, has suggested so many strange notions of its nature and affinities.

The helmet and cuirass are firmly united, and there is no trace of the jointed appendages, like pectoral fins, which characterize *Pterichthys*. The unprotected part of the trunk shows an ossification of the neural and hæmal spines, and of their appendages, the rays of a "dorsal" and "anal" fin ; and, by the analogy of *Cephalaspis*, the tail was most probably terminated by an unequal-lobed fin. The lower jaw is composed of two rami, loosely connected at the symphysis ; so that, when displaced, as they commonly are in crushed fossil specimens, they gave the notion of the fish being provided with laterally-working jaws, like those of the lobster. But, in reality, the jaw worked vertically upon a fixed upper jaw ; both jaws being provided with from ten to twelve teeth on each side, anchylosed to the bone.

An under-view of the cephalothoracic buckler of *Coccosteus*, according to Dr. Pander's restoration, is given in fig. 46, showing the sutures of most of the cephalic plates, and the external surface of the plates of the plastron. 9, *rostral plate ;* 7, *premedian ;* 5, *median ;* 8, *prelateral ;* 6, *lateral ;* 16 and 24, *the suborbital bone ;* 15, *preventromedian ;* behind the lozenge-shaped *ventromedian*, and on each side, are (22) the *preventrolateral* and (20) the *post-ventrolateral*. The same figures mark the above plates in the side view (fig. 45), with the addition of (12) the *dorsomedian* and (14) the *post-dorsomedian*.

The blank space between the neural (*n*) and hæmal (*h*) spines of the fossil endo-skeleton indicates the position of the

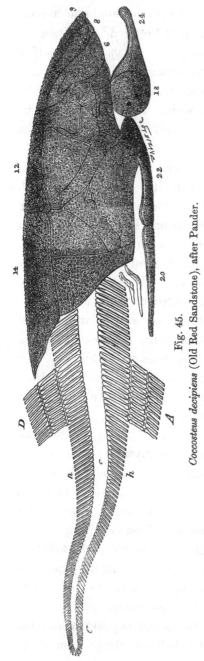

Fig. 45.

Coccosteus decipiens (Old Red Sandstone), after Pander.

soft "notochord" (*c*), which has been dissolved away. The cylindrical gelatinous body, so called (in Latin *chorda dorsalis*) pre-exists to the formation of the bony bodies of the vertebræ in all vertebrate animals ; and the development of those bodies seems never to have gone beyond this embryonal phase in any palæozoic fish ; such fishes are accordingly termed "notochordal," as retaining the notochord.

There are but two genera of existing fishes which manifest, when full grown, such a structure, associated with ossified peripheral elements of the vertebræ—viz., the *Protopterus* of certain rivers of Africa,* and the *Lepidosiren* of certain rivers of South America. Those fishes alone would, if fossilized, present the appearance of the vertebral column shown in fig. 45, and which characterizes all the oolitic fossil ganoid fishes (see figs.

* See Linnæan Transactions, vol. xviii. ; and Proceedings of the Linnæan Society, April 2, 1839.

54 and 55). This persistence in palæozoic and most mezozoic

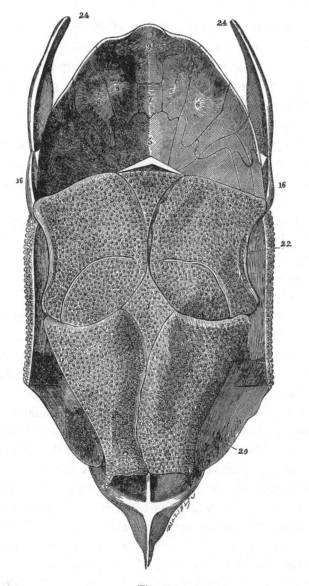

Fig. 46.
Cephalothoracic buckler, ventral aspect, *Coccosteus decipiens* (Devonian).

fishes of an embryonic vertebral character, transitory in nearly all existing fishes, significantly testifies to a principle of " progression."

The external "ganoid" surface of the buckler-plates of *Coccosteus* is ornamented with small hemispherical tubercles ; whence the generic name, signifying "berry-bone." The similarity of this ornamentation to that of the plates of the buckler in some tortoises, led to the belief, when the coccosteal plates were first found, of their being evidence of the chelonian genus *Trionyx* in Devonian beds. Passing notions also got into print of the crustaceous affinities of *Coccosteus ;* whence the trivial name of the type-species *decipiens*, or the " deceiving" *Coccosteus.*

Strange as seem the forms and structure of the placoganoid fishes of the " old red " period, there are not wanting existing species which throw much truer light on their nature than any existing *Chelonia* or *Crustacea.* The singular little family of " trunk-fishes " (*Ostracionidæ*) shows species in which the body is inclosed in a more or less quadrangular cuirass, composed of suturally-articulated ganoid plates, which are usually tuberculated on the external surface, and with the angles prolonged into spines in some species, like those of the helmet of *Cephalaspis.* The caudal part of the trunk protrudes from the back opening of the cuirass, as in *Coccosteus* and *Pterichthys*, and ossification of the endo-skeleton is incomplete. The species of this family are for the most part natives of seas of tropical or warm temperate latitudes.

In another family of existing fishes, called " Siluroids," there are species in which the broad cranial bones, connate with dermal ossifications, form a helmet to the head, whilst one or two dermal spine-bearing bones combine to form the part called " buckler " by Cuvier.* In the genus *Doras*, the lateral line is armed with bony ganoid plates ; and in *Callichthys*, these biserial plates are developed so as to incase the

* Histoire des Poissons, tom. xii.

whole body. But generally, as in *Pimelodus*, the hinder muscular part of the trunk is undefended, as in *Coccosteus*. The ganoid plates of the head and back shields are fretted with rows or ridges of confluent tubercles, radiating from the centre to the circumference of the plate, whilst the inner surface is smooth, as in *Coccosteus* (fig. 46) ; and, moreover, the dorsal plate in existing Siluroids sends down a median ridge from its inner surface, like that from the " dorso-median " plate in *Coccosteus*. The point of resemblance to be mainly noticed, however, is the contrast furnished by the powerful armature of the head and back with the unprotected nakedness of the posterior portions of the creature—a point specially noticeable in *Coccosteus*, and apparent also, though in a lesser degree, in some of the other genera of the old red, such as the *Pterichthyes* and *Asterolepides*. " From the snout of the *Coccosteus* down to the posterior termination of the dorsal plate the creature was cased in strong armour, the plates of which remain as freshly preserved in the ancient rocks of the country as those of the *Pimelodi* of the Ganges on the shelves of the Elgin museum ; but from the pointed termination of the plate immediately over the dorsal fin to the tail, comprising more than one-half the entire length of the animal, all seems to have been exposed, without the protection of even a scale ; and there survives in the better specimens only the internal skeleton of the fish and the ray-bones of the fins. It was armed, like a French dragoon, with a strong helmet and a short cuirass ; and so we find its remains in the state in which those of some of the soldiers of Napoleon's old guard, that had been committed unstripped to the earth, may be dug up in the future on the fatal field of Borodino, or along the banks of the Dwina or the Wap. The cuirass lies still attached to the helmet, but we only find the naked skeleton attached to the cuirass. The *Pterichthys* to its strong helmet and cuirass added a posterior armature of comparatively feeble scales, as if, while its upper parts were shielded with plate-armour, a lighter covering of ring or scale armour sufficed for the less vital parts beneath. In the *Asterolepis* the arrangement was somewhat similar, save that the plated cuirass was wanting. It was a strongly-helmed warrior in slight scale-armour ; for the disproportion between the strength of the plated head-piece and that of the scaly coat was still greater than in the *Pterichthys*. The occipital star-covered plates are, in some of the larger specimens, fully three-quarters of an inch in thickness, whereas the thickness of the delicately-fretted scales rarely exceeds a line.

" Why this disproportion between the strength of the armature in different parts of the same fish should have obtained, as in *Pterichthys* and *Asterolepis*, or why, while one portion of the animal was strongly armed, another portion should have been left, as in *Coccosteus*, wholly exposed, cannot of course be determined by the mere geologist. His rocks present him with but the fact of the disproportion, without accounting for it. But the natural history of existing fish, in which, as in the *Pimelodi*, there may be detected a similar peculiarity of armature, may perhaps throw some light on the mystery. In Hamilton's Fishes of the Ganges, the habitats of the various Indian species of *Pimelodi*, whether brackish estuaries, ponds, or rivers, are described, but not their characteristic instincts. Of the *Silurus*, however, a genus of the same great family, I read elsewhere that some of the species, such as the *Silurus Glanis*, being unwieldy in their motions, do not pursue their prey, which consists of small fishes, but lie concealed among the mud, and seize on the chance stragglers that come in their way. And of the *Pimelodus gulio*, a little strongly-helmed fish with a naked body, I was informed by Mr. Duff, on the authority of the gentleman who had presented the specimens to the Museum, that it burrowed in the holes of muddy banks, from which it shot out its armed head, and arrested, as they passed, the minute animals on which it preyed. The animal world is full of such compensatory defences ; there is a half-suit of armour given to shield half the body, and a wise instinct to protect the rest. Now it seems not improbable that the half-armed *Coccosteus*, a heavy fish, indifferently furnished with fins, may have burrowed, like the recent *Silurus Glanis* or *Pimelodus gulio*, in a thick mud, of the existence of which in vast quantity, during the times of the old red sandstone, the dark Caithness flagstones, the fœtid breccia of Strathpeffer, and the gray stratified clays of Cromarty, Moray, and Banff unequivocally testify ; and that it may have thus not only succeeded in capturing many of its light-winged contemporaries, which it would have vainly pursued in open sea, but may have been enabled also to present to its enemies, when assailed in its turn, only its armed portions, and to protect its unarmed parts in its burrow." *

Sub-Order 2.—LEPIDOGANOIDEI.

FAMILY I.—DIPTERIDÆ.

This family includes a few heterocercal fishes with a double anal as well as dorsal fin. The head is large and flattened ; the teeth subequal ; the scales perforated by small foramina ; the notochord persistent.

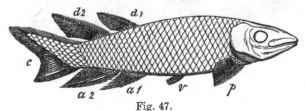

Fig. 47.

Dipterus macrolepidotus (Devonian).

In the genus *Dipterus* (fig. 47), the two dorsals, *d* 1, *d* 2, are opposite the two anals, *a* 1, *a* 2 : the ventrals, *v*, are in

* Hugh Miller, Rambles of a Geologist, p. 288.

K

advance of the first anal and first dorsal. The *Dipterus macrolepidotus* is characterized by the large size of its scales. Its remains are found in the old red sandstone of many localities of Scotland and England.

In the allied genus *Diplopterus* the vertical fins are opposite, but the dorsals are wider apart, and the teeth are larger and fewer. Four species have been recognised in the middle "old red" of Gamrie, Orkney, and Lethenbar, Ross-shire. Two species occur in the carboniferous series.

In the genus *Osteolepis* the vertical fins are alternate in position, the first dorsal being near the middle of the back. The teeth are sharp; not any of the species exceed a foot in length : they are all from the middle "old red."

FAMILY II.—ACANTHODII.

The species of this family are characterized by their very small scales : they are heterocercal and notochordal. There is a strong spine in front of each fin. The head is large ; the orbits approximate ; the mouth wide, formed chiefly by the maxillaries, and opening obliquely upwards, so that they

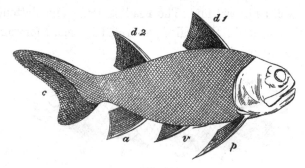

Fig. 48.
Diplacanthus striatus.

have somewhat the aspect of the *Uranoscopi.* They have many branchiostegal rays. The principal genera are from the old red sandstone, and are as follows :—*Cheiracanthus*, with a single dorsal situated in front of the anal ; *Acanthodes*, in which the dorsal is situated behind the anal ; and *Diplacanthus* (fig. 48), in which there are two dorsals.

The *Diplacanthus striatus* is found in the "old red" of Cromarty. In fig. 48, as in the other figures, *p* is the pectoral fin, *d* the dorsal, *v* the ventral, *a* the anal, and *c* the caudal. In this species the upper lobe of the caudal is much prolonged. The fin-spines in the *Acanthodii* were, like those of the recent dog-fish (*Spinax*), simply imbedded in the flesh, with their base, as it were, unfinished ; not provided, as in the Siluroids and other modern bony fishes, with a joint-structure.

Cheirolepis, with the minute scales of the family, has the dorsal behind the anal, but has no spine in any fin : the mouth is large, the teeth small and uniserial. Some species of the present family, *Acanthodes Bronnii, Ac. sulcatus*, existed in the seas of the carboniferous period.

FAMILY III.—CŒLACANTHI.

The species of this family are characterized by the hollowness of the rays or spines ; whence the name. The caudal fin has a peculiar structure, the vertebral column being continued into and beyond its middle part, supporting a kind of slender appendage between the two normal lobes. The species of the genus are most abundant in the Devonian and carboniferous formations ; but some occur in oolitic and even, if *Macropoma* be a true Cœlacanth, in cretaceous beds ; but all became extinct before the tertiary epoch.

Fig. 49.

Glyptolepis microlepidotus
(Devonian).

Glyptolepis had a heterocercal tail, with rounded scales, smooth externally, and with radiating compartments internally. The *G. microlepidotus*, of which a magnified view of the inner side of some

scales is given in fig. 49, occurs in the middle old red sandstone
of Scotland and England.

Phyllolepis is, as yet, known only by its large smooth or
concentrically furrowed scales, some of which are six inches
in diameter. *Ph. concentricus* occurs in the upper old red of
Clashbinnie; *Asterolepis* in the middle "old red" of Elgin;
Bothriolepis in the upper "old red" of Scotland and Russia;
and *Glyptopomus*, with the cranial bones sculptured externally,
in the upper "old red" of Dura Den.

FAMILY IV.—HOLOPTYCHIDÆ.

The type-genera of this family were first recognized and
characterized by the fossil scales, under the name *Holoptychius*
(Ag.), and by the fossil teeth, under the name *Rhizodus* (Ow.)
They include species which have left their remains in the
"old red" and the coal measures. They are nearly allied to
the Cœlacanthians, having, like them, but partially ossified
bones and spines, the interior of which retained their primitive
gristly state, and appear hollow in the fossils. The head was
defended by large externally sculptured and tuberculate ganoid
plates. The teeth consist of two kinds—small serial teeth,
and large laniary teeth—at long intervals; both kinds show-

ing the "labyrinthic" structure[*] at their base, which
is anchylosed to the jawbone.

The generic term *Rhizodus* is now retained for
the *Holoptychians* of the coal measures which have

Fig. 50.
Scale of *Holop-* more robust and obtuse serial teeth, and longer,
tychius nobil- sharper, and more slender laniaries, exemplified by
issimus (De-
vonian), half the *R. Hibberti.*[†] Species of true *Holoptychius*—
nat. size.   *e.g., H. giganteus* (Ag.), *H. nobilissimus* (Ag.), occur
in the old red sandstone. A noble specimen of the latter
species, 2 feet 6 inches in length, discovered in the old red
sandstone at Clashbinnie, near Perth, is now in the palæonto-

---

[*] Owen's "Odontography," Plate 63 B, fig. 1.   [†] Ib., Plates 35 and 36.

logical series of the British Museum. It is chiefly remarkable for the size and bold sculpturing of the ganoid scales (fig. 50).

Large fossil teeth, with the more complex " dendritic" disposition of the tissues, characterize a genus (*Dendrodus*), most probably of the Holóptychian family. The complexity is produced by numerous fissures radiating from a central mass of vasodentine, which more or less fills up the pulp-cavity of the seemingly simple conical teeth of this genus. Fig. 50 *a* is one of these fossil teeth of the natural size—*a*, a transverse section ; and fig. 50 *b*, a reduced view of a portion of the

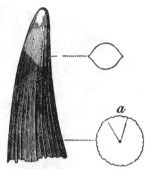

Fig. 50 *a*.

Tooth of *Dendrodus biporcatus* (nat. size).

same section enlarged twenty diameters. Thus magnified, a central pulp-cavity of relatively small size, and of an irregular lobulated form, is discerned, a portion of which is shown at *p ;* this is immediately surrounded by transverse sections of large cylindrical vascular or pulp canals of different sizes ; and beyond these there are smaller and more numerous medullary canals, which are processes of the central pulp-cavity. In the transverse section these processes are seen to be connected together by a net-work of smaller vascular canals belonging to a coarse osseous texture, into which the pulp has been converted, and this structure occupies the middle half of the section. All the vascular canals were filled up by the opaque matrix. From the circumference of the central net-work straight pulp-fissures radiate at pretty regular intervals to the periphery of the tooth ; most of these fissures divide once, rarely twice, in their course—the division taking place sometimes at their origin, in others at different distances from their terminations, and the branches diverge slightly as they proceed. Each of the above pulp-canals or fissures is continued

from a short process of the central structure, which is connected by a concave line with the adjoining process, so that
the whole periphery of the transverse section of the central

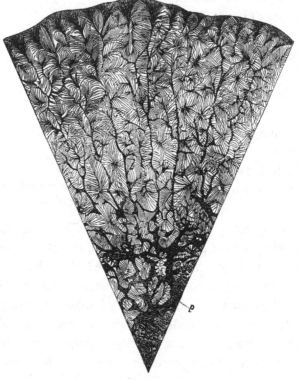

Fig. 50 *b*.
Magn. section of part of *Dendrodus biporcatus.*

coarse reticulo-vascular body of the tooth presents a crenate
outline.   From each ray and its primary dichotomous divisions short branches are sent off at brief intervals, generally
at right angles with the trunk, or slightly inclined towards the
periphery of the tooth.   These subdivide into a few short
ramifications like the branches of a shrub, and terminate in
irregular and somewhat angular dilations simulating leaves,
but which resolve themselves into radiating fasciculi of minute
dentinal tubes.   There are from fifteen to twenty-five or thirty-

six of these short and small lateral branches on each side of the medullary rays. The teeth of *Dendrodus* occur in the corn-stone beds of the " Old Red " at Scat-crag, near Elgin.

Such are some of the forms and structures of fishes that swam in the seas from which were deposited the sediment that has hardened into the " old red sandstones" of Great Britain, Russia, and other parts of the world. And in this process of consolidation the carcases of the fishes entombed in the primæval mud have had their share. For, just as a plaster-cast boiled in oil derives greater density and durability from that addition, so the oily and other azotized and ammoniacal principles of the decomposing fish operated upon the imme-diately surrounding sand so as to make it harder and more compact than the sediment not reached by the animal princi-ples. Accordingly it has happened that in the course of the upheaval and disturbance of old red strata, parts of it, broken up and exposed to the action of torrents, have been reduced to detritus, and washed away, with the exception of certain nodules, generally of a flattened elliptic form, which are harder than the surrounding sandstone. Such nodules form the bed of many a mountain stream in " old red sandstone " districts of Scotland. If one of these nodules be cleft by a smart and well-applied stroke of the hammer, the cause of its superior density will be seen in a more or less perfect speci-men of the fossilized remains of some animal, most commonly a fish.

But the placoganoid and lepidoganoid, heterocercal and notochordal, fishes of the Devonian epoch existed in such vast shoals in certain favourable inlets, that the whole mass of the sedimentary deposits has been affected by the decomposing remains of successive generations of those fishes. The De-vonian flagstones of Caithness are an instance. They owe their peculiar and valuable qualities of density, tenacity, and durability wholly to the dead fishes that rotted in their primitive constituent mud. From no other part of the world, perhaps, can a large flagstone be got, which a builder could set on its edge with assurance of its holding long together in that position. A great proportion of the county of Caithness formed, before its upheaval, the bottom of what may truly be termed a "piscina mirabilis." Yet there are minds, who, cognizant of the wonderful structures of the extinct Devonian fishes—of the evidence of design and adaptation in their structures—of the altered nature of the sediment surrounding them, and its dependence on the admixture of the decomposing and dissolved soft parts of the old fish—would deliberately reject the conclusions which healthy human reason must, as its Creator has constituted it, draw from such proofs of His operations. There are now individuals, one at least,* who prefer to try to make it be believed that God had recently, and at once, called into being all these phenomena; that the fossil bones, scales, and teeth, had never served their purpose—had never been recent—were never truly developed, but were created fossil; that the creatures they simulate never actually existed; that the superior hardness of the inclosing matrix was equally due to primary creation, not to any secondary cause; that the geological evidences of superposition, successive stratification, and upheaval were, equally with the palæontological evidences, an elaborate design to deceive and not instruct!

FAMILY V.—PALÆONISCIDÆ.

The Placoganoids, so richly represented in the Devonian epoch, disappear in the carboniferous one; the Lepidoganoids increase in number. In the present family they combine with rhomboid scales, a heterocercal tail, and jaws armed with numerous, minute, close-set, rather blunt teeth. The type-genus is *Palæoniscus* (fig. 51), species of which range throughout the carboniferous and Permian beds: it is characterized by

* See Omphalos, by P. H. Gosse, 8vo, 1858.

moderate-sized fins, the dorsal, *D*, being single, and opposite
the interval between the anal, *A*, and ventral, *V*, fins : each
fin has an anterior spine ; the fore-part of the head is obtuse.

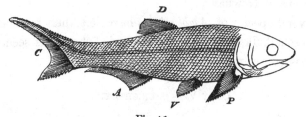

Fig. 51.
Palæoniscus (Permian).

In the *Palæonisci* from the coal formations at Burdie House,
near Edinburgh, the outer surface of the scales is striate and
punctate, *e.g.*, in *P. ornatissimus, P. striatus ;* but in the
Palæonisci of other British localities, and of the continental
and American coal formations, the scales are smooth, *e.g.*, in
P. fultus, from North America, *P. Duvernoyi* and *P. minutus*,
from the coal beds of Münster Appel.   In the *Palæonisci* from
the Permian copper schales and zechstein, the scales are striate
or punctate : the *Palæoniscus Freieslebeni* is the most common in
these beds, and was the first recognized species of the genus.   Of
this there are now forty known species, chiefly from carboniferous
and Permian eras : one from the Keuper beds at Rowington,
Warwickshire, appears to be the last
representative of the genus : it is
the *Palæoniscus superstes* of Egerton.

Fig. 52.
Scales of *Amblypterus striatus*
(Carboniferous).

   Amblypterus, with a geological
range like that of *Palæoniscus*, differs
in its shorter and deeper tail, and
larger body-fins, which are devoid of
anterior spines.   In fig. 52, *a* indi-
cates the outer surface of parts of two series of the rhomboidal
ganoid scales ; and *b* the inner surface of two scales, showing
the ridge produced at one end into a projecting peg, which fits

into a notch of the next scale, in the way that tiles are pegged together in the roof of a house. The species affording the above structure is the *Amblypterus striatus* from the coal-formations at Newhaven.

Several species of *Amblypterus* have left their remains in the muschelkalk, at which triassic period the genus seems to have passed away.

FAMILY VI.—SAURICHTHYIDÆ.

Magnificent species of heterocercal rhomb-scaled Ganoids, with large dispersed laniary teeth, sometimes of a size rivalling those of great Saurians, for which they have been mistaken, have left their remains in the coal strata at Carluke, near Glasgow, and other localities, and constitute the genus *Megalichthys* of Agassiz. The head is defended by strong ganoid plates, of a beautiful polish; the trunk-scales are usually granulate exteriorly. In this genus, as in the type of the family, the fulcra of the fin-rays are in two rows: all the known species of *Saurichthys* are triassic.

Pygopterus and *Acrolepis*, with fin-fulcra in a single row; *Eurynotus, Elonichthys, Plectrolepis, Graptolepis, Orognathus, Pododus, Acanthodes,* and *Diplopterus,* are carboniferous genera of Ganoids, with rhomboid scales. *Cœlacanthus, Isodus, Phyllolepis, Hoplopygus, Uronemus, Colonodus, Centrodus, Asterolepis, Psammosteus,* and *Osteoplax,* are genera of Ganoids with rounded scales, represented by species in carboniferous strata.

Of the above-named genera, *Acrolepis, Pygopterus, Palæoniscus,* and *Cœlacanthus,* continue to be represented in Permian beds; in which also are found species of the ganoid genera *Dorypterus, Holacanthodus,* and *Globulodus,* if the teeth on which the latter is based be not those of *Platysomus,* a pycnodont genus which is both Permian and carboniferous.

The formations of the mezozoic or secondary periods give

evidence of the full development of the ganoid order. In the lowest or "triassic" division this order is still represented by heterocercal and notochordal species belonging to some of the genera of the Permian period, as, *e. g., Cœlacanthus, Amblypterus,* and *Palœoniscus.* The genus *Placodus,* a supposed pycnodont fish of the muschelkalk, has been shown to be a conchivorous Saurian.*

In the oolitic division the heterocercal Ganoids are almost completely superseded by homocercal genera, which now, for the first time, appear on the stage of life ; but the ossification of the endo-skeleton is still incomplete. In the cretaceous series the Teleostian, or well-ossified, bony fishes, are numerous ; and here also first are seen fishes with the flexible "cycloid" or "ctenoid" scales, and of genera which continue to be represented by living species.

Of 33 genera of fishes in the lias, 4 only were represented in older strata, while the rest extend into the upper oolitic beds. Of these, 19 genera are Ganoids with rhomboid scales, and two (*Leptolepis* and *Gyrosteus*) have rounded scales.

FAMILY VII.—CATURIDÆ.

Homocercal rhombo-ganoids, with a short dorsal fin, and some of the teeth much larger than the rest and laniariform.

Genus CATURUS.—In this genus the jaws are armed with

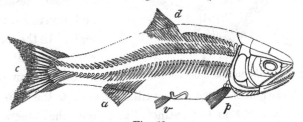

Fig. 53.
Caturus furcatus (Oolite, Solenhofen).

close-set, large, conical teeth ; the scales are delicate ; the fins

* Owen, in Phil. Trans. 1858, p. 169.

are of moderate size ; all the species are sub-homocercal * and
notochordal (fig. 53). The dorsal, *d,* is opposite the ventral, *v.*
One species of *Caturus* (*C. Bucklandi*) is from the lias ; but
the majority, like *C. furcatus,* are from the lithographic slates
of Solenhofen. The most recent known species (*C. similis*) is
from the chalk of Kent.

Pachycormus, Saurostomus, Sauropsis, Thryssonotus, and
Eugnathus, are liassic genera of the present family. It is
deemed by some Palæontologists to be represented at the
present day by the North American genus *Lepidosteus;* but
in this fish the notochord is converted into bony vertebral
bodies, united by ball-and-socket joints, and the tail is heterocercal.

FAMILY VIII.—PYCNODONTES.

The name of this group of ganoid fishes refers to the blunt
rounded form of the greater proportion of the teeth, especially
those attached to the palate and hind alveolar part of the
lower jaw : the few
anterior teeth are
small and sub-prehen-
sile ; but the whole
dentition bespeaks
fishes adapted to feed
on small testaceous
and crustaceous ani-
mals. In the modern
" Sea Breams " (Spa-
roids), with an analo-

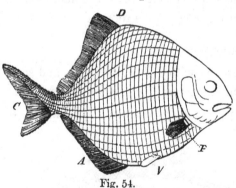

Fig. 54.

Platysomus gibbosus (zechstein of Mansfield).

gous dentition, the two premaxillaries oppose the two pre-
mandibulars, but in the extinct Pycnodonts the vomer, as in
Anarhichas, opposes its pavement of teeth to that of the two

---

* By this term is meant a symmetrical shape of the tail fin, with an unsymmetrical development of the supporting spines, the terminal vertebræ inclining
to the upper lobe.

closely approximated premandibular or dentary elements of the under jaw.

The Pycnodonts were for the most part deep-bodied fishes, symmetrically compressed from side to side. They were notochordal; a few of the earlier forms were heterocercal, but the majority of the family were homocercal.

The pycnodont type was first manifested in the carboniferous strata by the heterocercal genus *Platysomus*, and by the species *P. parvulus*, which has been found in that formation at Leeds : but this earliest pycnodont genus is chiefly represented by Permian species, of which *Platysomus gibbosus* (fig. 54) is a fine example.

In the lias, most beautiful fossil fishes of this group are found, which were referred by Bronn to the genus *Tetragonolepis*, and by Agassiz to the lepidoid sub-order. Sir P. Egerton has, however, shown that the dentition is truly "pycnodont," having a very close resemblance to that of *Microdon*, but with the masticatory apparatus smaller in proportion to the size of the fish. The scales, moreover, instead of being articulated by interlocking pegs and sockets, as in fig. 52, are joined in a peculiar way, which Sir P. Egerton describes as follows :— "Each scale bears upon its inner anterior margin a thick solid bony rib, extending upwards beyond the margin of the scale, and sliced off obliquely above and below, on opposite sides, for forming splices with the corresponding processes of the adjoining scales. These splices are so closely adjusted, that without a magnifying power, or an accidental dislocation, they are not perceptible. When *in situ*, and seen internally, these continuous lines decussate with the true vertebral apophyses, and cause the regular lozenge-shaped pattern so characteristic of the pycnodont family."* These decussating "pleurolepidal" lines are, however, confined to the space between the skull and the dorsal fin, as in fig. 55.

* Proceedings of the Geological Society, May 1853, p. 276.

The Pycnodonts so characterized are further distinguished from the closely-resembling lepidoid genus *Dapedius*, by having the small anterior teeth conical and single-pointed, instead of being bifurcate ; and although this character is subject to occasional variations, nevertheless, on taking a comprehensive view of all the dapedioid species, it seems to have been sufficiently constant to warrant the continuance of the separation of the group into the unicuspid and bicuspid species. And Sir P. Egerton has accordingly proposed to apply the generic terms *Æchmodus* (from αἰχμή, *a point*, and ὀδοὺς, *a tooth*),* for the unicuspid and pycnodont species, formerly termed *Tetragonolepis*, and to continue the name *Dapedius* for the bicuspid and unequivocally lepidoid homocercal deep-bodied Ganoids, many beautiful species of which are found in the lias.

Fig. 55.
Pycnodus rhombus (Upper Oolite).

Genus PYCNODUS (fig. 55).—The type-genus of this sub-order is characterized by the large size of the round flat crowned teeth, which cover the broad jaws as by a pavement of from three to five rows ;† at the fore-part of the jaws are two or more trenchant incisive teeth both above and

---

* Proceedings of the Geological Society, May 1854, p. 367.
† For the disposition of these teeth on the palate, see Owen's Odontography, vol. i., pl. 34, figs. 1 and 2 ; and for their microscopic structure, ibid, p. 71, pl. 33.

below. The oblique inner processes of the scales appear as distinct dermal ossicles decussating the neural spines in the space between the occiput and the dorsal fin (fig. 55).

The species of *Pycnodus* abound in the oolitic formations above the lias : the one figured (*P. rhombus*) is from a calcareous deposit, so charged with animal remains as to be fœtid, at Torre d'Orlando, near Naples. Species of *Pycnodus* (*P. cretaceous, e.g.*) occur in the chalk of Kent ; and one species (*P. toliapicus*) has left its remains in the eocene clay of Sheppy. Some teeth from German miocene have been referred to this genus ; but at this period, if not at the earlier tertiary one, *Pycnodus* became extinct.

FAMILY IX.—DAPEDIDÆ.

Notochordal rhombo-ganoids, with front teeth conical or bifurcate, back teeth obtuse, vertebral column and side scales continued into the upper lobe of an almost symmetrical tail-fin.

The type-genus, *Dapedius*, is a compressed deep-bodied fish, with a single dorsal, and a single series of fin-fulcra ; the front teeth are commonly notched. All the species are from liassic strata. *Amblyurus*, with a similar form, and also liassic, has a very narrow anal, and a wide mouth with small pointed teeth. *Semionotus* and *Pholidophorus* are long-bodied fishes, the species of which range from the lias upwards to the Purbecks (*Pholidophorus ornatus*), and to the chalk (*Semionotus Bergeri*).

FAMILY X.—LEPIDOTIDÆ.

Homocercal rhombo-ganoids, with obtuse teeth and well ossified vertebræ.

The type-genus of this family, *Lepidotus*, is remarkable for the density and polish of its full-sized imbricated rhomboid scales; it has a short dorsal fin opposite the anal, and has two rows of fulcra to the anterior rays of all the fins. The species range from the lias to the chalk; one species, indeed (*Lepidotus Maximiliani*), lingers, after the commencement of the tertiary period, in the "calcaire grossier" of Paris.

In *Nothosomus* and *Ophiopsis* the fin-fulcra are in a single row, and the dorsal fin is very long. *Notagogus* and *Propterus* have the dorsal fin almost cleft into two.

Family XI.—Leptolepidæ.

The Ganoids of this family are homocercal, and have rounded scales. In the type-genus (*Leptolepis*, fig. 56), the

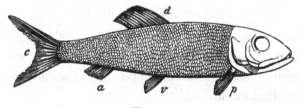

Fig. 56.

Leptolepis sprattiformis (Oolite, Solenhofen).

scales are extremely thin, yet a fine layer of ganoin may be discovered on them. The teeth are minute and *en brosse*, with two of larger size in front of the mouth. It has not been determined whether the notochord is ossified; but traces of distinct vertebral bodies appear to the writer to be discernible in some specimens. Species of *Leptolepis* range from the lias to the calcareous slates of Eichstadt. They are very common in the lithographic slates of Solenhofen and Pappenheim.

Family XII.—Macropomidæ.

Genus Macropoma.—Fine specimens of homocercal ganoid fishes, with rounded scales, sculptured externally, as in fig. 57, have been discovered in the chalk formations of Kent and Sussex. They have been referred by Agassiz to the genus called *Macropoma*, significative of the large size of the gill-cover, and to the cœlacanthal family. Casts of the "interior" of the alimentary canal, showing impressions of a broad spiral valve, are preserved in certain specimens in the British Museum. One species (*M. Egertoni*) is from the Speeton clay; the other (*M. Mantelli*) from the chalk.

Fig. 57.
Macropoma
Mantelli
(Chalk).

FAMILY XIII.—STURIONIDÆ.

The family *Sturionidæ*, represented by the sturgeons of the present seas, makes its first appearance in the lias, under the generic form of *Chondrosteus*, which has recently received a full description and illustration in a memoir communicated by Sir P. Egerton to the Royal Society of London.[*] In this it is shown "that *Chondrosteus*, though essentially sturionian, yet evidences a transitional form between the sturgeons and more typical Ganoids; that its food was similar to that of the existing members of the family, but that it was procured in a tranquil sea, rather than in the tumultuous waters frequented by sturgeons at the present time."

In the tertiary division of geological time the ganoid order rapidly diminishes, and its place is taken by fishes with better ossified internal skeletons, and with thinner, more flexible, and usually soluble scales. The gills are supported on bony arches, and are protected by branchiostegal rays, and by an operculum or gill-cover. The aortic bulb is provided with but two valves; and the optic nerves decussate. For this group, including the majority of existing fishes, and of those which made their appearance during the tertiary period, Müller proposed the name "Teleostei," which almost corresponds with the "osseous fishes" of Cuvier.

[*] Proceedings of the Royal Society, April 20, 1858.

ORDER IV.—ACANTHOPTERI.

Char.—Endo-skeleton ossified ; fins with one or more of the
first rays unjointed or inflexible spines ; ventrals in most
beneath or in advance of the pectorals ; swim-bladder
without air-duct.

Sub-Order 1.—CTENOIDEI.

Exo-skeleton as ctenoid scales (fig. 58).

Of this sub-order may be given two genera, both of which
are now extinct.
One (*Semiopho-*
rus) belongs to
the chetodont

Fig. 58.
Scale of *Perca*
(Recent).

family ; the
other (*Smerdis*)
to the Percoids.
The genus
Semiophorus,
Ag. (fig. 59), is
represented ex-
clusively by ex-
tinct species pe-
culiar to the
tertiary deposits
at Monte Bolca.
It is character-
ized by the ex-

Fig. 59.
Semiophorus velicans (Monte Bolca).

treme height or prolongation of the anterior part of the dorsal fin, D, and for the correlative elongation of the slender pointed ventral fins. The anal fin, A, is much shorter than the dorsal. Owing to the soluble nature of the scales, and to the well-ossified skeleton, the fossils of this, as of most other tertiary fishes, are exemplified by the vertebral column and skull more than by the skin.

Genus SMERDIS.—The species composing this genus are of small size, and are wholly extinct ; they likewise are chiefly

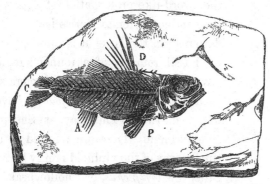

Fig. 60.
Smerdis minutus (Gypsum of Provence).

met with in the tertiary ichthyolite beds of Monte Bolca ; but some (*e.g.*, the *Smerdis minutus*, fig. 60) are from eocene deposits in France. In all the species the first suborbital or lacrymal bone is strongly dentate, as is also the preoperculum ; but this has no spine at the angle. The operculum terminates behind by a rounded prominence. There are two dorsals. The scales are minute, but are occasionally preserved.

Sub-Order 2.—CYCLOIDEI.

This sub-order includes the teleostian fishes with undivided and unjointed spines at the fore part of the dorsal, and with smooth flexible circular or elliptical scales (fig. 61). It is not represented by any species of older date than the cretaceous epoch ; and both here and in the eocene tertiaries by extinct

species, mostly of extinct genera.   It is most
richly represented at the pre-
sent day by the Sphyrenoid,
Scomberoid, and Xiphioid
families.

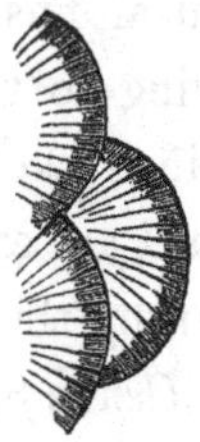

There are two kinds of existing sword-fish, *Xiphias* and
Histiophorus ; in the former
the sword-like prolongation of
the confluent premaxillaries is
flattened, in the latter it is rounded.

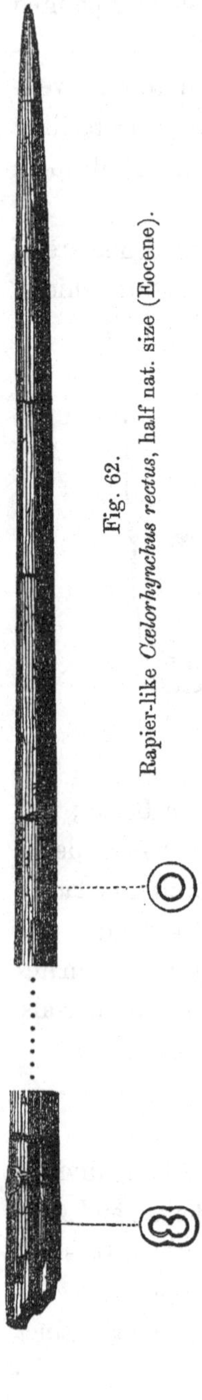

Fig. 61.
Scales of a Scom-
beroid fish.

Fossil remains of a rounded rapier-like
" sword," but much longer and more slender
than in the existing *Histiophorus,* have been
found in the eocene clay at Sheppy and
Bracklesham.   They are referred to an extinct genus of the xiphioid family by Agassiz,
called *Cœlorhynchus,* or " hollowbeak."   The
most perfect specimen hitherto found is
figured in fig. 62, of half the natural size.   It
forms part of the instructive collection of
Captain Le Hon at Brussels.   The upper
transverse section shows the single cavity at
the middle of the rostrum ; and the lower
section shows the double or divided cavity
near its base.

ORDER V.—ANACANTHINI.

Char.—Endo-skeleton ossified ; exo-skeleton
in some as cycloid, in others as ctenoid
scales ; fins supported by flexible or
jointed rays ; ventrals beneath the
pectorals, or none ; swim-bladder without air-duct.

Fig. 62.

Rapier-like *Cœlorhynchus rectus*, half nat. size (Eocene).

FAMILY.—PLEURONECTIDÆ.
(*Flat-Fishes.*)

In this family the symmetrical form is lost, and both eyes are on one side of the head. Species of still existing genera of this much-modified family have been found in tertiary deposits. The little turbot (*Rhombusminimus, e. g.,* fig. 63) occurs in the tertiary deposits of Monte Bolca. An equally extinct species of sole (*Solea antiqua*) has been found in tertiary marls near Ulm.

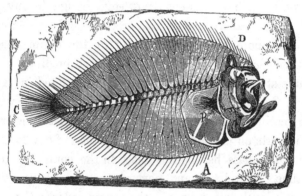

Fig 63.
Rhombus minimus (Monte Bolca).

Fossil fishes of the cod, mullet, carp, salmon, and herring genera, are found in the tertiary formations, but are distinct from all known species. The Ganoids in these formations are reduced to the genera *Lepidosteus* and *Acipenser;* but may have been represented by the palates with crushing teeth, from the Sheppy clay, to which the names *Pisodus**[*] and *Phyllodus*[†] have been given.

With respect to the fishes of the tertiary period, "they are so nearly related," says Agassiz, "to existing forms, that it is often difficult, considering the enormous number (above

---

[*] See Owen's Odontography, p. 138, pl. 47, fig. 3.
[†] Ibid, p. 139, pl. 47, figs. 1 and 2.

8000) of living species, and the imperfect state of preservation
of the fossils, to determine exactly their specific relations.
In general I may say that I have not yet found a single
species which was perfectly identical with any marine existing fish, except the little Capelin (*Mallotus villosus*), which
is found in the nodules of clay of unknown geological age in
Greenland." These nodules are mostly very recent, and exemplify the operation of the dissolving soft parts of the fish
in consolidating the surrounding matrix.

We cannot, from present knowledge, assign to any past
period of the earth's history a characteristic derived from a
fuller and more varied development of the entire class of
fishes than has since been manifested, nor predicate of the
present state of the class that it has degenerated in regard
either to the number, bulk, powers, or range of modifications
of the piscine type. A retrospect of the genetic history of
fishes imparts an idea rather of mutation than of development,
to which the class has been subject in the course of geological time. Certain groups, now on the wane, have existed
in plenary development, as, *e. g.*, the ganoid order in the mezozoic period, and the cestraciont form of Plagiostomes in both
palæozoic and mezozoic times.

As to the variety of the forms of fishes, seeing that the
earth yields no indisputable evidence of Ctenoids or Cycloids
anterior to the cretaceous epoch, yet still retains living representatives of both Ganoids and Placoids, the present would
appear to be the culminating period in the development of
fishes, in respect of the number of ordinal forms or modifications of the class. It represents, however, rather a period of
mutation of the piscine character, depending upon the progressive assumption of a more special piscine type, and progressive departure from a more general vertebrate type. The
Scomberoids, as fishes, are at the head of the piscine modification of the vertebrate type. But as the retention of general

vertebrate characters implies closer affinity with the proximate cold-blooded class, so a higher character of organization may be predicated of the palæozoic Placoids and Ganoids than of the Ctenoids and Cycloids forming the great bulk of the class at the present day. The comparative anatomist dissecting a shark, a Polypterus, or a Lepidosteus, would point to the structures of the brain, heart, generative organs, and in the last two genera to the air bladder, as being of a higher or a more reptilian character than the corresponding parts would present in most other fishes. But the palæontologist would point to the persistent notochord, and to the heterocercal tail in palæozoic and many mezozoic fishes, as evidence of an " arrest of development," or of a retention of embryonic characters in those primæval fishes.

No class of animals is more valuable in its application to the great point now mooted by the Uniformitarians and Progressionists of the present day than that of fishes ; for they are exempt from the attack of the Uniformitarian on the score of the defective nature of negative evidence, to which attack conclusions from the known genetic history of air-breathing animals are open. Many creatures living on land may never be carried out to sea ; but marine deposits may be expected to yield adequate grounds for general conclusions as to the character of the vertebrate animals that swarmed in the seas precipitating such deposits.

One other conclusion may be drawn from a general retrospect of the mutations in the forms of the fishes at different epochs of the earth's history,—viz., that those species, such as the nutritious cod, the savoury herring, the rich-flavoured salmon, and the succulent turbot, have greatly predominated at the period immediately preceding and accompanying the advent of man ; and that they have superseded species which, to judge by the bony Garpikes (*Lepidosteus*), were much less fitted to afford mankind a sapid and wholesome food.

ICHNOLOGY.*

In entering upon the genetic history of the class of reptiles, we have to inquire, as in that of fishes, in what period of the earth's history the class was introduced, and under what forms ; at what period it attained its plenary development, in regard to the size, grade of structure, number and diversities of its representatives ; and the relations which the existing members of the class bear to its past condition. Fifteen years ago, the oldest known reptilian remains were those of the so-called "Thuringian Monitor," from the Permian copper-slates of Germany. Five years ago, the batrachian *Apateon*, or *Archegosaurus* was discovered in a Bavarian coal-field ; and about the same period, footprints in carboniferous sandstones of North America, had been recognized as evidence of the commencement of reptilian existence at that period of the earth's history. Air-breathing ambulatory animals may leave other evidence of their former presence upon earth than their fossilized remains.

There are several circumstances under which impressions made on a part of the earth's surface, soft enough to admit them, may be preserved after the impressing body has perished. When a shell sinks into sand or mud, which in course of time becomes hardened into stone, and when the shell is removed by any solvent that may have filtered through the matrix, its place may become occupied by crystalline or other mineral matter, and the evidence of the shell be thus preserved by a cast, for which the cavity made by the shell has served as a mould. If the shell has sunk with its animal within it, the plastic matrix may enter the dwelling-chamber as far as the retracted soft parts will permit ; and as these slowly melt away, their place may become occupied by crystallized deposits of

* Ιχνος, *a footstep*, and λόγος.

any silicious, calcareous, or other crystallizable matter that may have been held in solution by water percolating the matrix, and such crystalline deposit may receive and retain some colour from the soft parts of which it thus becomes a cast.

Evidences of soft-bodied animals, such as *Actiniæ* and *Medusæ*, and of the excremental droppings of higher animals, have been thus preserved. Fossil remains, as they are called, of soft plants, such as sea-weeds, reeds, calamites, and the like, are usually casts in matrix made naturally after the plant itself has wholly perished.

Even where the impressing force or body has been removed directly or shortly after it has made the pressure, evidence of it may be preserved. A superficial film of clay, tenacious enough to resist the escape of a bubble of gas, may retain, when petrified, the circular trace left by the collapse of the burst vesicle. The lightning flash records its course by the vitrified tube it may have constructed out of the sandy particles melted in its swift passage through the earth. The hailstone, the ripple wave, the rain-drop, even the wind that bore it along and drove it slanting on the sand, have been registered in casts of the cavities which they originally made on the soft sea-beach ; and the evidence of these and other meteoric actions, as sun-cracks and frost-marks, so written on imperishable stone, have come down to us from times incalculably remote. Every form of animal life that, writhing, crawling, walking, running, hopping, or leaping, could leave a track, depression, or foot print, behind it, might thereby leave similar lasting evidence of its existence, and also to some extent of its nature.

The interpretation of such evidences of ancient life has much exercised the sagacity of naturalists since Dr. Duncan, in 1828, first inferred the existence of tortoises at the period of the deposition of certain sandstones in Dumfriesshire, from the impressions left on those sandstones, and the casts after-

wards formed in those impressions. The faculty of interpreting
has been still more racked by similar evidences of more extraordinary footprints, probably of large batrachian reptiles, first
noticed in 1834 at Hildberghausen in Saxony, in sandstones
of the same geological age as those in Scotland.

The vast number and variety of such impressions, due
either to physical or meteoric forces, to dead organic bodies,
parts or products, or to the transitory actions of living
beings, have at length raised up a distinct branch of palæontological research, to which the term "Ichnology" has been given.

In this class of evidences the impressions called "protichnites"* (fig. 64), left upon the "Potsdam sandstones"† of the
older Silurian age in Canada, are the most ancient ; but the
footprints of birds surpass all others in regard to their number, distinctness, and variety of sorts.

But how, it may be asked, are such footprints preserved ?
A common mode may be witnessed daily on those shores
where the tide runs high, and the sea-bottom is well adapted
to receive and retain the impressions made upon it at low-
water.

Dr. Gould of Boston, U. S., first called the attention of naturalists to this interesting operation on the shores of the Bay of
Fundy, where the tide is said to rise in some places seventy
feet in height. The particles deposited by that immense tidal
wave are derived from the destruction of previously existing
rocks, and consist of silicious (flinty) and micaceous (talcky)
particles, cemented together by calcareous (limy) or argillaceous
(clayey) paste, containing salts of soda, especially the muriate
(common salt), and coloured with various shades of the oxide or
rust of iron, of which the red oxide predominates. The perfection
of the surface for receiving and retaining an impression depends

* See Owen, "Description of the Impressions and Footprints of the Protichnites from the Potsdam Sandstone of Canada," Quarterly Journal of the
Geological Society, 1852, p. 214.   † Logan, ibid, p. 2.

much upon the micaceous element. Vast are the numbers of
wading and sea birds that course to and fro over the extensive
tract of plastic red surface left dry by the far retreat of the
tide in the Bay of Fundy. During the period that elapses
between one spring tide and the next, the highest part of the
tidal deposit is exposed long enough to receive and retain many
impressions ; even during the hours of hot sunshine, to which,
in the summer months, this so-trodden tract is left exposed,
the layer last deposited becomes baked hard and dry, and
before the returning tidal wave, turbid with the same commi-
nuted materials of a second stratum, has power to break up the
preceding one, the impressions left on that stratum have
received the deposit. A cast is thus taken of the mould previously made, and the sediment superimposed by each succeeding tide, tends more and more surely to fix it in its place.
Then, let ages pass away, and the petrifying influences conso-
lidate the sand layers into a fissile rock : it will split in the
way it was formed, and the cleavage will expose the old moulds
on one surface and the casts on the other.

Another condition for fixing the impressions on a sandy
shore is the following :—When an extensive level tract is
left dry by the retreating tide, as at the estuary of the small
rivers entering the Bay of Morecombe, on the Lancashire
coast, those rivers occasionally overflow the sands at low-water,
and deposit in the footprints made previous to such overflow
the fine mud which sudden heavy rains have brought down
from the surrounding hills. Again, those sudden " freshets,"
as they are locally called, sometimes as quickly subside, and
the thin layer of argillaceous mud is left dry on the sand
before the returning tide. Such layer of mud readily receives
and retains the footprints of the many birds that course over
the flat expanse ; and as the tide returns, it deposits in such
footprints a layer of the fine sand which the rising waters
hold in suspension.

The best-defined footprints in the new red sandstone quarries at Stourton, on the Cheshire coast, are found where strata of sandstone are separated by a thin layer of argillaceous stone, which, when exposed, soon breaks up and crumbles away. This layer has, however, received the impressions when it was plastic, and the superincumbent deposit of sandstone retains those impressions in relief upon its under surface. The observations which have just been recorded, of the circumstances that produce an interposition of a thin layer of claystone between thicker beds of sandstone, and which circumstances the writer has witnessed in the Bay of Morecombe, explain the formation and the preservation of the best "ichnites" of the labyrinthodont and other reptiles in the new red sandstone of Stourton.

There is a third condition under which impressions, and casts of impressions, on a sandy beach may be preserved. On a dry windy day clouds of fine sand are drifted along the surface exposed at low-water, are spread lightly over all its little inequalities, and fill up every impression that may have been made on it since it was left bare by the retreating waves. On the return of the tide, the fine sand filling the impressions is moistened, and more wet fine sand is added to it; and a cast is thus fixed in the moulds, to be more and more firmly fixed by each deposition from successive tidal waves.

Thus may be witnessed the actual conditions and the circumstances daily occurring that tend to preserve footprints and other impressions made on the sea-shore, and which have operated in past time to similarly preserve the impressions then made on tracts alternately exposed and covered by the tidal wave. The merit of having first discerned the nature and cause of the numerous small hemispheric pits and tubercular casts in relief on the surface of certain sandstone slabs, is due to John Cunningham, Esq., F.G.S., architect, of Liverpool.* Since that light was thrown on their nature, they have been recognized under various modifications, as impressions of soft rain, of the big-dropped thunder-shower, of rain driven obliquely by the gale, and making impressions with the side of the cup highest opposite the point whence the wind blew, of frozen rain or hail, etc. Dr. Deane, in 1845, after witnessing the first exposure and raising of the red sand slabs, near Greenfield, Mass., U. S., writes, "They were characters fresh as upon the morning when they were impressed ; 'on that morning gentle showers watered the earth,'" etc. Whenever a stratum is proved to be a "sedimentary" one—*i. e.*, to be due to the precipitation of its constituent particles from water, in which they had been previously suspended—we have evidence of some expanse of water,—proof, in fact, of the existence of that element, with all its properties of condensation by cold, and expansion and vaporization by heat and exposure. Evaporation makes the raw material of rain. No wonder, then, that impressions of rain-drops should be seen on the oldest sedimentary rocks. Conditions are coordinated in meteoric as in organic phenomena ; one being given, the rest may be deduced.

The oldest rocks in which rain-drop impressions have been observed are those of the Cambrian age at Longmynd, Wales, described and figured by Mr. Salter.† Many of the micaceous flags of the same formation are covered with ripple, or current marks. They show borings of worms, and a trace of a trilobite (*Palæopyge*) nearly allied to the *Dikelocephalus* —the oldest known trilobite of America (Lower Silurian or Cambrian at St. Croix, Minnesota).

It is in "Potsdam sandstones" of the same geological antiquity that the impressions have been discovered which the

---

* Communicated by Dr. Buckland to the meeting of the British Association at Newcastle, 1838; and subsequently by Mr. Cunningham to the Geol. Soc. (Proc. of the Geol. Soc., vol. iii., 1839, p. 99.)

† Quar. Jour. of the Geol. Soc., vol. xii., 1856, p. 250, pl. iv., fig. 4.

writer has interpreted to be those of a large entomostracous Crustacean ;* in evidence of which the following sample, applicable to a single species, may be given, in illustration of the ichnologist's mode of work.

PROTICHNITES.

Protichnites septem-notatus (fig. 64).

The subject so named consists of a series of well-defined impressions, continued in regular succession along an extent of 4 feet ; and traceable with an inferior degree of definition, along a further extent of upwards of 2 feet.

In the extent of 4 feet there are thirty successive groups of footprints on each side of a median furrow, which is alternately deep and shallow along pretty regular spaces of about $2\frac{1}{2}$ inches in extent. The number of prints is not the same in each group ; where they are best marked, as in fig. 64, 1 L, we see 3 prints in one group, a, a', a'', 2 prints in the next, b, b', and 2 in the third, c, c', which is followed by a repetition of the group of 3 prints, a, a', a'', making the numbers in the three successive groups 3, 2, 2 ; the three groups of impressions being recognizably repeated in succession along the whole series of tracks on both sides of the median groove.

The principal footprints are disposed in pairs, placed with different degrees of obliquity, in each of the three groups towards the median track ; the innermost print in the second, B, and third, C, pairs, which are best marked, being usually rather more than half the size of the outer print, b' and c'.

The two footprints of the same pair are a little further apart from each other, in the three succeeding pairs, as at a', a'', b, b', c, c', especially in the second and third groups of each set ; the two forming the pair a', a'', again approximating in the next series, and the pairs b, b' and c, c', diverging in the same direction and degree ; and this alternate approximation

* Quarterly Journal of the Geological Society, vol. viii., p. 214, 1852.

and divergence is repeated throughout the entire series of the
present tracks.

But what strikes the ichnologist, heretofore conversant
chiefly with the footprints of bipeds or quadrupeds, is the
occurrence in the present series of the third impression *a*,

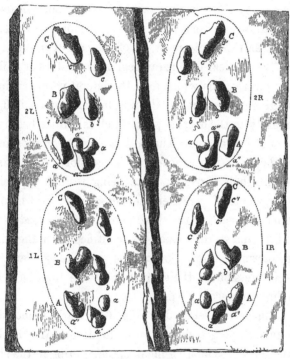

Fig. 64.
Protichnites 7-notatus (Cambrian).

which complicates the most approximated pair A, being placed
in front and a little to the inner side of the hindmost impression, *a″*, of that pair. The superadded impression *a*, is about
the same size as the innermost in each pair, the average
diameter of that impression being 5 lines.

Taking this view of the impressions, it appears that whilst
the innermost in each pair, *a′, b, c*, are of equal size, the outermost, *a″, b′, c′*, 1 L, progressively increase in size, from the most

approximated to the most divergent of the three pairs ; that
of the first, a'', being narrow in proportion to its length, that
of the second b', as broad as long, and the outermost, c', c'', of
the third pair being oblong, but larger than that in the first
pair.   In some places where the most approximated pair of
impressions, a', a'', are deeply marked, they are complicated by
a fourth shallow and very small pit, a''', 2 L, midway between
the third, a, and the outermost, a'', of the pair of impressions.

There are no clear or unequivocal marks of toes or nails on
any of the impressions which form the lateral pairs or triplets.
Their margins are not sharply defined, but are rounded off, and
sink gradually to the deepest part, which is a little behind the
middle of the depression.   There is a slight variation in the
form and depth of the answerable impressions, but not such
as to prevent their correspondence being readily appreciable
through the whole of the extent here described ; that is to say,
the innermost of each of the three pairs here described as first,
A, second, B, and third, C, may be identified with the corresponding innermost impression on the opposite side, and with
the same impression of the same pair in the three preceding
and the three succeeding pairs.

The impressions selected for fig. 64 clearly demonstrate
that the animal, progressing in an undulating course, made at
each action of its locomotive members, answering to the single
step of the biped and the double step of the quadruped, not
fewer than, in *Protichnites* 7-*notatus*, fourteen impressions,
seven on the right and seven on the left ; and in *Protichnites*
8-*notatus*, sixteen impressions, eight on the right and eight on
the left ; these seven and eight impressions respectively being
arranged in three groups—viz., in *Protichnites* 7-*notatus*, three,
two, and two ; in *Protichnites* 8-*notatus*, three, two, and three
—the groups being re-impressed, in successive series, so
similarly and so regularly as to admit of no doubt that they
were made by repeated applications of the same impressing

instruments, capable of being moved so far in advance as to clear the previous impressions and make a series of new ones at the same distance from them as the sets of impressions in the series are from each other. What then was the nature of these instruments? To this four replies may be given, or hypotheses suggested :—They were made either, first, as in the case of quadrupedal impressions, each by his own limb, which would give seven and eight pairs of limbs to the two species respectively ; or, secondly, certain pairs of the limbs were bifurcate, as in some insects and crustaceans, another pair or pairs being trifurcate at their extremities ; and each group of impressions was made by a single so subdivided limb, in which case we have evidence of a remarkably broad and short, and, as regards ambulatory legs, hexapod creature ; or, thirdly, three pairs of limbs were bifurcate, and the supplementary pits were made by small superadded limbs, as in some crustaceans ; or, fourthly, a single broad fin-like member, divided at its impressing border into seven or into eight obtuse points, so arranged as to leave the definite pattern described, must have made the series of three groups by successive applications to the sand.

The latter hypothesis appears to be the least probable,—first, as being most remote from any known analogy ; and, secondly, because there are occasional varieties in the groups of footprints which would hardly accord with impressions left by one definitely subdivided instrument or member. Thus in the group of impressions marked 1 L in fig. 64, the outer impression, c', is single, but in the preceding set it is divided ; whilst the impressions, a, a', are confluent in that set, and are separate in 1 L. The same variety occurs in the outer pair, c', c'', in *Protichnites 8-notatus*.

Yet, with respect to the hypothesis that each impression was made by its own independent limb, there is much difficulty in conceiving how seven or eight pairs of jointed limbs could be aggregated in so short a space of the sides of one animal. So that the most probable conception is, that the creatures which have left these tracks and impressions on the most ancient of known sea-shores belonged to a crustaceous genus,—either with three pairs of limbs employed in locomotion, and severally divided to accord with the number of prints in each of the three groups,—or bifurcated merely, the supplementary and usually smaller impressions being made by a small and simple fourth, or fourth and fifth pair of extremities.

The great entomostracous king-crab (*Limulus*) which has the small anterior pair of limbs near the middle line, and the next four lateral pairs of limbs bifurcate at the free extremity, the last pair of lateral limbs with four lamelliform appendages, and a long and slender hard tail, comes nearest to the above idea of the kind of animal which has left the impressions on the Potsdam sandstone.

The shape of the pits, so clearly shown in the ice-rubbed slabs, impressed by *Protichnites* 8-*notatus*, accords best with the hard, subobtuse, and subangular terminations of a crustaceous ambulatory limb, such as may be seen in the blunted legs of a large *Palinurus* or *Birgus*; and it is evident that the animal of the Potsdam sandstone moved directly forwards after the manner of the *Macroura* and *Xiphosura*, and not sideways, like the brachyurous Crustaceans.

The appearances in the slab impressed by the *Protichnites multi-notatus* favour the view of the median track having been formed by a caudal appendage, rather than by a prominent part of the under surface of the trunk.

The imagination is baffled in the attempt to realize the extent of time past since the period when the creatures were in being that moved upon the sandy shores of that most ancient Silurian sea : and we know that, with the exception of certain microscopic forms of life, all the actual species of animals came into being at a period geologically very recent in comparison with the Silurian epoch.

Fig. 68.
Foot-prints of *Labyrinthodon*
(Cheirotherium).

The deviations from the living exemplars of animal types usually become greater as we descend into the depths of time past; of this the Archegosaur and Ichthyosaur are instances in the reptilian class, and the *Pterichthys* and *Coccosteus* in that of fishes. If the vertebrate type has undergone such inconceivable modifications during the Secondary and Devonian periods, what may not have been the modifications of the articulate type during a period probably more remote from the secondary period than this is from the present time ? In all probability no living form of animal bears such a resemblance to that which the Potsdam footprints indicate as to afford an exact illustration of the shape and number of the instruments, and of the mode of locomotion, of the Silurian *Protichnites*.

Since the foregoing interpretation of the Silurian Ichnites of North America. was published, similar impressions have been observed in rocks of the like high antiquity in Scotland, as at Binks, Eskdale, which have received the name of *Protichnites Scoticus.**

AMPHIBICHNITES.

Genus CHEIROTHERIUM.—Fig. 68 gives a reduced view of a portion of new red sandstone, with three pairs of footprints in relief : the first and third of the left, the second of the right, side. Consecutive impressions of such prints have been traced for many steps in succession in quarries of that formation in Warwickshire and Cheshire, more especially at a quarry of a whitish quartzose sandstone at Storton Hill, a few miles from Liverpool. The footmarks are partly concave and partly in relief ; the former are seen upon the upper surface of the sandstone slabs, but those in relief are only upon the lower surfaces, being in fact natural casts, formed on the subjacent footprints as in moulds. The impressions of the

* Harkness and Salter "On the Lowest Rocks of Eskdale," Quarterly Journal of the Geological Society, vol. xii., pp. 238, 243, fig. 2.

hind foot are generally 8 inches in length and 5 inches in
width ; near each large footstep, and at a regular distance—
about an inch and a half—before it, a smaller print of the fore
foot, 4 inches long and 3 inches wide, occurs. The footsteps
follow each other in pairs, each pair in the same line, at intervals
of about 14 inches from pair to pair. The large as well as the
small steps show the thumb-like outermost toe alternately on
the right and left side, each step making a print of five toes.

Footprints of corresponding form, but of smaller size, have
been discovered in the quarry at Storton Hill, imprinted on
five thin beds of clay, lying one upon another in the same
quarry, and separated by beds of sandstone. From the lower
surface of the sandstone layers the solid casts of each impression project in high relief, and afford models of the feet, toes,
and claws of the animals which trod on the clay.

Similar footprints were first observed in Saxony, at the
village of Hessburgh, near Hillburghausen, in several quarries
of a grey quartzose sandstone, alternating with beds of red
sandstone, and of the same geological age as the sandstones of
England that had been trodden by the same strange animal.
The German geologist who first described them (1834) proposed the name of *Cheirotherium* (*cheir*, the hand, *therion*,
beast) for the great unknown animal that had left the footprints, in consequence of the resemblance, both of the fore and
hind feet, to the impression of a human hand ; and Dr. Kaup
conjectured that the animal might be a large species of the
opossum kind ; but in Didelphys the thumb is on the inner
side of the hind-foot. The fossil skulls, jaws, teeth, and a few
other bones, in the sandstones exhibiting the footprints in
question, and corresponding in size with those impressions,
belong to labyrinthodont or huge extinct batrachian reptiles.

The impressions of the *Cheirotherium* resemble those of
the footprints of a salamander, in having the short outer toe
of the hind foot projecting at a right angle to the line of the

mid toe, but are not identical with those of any known
Batrachian or other reptile. They show a papillose integument as in some mammals, but also like that on the sole of
certain Geckos, and which may be another mark of sauroid
departure from the modern batrachian type. The proximity
of the right and left prints to the median line indicates a
narrower form of body, or its greater elevation upon longer and
more vertical limbs, than in tailless Batrachia. In the attempt
to solve the difficult problem of the nature of the animal which
has impressed the new red sandstone with the cheirotherian
footprints, we cannot overlook the fact, that we have in the
Labyrinthodons also batrachoid reptiles, differing as remarkably
from all known Batrachia, and from all other reptiles, in the
structure of their teeth; both the footsteps and the fossils are,
moreover, peculiar to the new red sandstone; the different
size of the footprints referred to different species of *Cheiro-
theria* correspond with the different size of ascertained species
of *Labyrinthodon;* and the present facts best support the
hypothesis, that the footprints called "cheirotherian," are
those of labyrinthodont reptiles.

Genus OTOZOUM.—The footprints in the red sandstones,
probably of liassic age, in Connecticut, described by Prof.
Hitchcock under the above name, equalled in size the largest
of those of the *Cheirotherium* (*Ch. Hercules*), but the hind
foot had but four toes, whilst the fore foot had five toes. It
would seem that the hind foot, which was larger than the fore
foot, obliterated the print of that foot, by being placed upon it
in walking. In the few instances of the fore foot print the toes
are turned outward, and the fourth and fifth seem to have been
connate at their base. An impression of a web has been clearly
discerned in the hind foot. Only one toe on this foot shows a
claw, the rest are terminated by "pellets," as in the *Batrachia,*
to which family Dr. Hitchcock refers these footprints, though
with a surmise of the possibility of their marsupial nature.[*]

* Ichnology of Massachusetts, 4to, 1858, p. 123.

Genus BATRACHOPUS (*Batrachopus primævus*, King.)—In 1844, Dr. King of Greensburg, Pennsylvania, discovered fossil footmarks, which he announced as being those of a reptile, in the sandstone of the coal measures, near that town. No reptilian footprints had previously been found lower in the series than the New Red sandstone. Dr. King states the impressions to be " near 800 feet beneath the topmost stratum of the coal formation."

Sir C. Lyell, in *Silliman's Journal*, July 1846, describes his visit to Greensburg, where he examined these footmarks, and confirmed Dr. King's description of them. He considered them to be allied to the labyrinthodont footprints which have been referred to the genus *Cheirotherium*. He says—" They consist, as before stated, of the tracks of a large reptilian quadruped, in a sandstone in the middle of the carboniferous series, a fact full of novelty and interest ; for here in Pennsylvania, for the first time, we meet with evidence of the existence of air-breathing quadrupeds capable of roaming in those forests where the Sigillaria, Lepidodendron, Caulopteris, Calamites, ferns, and other plants flourished."

These footmarks were first observed standing out in relief from the lower surface of slabs of sandstone resting on thin layers of fine unctuous clay, which also exhibited the cracks due to shrinking and drying. Now these cracks, where they traversed the footprints, had produced distortion in them, for the mud must have been soft when the animal walked over it and left the impressions ; whereas, when it afterwards dried up and shrunk, it would be too hard to receive such indentations, and could only affect them in the way of subsequent dislocation.

No less than twenty-three footsteps, the greater part so arranged as to imply that they were made successively by the same animal, were observed in the same quarry.

Everywhere there was a double row of tracks, and in each row they occur in pairs, each pair consisting of a hind and fore foot, and each being at nearly equal distances from the next pair. The hind foot-print is about one-third larger than the fore foot-print: it has five toes, but the front one only four ; some of them exhibit a stunted rudiment of the innermost toe or "pollex," which is the undeveloped one. The outermost toe in the hind footprint is shorter and rather thicker than the rest, and stands out like a thumb on the wrong side of the hand.

With this general resemblance to the footprints of Labyrinthodon, from the new red sandstones of Europe, there are well-marked distinctions. In the first place, the right and left series of impressions are wider apart, indicative of a broader-bodied animal. The front print in *Batrachopus* has only four well-developed toes instead of five, as in *Labyrinthodon ;* it is also proportionably larger,—the fore foot in *Labyrinthodon* being less than half the size of the hind foot. The distance between the fore and hind print of each pair, and of one such pair from the next on the same side, is nearly the same in *Batrachopus* and *Labyrinthodon.*

Genus SAUROPUS, Rogers.—Very similar footprints were discovered and described by Mr. Isaac Lea in a formation of red shales, at the base of the coal measures at Pottsville, 78 miles N.E. of Philadelphia. These are of older date than the preceding, inasmuch as a thickness of 1700 feet of strata intervenes between the footprints at Greensfield and the Pottsville impressions.

Professor H. D. Rogers, in 1851, announced his discovery in the same red shales, between the Devonian and carboniferous series, of three species of four-footed animals, which he deems to have been rather saurian than batrachian, seeing that each foot was five-toed ; one species, the largest of the three, presented a diameter for each footprint of about two inches, and showed the fore and hind feet to be nearly equal in dimensions. It exhibits a length of stride of about nine inches, and a breadth between the right and left footsteps of nearly four inches. The

impressions of the hind feet are but little in the rear of the
fore feet. With these footmarks were associated shrinkage
cracks, such as are caused by the sun's heat upon mud, and
rain-drop pittings, with the signs of the trickling of water on
a wet beach,—all confirming the conclusions derived from the
footprints, that the quadrupeds belonged to air-breathers, and
not to a class of animals living in and breathing water.

CLASS II.—REPTILIA.

Order I.—GANOCEPHALA.*

The name of this order has reference to the sculptured and
externally polished or " ganoid " bony plates with which
the entire head was defended. These plates include the
" post-orbital " and " super-temporal " ones, which roof
over the temporal fossæ. There are no occipital condyles. The teeth have converging inflected folds of
cement at their basal half. The notochord is persistent;
the vertebral arches and peripheral elements are ossified;
the pleurapophyses are short and straight. There are
pectoral and pelvic limbs, which are natatory and very
small; large median and lateral " throat-plates;" scales
small, narrow, sub-ganoid; traces of branchial arches.
The above combination of characters gives the value of
an ordinal group in the cold-blooded Vertebrata.

Genus APATEON, Von M.; ARCHEGOSAURUS,† Goldf.

The extinct animals which manifest the above ordinal
characters were first indicated by certain fossils, discovered in
the sphærosideritic clay-slate forming the upper member of the
Bavarian coal measures; and also in splitting spheroidal con-
cretions from the coal-field of Saarsbruck, near Treves. They
were originally referred to the class of fishes (*Pygopterus Lucius,*
Agassiz) : but a specimen from the Brandschiefer of Münster-

---

* Γανος, lustre; κεφαλη, head.
† Απατεων, *a cheat;* αρχηγος, *beginning;* σαυρος, *lizard.*

Appel presented characters which were recognized by Dr. Gergens to be those of a salamandroid reptile.* Dr. Gergens placed his "salamander" in the hands of H. von Meyer for description, who communicated the result of his examination in a later number of the under-cited journal.† In this notice the author states that the salamandroid affinities of the fossil in question, for which he proposes the name of *Apateon pedestris*, "are by no means demonstrated."‡ "Its head might be that of a fish as well as that of a lizard, or of a Batrachian." "There is no trace of bones or limbs." M. von Meyer concludes by stating that, in order to test

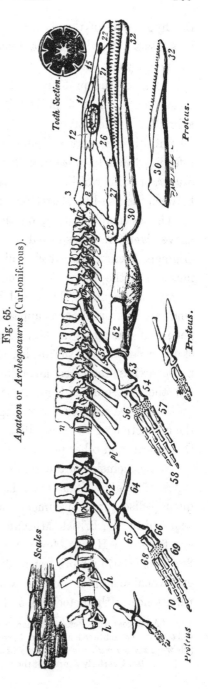

Fig. 65. *Apateon* or *Archegosaurus* (Carboniferous).

* Mainz, Oktober 1843. "In dem Brandschiefer von *Münsterappel* in *Rhein-Baiern* habe ich in vorigen Jahre einen Salamander aufgefunden. Gehört dieser Schiefer der Kohlen-formation? in diesem falle wäre der Fund auch in anderen Hinsicht interessant." (Leonhard und Bronn, *Neues Jahrbuch für Mineralogie*, etc., 1844, p. 49.)

† Ibid, 1844, p. 336.

‡ "Ob das—*Apateon pedestris*— ein Salamander-artiges Geschöpf war, est keinesweg ausgemacht." (Ibid.)

the hypothesis of the *Apateon* being a fossil fish, he has sent to Agassiz a drawing with a description of it.

Three years later, better preserved and more instructive specimens of the problematical fossil were obtained by Professor von Dechen from the Bavarian coal-fields, and were submitted to the examination of Professor Goldfuss of Bonn : he published a quarto Memoir on them, with good figures, referring them to a saurian genus which he calls *Archegosaurus*, or primæval lizard, deeming it to be a transitional type between the fish-like Batrachia and the lizards and crocodiles.*

The estimable author, on the occasion of publishing the above Memoir, transmitted to me excellent casts of the originals therein described and figured. They are in the museum of the Royal College of Surgeons, and are described in my Catalogue of the Fossil Reptiles, 4to, 1854, p. 117.† One of the specimens appeared to present evidence of persistent branchial arches. The osseous structure of the skull, especially of the orbits, through the completed zygomatic arches, indicated an affinity to the Labyrinthodonts ; but the vertebræ and numerous very short ribs, with the indications of stunted swimming limbs, impressed me with the conviction of the near alliance of the *Archegosaurus* with the *Proteus* and other perennibranchiate reptiles.

This conclusion of the affinity of *Archegosaurus* to existing types of the reptilian class has been confirmed by subsequently-discovered specimens, some of which have been acquired by the British Museum, others have been described and figured by H. von Meyer in his *Palæontographica* (Bd. vi., 2to Lief. 1857) ; more especially by his discovery of the embryonal condition of the vertebral column—*i. e.*, of the persistence of the notochord, and the restriction of ossification

* "Archegosaurus, Fossile-Saurier aus dem Stein kohlengebirge die den Uebergang der Ichthyoden zu den Lacerten und Krokodilen bilden," p. 3. (*Beitrage zur vorweltlichen Fauna des Steinkohlengebirges*, 4to, 1847).

† See also, Quarterly Journal of the Geological Society, vol. iv., 1848.

to the arches and peripheral vertebral elements.* In this structure the old carboniferous reptile resembled the existing *Lepidosiren*, and affords further ground for regarding that remarkable existing animal as one which obliterates the line of demarcation between the fishes and the reptiles.

Coincident with this non-ossified state of the basis of the vertebral bodies of the trunk (fig. 65, *c*), is the absence of the ossified occipital condyles which characterize the skull in better developed *Batrachia*. The fore part of the notochord has extended into the basi-sphenoid region, and its capsule has connected it by ligament to the broad flat ossifications of expansions of the same capsule, forming the basi-occipital or basi-sphenoid plate. In fig. 65 are represented the chief modifications of the vertebræ, as shown in the neck, thorax, abdomen, sacrum, and tail. The vertebræ of the trunk in the fully-developed full-sized animal present the following stage of ossification :—

The neurapophyses (fig. 65, *n*) coalesce at top to form the arch, from the summit of which was developed a compressed, sub-quadrate, moderately high spine, with the truncate or slightly convex summit expanded in the fore-and-aft direction so as to touch the contiguous spines in the back ; the spines are distinct in the tail. The sides of the base of the neural arch are thickened and extended outwards into diapophyses, having a convex articular surface for the attachment of the rib, *pl ;* the fore-part is slightly produced at each angle into a zygapophysis looking upwards and a little forwards ; the hinder part was much produced backwards, supporting two-thirds of the neural spine, and each angle developed into a zygapophysis, with a surface of opposite aspects to the anterior one. In the capsule of the notochord three bony plates were developed, one on the ventral surface, and one on each side,

---

* Reptilien aus der Steinkohlen Formation in Deutchland, Sechster Band, p. 61.

at or near the back part of the diapophysis. These bony plates may be termed cortical parts of the centrum, in the same sense in which that term is applied to the element which is called "body of the atlas" in man and Mammalia, and "sub-vertebral wedge-bone" at the fore-part of the neck in Enaliosauria.

As such neural or inferior cortical elements co-exist with seemingly complete centrums in the *Ichthyosaurus*, thus affording ground for deeming them essentially distinct from a true centrum, the term "hypopophysis" has been proposed for such independent inferior ossifications in and from the notochordal capsule ; and by that term may be signified the sub-notochordal plates in *Archegosaurus*, which co-exist with proper hæmapophyses (*h*) in the tail. In the trunk they are flat, subquadrate, oblong bodies, with the angles rounded off ; in the tail they bend upwards by the extension of the ossification from the under to the side parts of the notochordal capsule ; sometimes touching the lateral cortical plates. These serve to strengthen the notochord and support the intervertebral nerve in its outward passage. The ribs (*pl*) are short, almost straight, expanded and flattened at the ends, round and slender at the middle. They are developed throughout the trunk and along part of the tail, co-existing there with the hæmal arches, as in the Menopome.* The hæmal arches (*h*), which are at first open at their base, become closed by extension of ossification inwards from each produced angle, converting the notch into a foramen. This forms a wide oval, the apex being produced into a long spine ; but towards the end of the tail the spine becomes shortened, and the hæmal arch reduced to a mere flattened ring.

The size of the canal for the protection of the caudal bloodvessels indicates the powerful muscular actions of that part,

* "Principal Forms of the Skeleton," Orr's Circle of the Sciences, p. 187, fig. 11.

as the produced spines from both neural and hæmal arches bespeak the provision made for muscular attachments, and the vertical development of the caudal swimming organ.

The skull of the *Archegosaurus* appears to have retained much of its primary cartilage internally, and ossification to have been chiefly active at the surface ; where, as in the combined dermo-neural ossifications of the skull in the sturgeons and salamandroid fishes—*e. g.*, *Polypterus, Amia, Lepidosteus*—these ossifications have started from centres more numerous than those of the true vertebral system in the skull of saurian reptiles. This gives the character of the present extinct order of *Batrachia.*

The skull is much flattened or depressed, triangular, with rounded angles, and the front one more or less produced according to the species ; and in some species according to the age of the individual. The base is concave ; the sides nearly straight, or slightly concave. The basi-occipital appears to have retained its primordial soft, unossified state. Of the ex-occipitals, in a distinctly ossified state, no clear view has yet been had. The super-occipital (fig. 65, 4), is represented, as in the salamandroid fishes, by a pair of flat bones, more probably developed in the epicranial membrane and integument than in the cartilaginous protocranium. The pair of bones external to these, and forming the prominent angles of the occipital region, represent the "par-occipitals." The lower peripheral surface of the basi-sphenoidal cartilage is ossified with a concave border towards the notochord behind, to the capsule of which it seems to have been attached. The alisphenoids were doubtless cartilaginous, and the protocranium there unaltered, as it was apparently in the ex-occipital region. The peripheral ossifications above representing the "parietal" (7), form a pair of oblong flat bones, with the "foramen parietale" in the mid-suture. External to these, and wedged between the parietals, the super- and par-occipitals, are the pair of bones answering to the "mastoids" (8). They give attachment externally and below to the tympanic (28), and to a subsidiary bony plate, holding the position of that development of the mastoid and squamosal, which roofs over the temporal fossa in the *Chelonia* : it may be termed " supra-squamosal" (the bone between 8 and 27 in fig. 65). The frontal bones (11), divided by a mid-suture, like the parietals, increase in length, and are continued far in advance of the orbits.   The bone (12) which occupies the position of the post-frontal in *Chelonia* is ossified from two centres, one articulating with the mastoid (8), the other, which is external to it, with the supra-squamosal. This other bone may be termed the "post-orbital," as proposed by Von Meyer.   The post-frontal extends forward above the orbit to meet the pre-frontal, separating the frontal (11) from the orbit, as in the sturgeon (*Acipenser*), *Polypterus*, and *Lepidosteus*, and also in some *Chelones.* The pre-frontal extends far forward, terminating in a point between the nasal (15) and lacrymal. The nasals (15), divided by the median suture, extend to the external nostrils, their prolongation varying with the species and age of the individual.

Thus far the ossification of the superficies of the skull of *Archegosaurus* closely conforms to that of the salamandroid ganoid fishes above cited ; and the homologous bones are determinable without doubt. The lacrymal bone obviously answers to the front large suborbital scale-bone in fishes ; its large size and forward extension in *Archegosaurus* is a mark of that affinity.

The upper jaw consists of pre-maxillary (22), maxillary (21), and palatine bones. The pre-maxillaries are divided by a median suture, as in *Lepidosteus* and *Crocodilus*, and are short bones, the breadth exceeding the length in *A. latirostris,* and also in the young of *A. Decheni ;* but in the old animal opposite proportions prevail. In *A. Decheni* each pre-maxillary contains eight teeth ; in *A. latirostris* not less than eleven.

The maxillary (21) which extends from the pre-maxillary to beneath and beyond the orbit, presents a great length, varied according to species and age; it is of small vertical extent, and terminates in a point, which reaches the tympanic. Anteriorly it unites with the pre-maxillary, and enters into the formation of the back boundary of the nostril; mesially it unites above with the lacrymal and suborbital, and below forms the outer boundary of the choanal aperture, joining the vomer anteriorly, and the palatine posteriorly. The palatine is a long narrow bone, rather expanded at both extremities; it forms anteriorly the hinder border of the choanal aperture, and mesially throughout a great part of its extent the outer boundary of the great palatal vacuity. It supports a row of teeth, of which one or two at the fore part are of large size.

Between the orbit and the maxillary extends an oblong flat bone (26), forming the lower or outer border of the orbit, uniting with the pre-frontal and lacrymal anteriorly, with the maxillary below, and with the tympanic (9) and another bone behind. In this position, and in its connections, it agrees with the malar of the crocodile, and also with the suborbital bone or bones of fishes. The latter are unequivocally muco-dermal bones, and may not be the homologues of the endo-skeletal malar bone of saurians, birds, and mammals. To which of the bones, therefore,—suborbital or malar,—the one in question of the *Archegosaurus* answers, may be doubtful. The writer inclines to view it as a dermal ossification, and to conclude that, as in the higher *Batrachia*, the true malar and the zygomatic arch are not developed. Admitting the doubt on this point, the bone (26) may be termed the "suborbital."

With regard to the next bone (27), the same question, whether it answers to the squamosal in the crocodile, or whether it is a dermal ossification, applies. If a homology with a determinate endo-skeletal bone in the crocodile and higher vertebrates were to be predicated, it would be the "squamosal." Above 27, between it and 8, is the "supra-squamosal." Essentially it indicates the tendency to excessive dermal ossification of the skull, like that which extends into the superficial temporal fascia from the squamosal and mastoid in the *Chelonia;* this separate ossification in *Archegosaurus* roofs over the temporal fossa. It is the homologue of the supernumerary surface-bone called "supersquamosal" in the Labyrinthodonts; and both this and the "postorbital" corre-spond in position with the posterior suborbital scale-bones in *Amia* and *Lepidosteus.*

The hinder angles of the skull are formed by the tympanic; in young individuals the tympanic does not extend backward beyond the par-occipital, but as age advances it projects further backward. It appears to abut internally against the pterygoid.

The two rami of the mandible were loosely united at a short symphysis, not exceeding the breadth or depth of the jaw at that point; the depth gradually augments to near the articular end, but never exceeds a sixth, and is usually only an eighth of the length of the jaw, no definite coronoid process being developed; the upper and lower borders are nearly straight as far as the deepest part. The lower border behind this part rises rather abruptly to an angle, which is just below the articular pit. The angular element (30) presents a con-vexity answering to the point of ossification whence some faint ridges radiate upon its outer surface. The dentary (32), if it does not form the articular surface, begins very near it, and each ramus appears to be composed of these two bones. The dentary developes the coronoid rising. Neither articular nor splenial element has been clearly demonstrated. If an articu-lar element has existed, it has been very small.

From fishes the lower jaw of *Archegosaurus* differs in the great length or forward extension of the angular piece (30); but it resembles the piscine type in the simplicity of its com-position. The angular piece is, however, longer in the Ganoids

—*e. g.*, *Amia*, *Polypterus*, *Lepidosteus*,—than in other fishes ; in *Lepidosiren* its proportions are almost those of the *Archegosaurus;* and it offers similar proportions in the mandible of the *Axolotl* and *Proteus* (fig. 65).

The teeth in *Archegosaurus* have the simple conical pointed shape. They are implanted in the premaxillary, maxillary, mandibular, and vomerine bone, and in a single row in each. In the short premaxillaries there are from 8 (*A. Decheni*) to 12 (*A. latirostris*) ; they are rather larger than the maxillary teeth. These follow in an unbroken series to beneath and beyond the orbit, and are about 30 in number ; but their interspaces are such as would lodge double that number in the same extent of alveolar border. The vomerine teeth are in a single row, parallel with and near to the maxillary row ; one or two behind the choane are much larger than the rest, which resemble the maxillary teeth in size. The mandibular teeth extend backward to the coronoid rising, and decrease in size, the front ones being the largest. Each tooth is implanted by a simple base in a shallow cup-shaped socket, with a slightly raised border, to which the circumference of the tooth becomes anchylosed. The tooth is loosened by absorption and shed to make way for a successor. These are developed on the inner, hind, and fore part of the base of the old tooth. The teeth are usually shed alternately. They consist of osteodentine, dentine, and cement. The first substance occupies the centre ; the last covers the superficies of the tooth, but is introduced into its substance by many concentric folds extending along the basal half. These folds are indicated by fine longitudinal, straight striæ along that half of the crown. The section of the tooth at that part (see fig. 65, tooth-section) gives the same structure which is shown by a like section of a tooth of the *Lepidosteus oxyurus*.[*]

The same principle of dental structure is exemplified in

[*] Wyman, American Journal of the Natural Sciences, Oct. 1843.

N

the teeth of most of the ganoid fishes of the carboniferous and Devonian systems, and is carried out to a great and beautiful degree of complication in the "old red" Dendrodonts.

The repetition of this structure in the teeth of one of the earliest genera of *Reptilia*, associated with the defect of ossification of the endo-skeleton and the excess of ossification in the exo-skeleton of the head and nape, instructively illustrates the true affinities and low position in the reptilian class of the so-called *Archegosauri*.

Resting upon and protected by the throat-plate in the middle line, there is a longish slender bone, which must belong to the median series of the hyoid system, either basi- or uro-hyal; it is most probably homologous with the uro-hyal of *Amphiuma* and other Perennibranchiates. That two pairs of slender bones projected outward and backward from the median series, is shown by more than one specimen of *Archegosaurus* in the British Museum. The anterior pair is the longest; these are situated as if they had been attached, one to each side of the broad "throat-plate," which may have represented a basi-hyal. The anterior pair are homologous with the corresponding longer pair of appendages to the broad basi-hyal of *Amphiuma*, and are cerato-hyals. The shorter posterior pair answer to the branchi-hyals in *Amphiuma* and other Perennibranchs. There is no such pair in the hyoidean arch of any known Saurian.

External to the ends of the above lateral elements of the hyoid apparatus, feeble traces of arched series of bony nuclei were detected by Goldfuss, and interpreted by him as remains of partially ossified branchial arches. In all those specimens possessing them they present the outline of two or three arches in dots, or slightly curved series of dots or points. In the small relative size of these indications of branchial arches, the *Archegosaurus* agrees with the *Amphiuma*.

No doubt, in the fully-grown *Archegosaurus*, the lungs

would be equal to the performance of the required amount of respiration ; but the retention of such traces of the embryonal water-breathing system in the adult leads to the inference that the animal must have affected a watery medium of existence for as great a proportion of its time as is observed to be the case in the existing perennibranchiate reptiles ; in which, notwithstanding the degree of development of the lungs, the respiratory function seems to be mainly performed by the gills.

The additional marks of affinity to fishes which the *Archegosaurus* presents in its persistent notochord, cartilaginous basi-occipital, dermal ossifications on the head, and minute body-scales (fig. 65, scales), remove it further from the saurian reptiles, and exhibit it more strongly in the light of an osculant form between the Batrachians and the Ganoids.

The under surface of the body between the head and trunk is defended by broad bony plates, three in number. One is median and symmetrical, of an elongate lozenge shape, with the angles rounded off ; slightly convex externally, a little produced along the middle of the anterior half into something like a low *quasi*-keel. The outer surface is sculptured by radiating furrows, except at so much of the marginal part as is overlapped by the lateral pieces, and by the scapular arch. The lateral throat-plates are attached to the anterior half of the sides of the median one, are shaped like beetles' elytra, and converge forwards. Their centre of ossification is towards their outer and back part, from which the external ridges and grooves radiate towards the inner border.

Von Meyer[*] compares these dermal shields to the ento- and epi-sternal elements of the plastron of Chelonia ; their truer homology seems to the writer to be with the median and lateral large throat-plates or scales of *Megalichthys* and *Sudis*

---

[*] " Die kehlbrust platten konnte man der unpaarigen Platte und dem ersten Platten-paar im Bauchpanzer der Schildkröter vergleichen." (*Op. cit.*, p. 100.)

gigas. The ento-sternal element is the only endo-skeletal piece uncombined with a dermal ossification in most Chelonia. The epi-sternal, like the hyo- and hypo-sternals, appear to be abdominal ribs, with superadded dermal ossifications in *Chelonia.*

The scapulæ (fig. 65, 51) are instructively exhibited in the very young specimen of the *Archegosaurus* figured in t. xiv., fig. 4, of Von Meyer's treatise. The coracoids being doubtless wholly cartilaginous at that stage, are not discernible in the specimen referred to. The upper slender end of the scapula is opposite the side of the vertebral column, about the fifth neurapophysis from the head, and it curves gently downward and forward, expanding at its humeral end. This expansion is more sudden in the fully-developed animal, giving the bone the shape of a rudder, and the direction of the scapula is changed. At least in the specimens (the great majority) in which the skeleton is seen from above, the slender dorsal end of the scapula is seen overlying, or near the hinder border of the lateral throat-plate, and it extends outward and backward to its expanded humeral end. The coracoids (52) are represented by a pair of flat reniform plates, with the convex border turned forward, the concave one backward ; they seem to have overlapped the smooth margins of the posterior half of the median throat-plate. It is most probable that, as in *Amphiuma,* a portion of the broad coracoid remained in the cartilaginous state, and that the full reniform plate answers to the ossified part of that coracoid which it resembles in shape and relative position. The position of the slender scapulæ, styliform and rib-like, as in the Perennibranchiates, is instructively shown in t. xviii., figs. 1 and 2, of M. von Meyer's treatise. The coracoids, as in *Amphiuma,* form the chief part of the articular cavity for the humerus.

The perennibranchiate affinities of *Archegosaurus* are shown as clearly by the scapular as by the hyoidean arch. The fore-limb does not exceed half the length of the head. The humerus

(53) is a short thick bone, slightly constricted at the middle, expanded and rounded at both ends, the proximal one being the largest. For some time the bone is hollow and open at each end; when ossification finally closes the terminal apertures, it shows that the ends were connected to the coracoid and to the fore-arm by interposed ligamentous matter,—not, as in true Saurians, by a synovial joint. Of the two bones of the fore-arm the ulna is a little longer and larger than the radius (54). Both bones present the simplest primitive form, gently constricted in the middle, with the proximal ends a little concave, the distal ones a little convex. The space between the antibrachium and the metacarpus plainly bespeaks the mass of cartilage representing, as in *Amphiuma*, the carpal segment (56) in *Archegosaurus*. No trace of a carpal bone is found save in the largest and oldest examples, in which five or six small roundish ossicles are aggregated near the ulnar side of the carpus. Four digits are present; and considering the pollex to be, as usual, wanting, the second digit answering to the medius of pentadactyle feet, is the largest, and includes at least four phalanges (58); these, with the metacarpals (57), are long, slender, terminally expanded, and truncate. They obviously supported a longish, narrow, pointed paddle. The outermost or little finger was the shortest, and has the shortest metacarpal and first phalanx.

It is true that in *Mystriosaurus* the fore limbs are relatively almost as short as in *Archegosaurus*; and the oolitic crocodile recalls the arrest of development of the same limbs in the marsupial Potoroos; but in *Archegosaurus*, not only is the small size of the fore limbs, but also their type of structure, especially that of their scapular arch, closely in accordance with that in the *Perennibranchiata*, as shown in the tridactyle fore-limb of the *Proteus anguinus*, of which a figure is added to that of the *Archegosaurus* in fig. 65.

The ilium (62), like the scapula, is expanded at its articular or femoral end. It is less long and slender; one border is straight, the other concave, by the expansion toward that border of the femoral end. Two shorter bones on each side complete the pelvis below. One is of a simple form, straight, thicker in proportion to its length than in the ilium : it may be ischium.

The other bone is shown, with its fellow, in t. xiii., fig. 6, and xviii., figs. 8 and 9, of Von Meyer's treatise. That author compares the pair of bones to the *Aptychus* in shape ; they may be the pubic bones. On this hypothesis, they are restored to their true position at 64 (pubis) in fig. 65. The femur (65) is slightly expanded, and truncate at both ends ; it is not longer than the ilium. The tibia (66) and fibula are separate bones; like those of the fore-arm ; the margins, which are turned toward each other, are most concave. They are rather more than half the length of the femur.

The foot-bones are separated by a fibro-cartilaginous tarsal mass (68) from those of the leg. The form of the phalanges, expanded and truncate at both ends, bespeaks their simple ligamentous joints, and that they supported, like the fore-limb, a fin or limb adapted simply for swimming. The argument for the saurian affinities of *Archegosaurus*, based by V. Meyer on the short fore-limbs of *Mystriosaurus*, already invalidated by the difference of structure, is controverted by the fact, that the hind limbs of *Archegosaurus*, like those of the Perenni-branchs, are not only as simple in structure, but also as short, as the fore-limbs.

Genus DENDRERPETON.—In 1852 Sir Charles Lyell and Mr. Dawson, in the course of their investigations of the coal strata of Nova Scotia, remarkable for the erect fossil trees in certain parts, discovered in the hollow of the trunk of one of these trees (*Sigillaria*, 2 feet in diameter), which was wholly converted into coal, some small bones, which Professor Wyman of Boston surmised to have belonged to a batrachian reptile.

By the professor's advice they were brought to England and submitted to the writer, who has described and figured them* as batrachian, under the name *Dendrerpeton Acadianum*, and with close affinities, from the plicated structure of the teeth, the sculpturing of some broad cranial plates, and the structure and proportions of certain limb-bones, to the genus *Archegosaurus*. The subsequent discovery of carinate scales with bones of the *Dendrerpeton* adds to the probability of its appertaining to the Ganocephalous order.

Genus RANICEPS.—In about the centre of the great carboniferous basin of Ohio, United States, at the mouth of the "yellow creek," is a seam of coal 8 feet in thickness, the lower four inches of which is "cannel coal." In this has been found the skull, part of the vertebral column, scapular arch, and fore limbs of a reptile referred by Professor Wyman† to the batrachian sub-class, under the name of *Raniceps*. Two closely-allied fossils, also referred to *Batrachia*, have been found in the same formation and locality.

Order II.—LABYRINTHODONTIA.

Head defended, as in the Ganocephala, by a continuous casque of externally sculptured and unusually hard and polished osseous plates, including the supplementary "post-orbital" and "super-temporal" bones, but leaving a "foramen parietale." Two occipital condyles. Vomer divided and dentigerous. Two nostrils. Vertebral bodies, as well as arches, ossified, biconcave. Pleurapophyses of the trunk, long and bent. Teeth rendered complex by undulation and side branches of the converging folds of cement, whence the name of the order.

The reptiles presenting the above characters have been divided into genera, according to minor modifications exempli-

---

* Quarterly Journal of the Geological Society, vol. ix., 1853.
† American Journal of Science and Arts, March 1857.

fied by the form and proportions of the skull, and by the
relative position and size of the orbital, nasal, and temporal
cavities.

Genus BAPHETES, Ow.

Sp. *Baphetes planiceps.*—In January 1854 the writer communicated to the Geological Society of London a description
of part of a fossil cranium of an animal, from the Pictou coal,
Nova Scotia, measuring 7 inches across the orbit.   From the
characters then specified, the fossil was determined to be the
fore part of a skull of a sauroid Batrachian of the extinct family
of the Labyrinthodonts.   It agreed with them in the number,
size, and disposition of the teeth ; in the proportions and mode
of connection of the premaxillaries, maxillaries, nasals, pre-
frontals and frontals ; and in the resultant peculiarly broad
and depressed character of the skull, the bones of which also
present the same well-marked external sculpturing as in the
Labyrinthodonts : and amongst the genera that have been
established in that family, the form of the end of the muzzle,
or upper jaw, in the Pictou coal specimen, best accorded with
that in the *Capitosaurus* and *Metopias* of Von Meyer and
Burmeister.   But the orbits had been evidently larger and
of a different form than in the reptiles so called ; and, for the
convenience of distinction and reference, it was proposed to
name the fossil *Baphetes planiceps* (βάπτω, *I dip* or *dive*), in
reference to the depth of its position and the shape of its head.

Being thus introduced at the carboniferous period to the
labyrinthodont order, which attained its full development in
the triassic period, the more decisive evidences and typical
illustrations of that extinct group of reptiles will next be
described.

Genus LABYRINTHODON, Ow.

At the period of the deposition of the new red sandstone,
in the present counties of Warwick and Cheshire, the shores
of the ancient sea, which were then formed by that sandy

deposit, were trodden by reptiles having the essential bony
characters of the modern *Batrachia*, but combining these with
other bony characters of crocodiles, lizards, and ganoid fishes ;
and exhibiting all under a bulk which, as made manifest by
the fossils and footprints, rivalled that of the largest crocodiles
of the present day. The form of the largest Labyrinthodonts,
if we may judge by the great breadth and flatness of the skull,
and the proportions of certain bones, seems to have been
something between that of the toad or land-salamander.

 The smooth-skinned Batrachians have no fixed type of external form like the existing higher orders of reptiles, but some, as
the broad and flat-bodied toads and frogs, most resemble the
Chelonians, especially the soft-skinned mud-tortoises (*Trionyx*);
other Batrachians, as the *Cœciliœ*, resemble Ophidians ; a third
group, as the newts and salamanders, represent the Lacertians ;
and among the perennibranchiate reptiles there are species
(*Siren*) which combine with external gills the mutilated condition of the apodal fishes.

Thus it will be perceived that, even if the entire skeleton
of a Labyrinthodont had been
obtained, there is no fixed or
characteristic general outward
form in the Batrachian order
whereby its affinity to that group
could have been determined.
The common characters by
which the Batrachians, so diversified in other respects, are
naturally associated into one
group or sub-class of reptiles,
besides being taken from the
condition of the circulating and
generative systems, and other
perishable parts, are manifested in modifications of the skeleton,

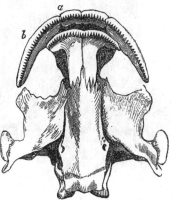

Fig. 65 *a*.

Cranium and upper jaw and teeth of
the Menopome (*Menopoma
alleghanniense*).

and principally in the skull. This is joined to the atlas by the medium of two tubercles, developed exclusively from the ex-occipitals ; the bony palate is formed chiefly by two broad and flat bones (fig. 65 *a, c*), called " vomerine," and generally supporting teeth. It is only in the Batrachians among existing reptiles that examples are found of two or more rows of teeth on the same bone, especially on the lower jaw (*Cæcilia, Siren*). With regard to vertebral characters, no such absolute batrachian modifications can be adduced as those above cited from the anatomy of the cranium. Some Batrachians have the vertebræ united by ball-and-socket joints, as in most recent reptiles ; others by biconcave joints, as in a few recent and most extinct Saurians. Some species have ribs, others want those appendages ; the possession of ribs, therefore, even if longer than those of the *Cæcilia*, by a fossil reptile combining all the essential batrachian characters of the skull, would not be sufficient ground for pronouncing such reptile to be a Saurian. Much less could its saurian nature be pronounced from the circumstance of its possessing large conical striated teeth : the ordi-

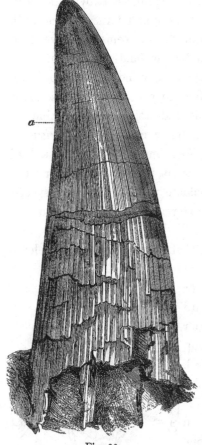

Fig. 66.

Canine tooth of the *Labyrinthodon Jagaeri* (nat. size.)

nary characters of size, form, number, and even of presence
or absence of teeth, vary much in existing Batrachians ; and
the location of teeth on the vomerine bones is the only dental
character in which they differ from all other orders of reptiles.

The writer's acquaintance with the remarkable fossils
under consideration was begun by the examination, in 1840,
of portions of teeth from the new red sandstone of Coton End
quarry, Warwickshire. The external characters of these teeth
corresponded with those (fig. 66) which had previously been
discovered by Professor Jaeger in the German Keuper formation in Wirtemberg, and on which the genus *Mastodonsaurus*
had been founded.

The results of a microscopic examination of the teeth of the
Mastodonsaurus from the German Keuper, and of those from
the new red sandstone of Warwickshire, proved that the teeth
from both localities possessed in common a very remarkable
and complicated structure (fig. 67), to the principle of which,
—viz., the convergence of numerous inflected folds of the
external layer of cement towards the pulp-cavity,—a very
slight approach was made in the fang of the tooth of the
Ichthyosaurus, whilst a closer approximation to the labyrinthic
structure in question was made by the teeth of several species
of ganoid fishes, and by those of *Archegosaurus.*

Thus, inasmuch as the extinct animals in question manifested in the intimate structure of their teeth an affinity to
fishes, it might be expected that, if they actually belonged to
the class of reptiles, the rest of their structure would manifest
the characters of the lowest order,—viz., the *Batrachia,* the
existing members of which pass, though not by the dental character alluded to, yet by so many other remarkable degrada-
tions of structure, towards fishes.

In the same formation in Wirtemberg from which the labyrinthic teeth of the so-called *Mastodonsaurus* had been derived, a
fragment of the posterior portion of the skull has been obtained,

showing the development of a separate condyle on each ex-
occipital bone ; whence Professor Jaeger, recognizing the
identity of this structure with the batrachian character above
mentioned, founded upon the fossil a new genus of *Batrachia*
which he called "*Salamandroïdes giganteus.*" Subsequent discoveries, however, satisfied the Professor that the bi-condylous
fragment of skull, representing the genus *Salamandroides*,
belonged to the same reptile as the teeth on which he had
founded the genus *Mastodonsaurus*. The following fossils,
from the new red sandstone of Warwickshire, gave additional
proof of the batrachoid nature of the genus to which those

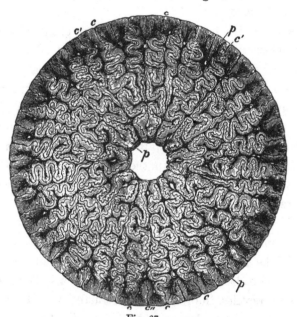

Fig. 67.

Transverse section of a tooth of the *Labyrinthodon* (magn.)

fossils belong, with the establishment of five distinct species,
one of which is most probably identical with the *Mastodonsaurus salamandroïdes* of Professor Jaeger. In reference to
the generic denomination *Mastodonsaurus*, it unavoidably recalls the idea of the mammalian genus *Mastodon*, or else a mam-

milloid form of tooth, whereas all the teeth of the reptile so called are originally, and most of them are permanently, of a cuspidate and not of a mammilloid form ; secondly, because the second element of the word, *saurus*, indicates the genus to belong to the saurian or lacertian order of reptiles.   For these reasons, the writer has proposed to designate the genus in question *Labyrinthodon*, in allusion to the peculiar and characteristic structure of the teeth (fig. 67).

The specimens from British localities are referable to five species—viz., 1. *Labyrinthodon salamandroïdes ;* 2. *L. leptognathus ;* 3. *L. pachygnathus ;* 4. *L. ventricosus ;* and 5. *L. scutulatus ;* and we shall here briefly notice the characters exhibited by the bones assignable to the second, third, and fifth species.

Labyrinthodon leptognathus.—The remains of this species consist of fragments of the upper and lower jaws, two vertebræ, and a sternum.   They were found in the new red sandstone quarries at Coton End near Warwick.

A dorsal vertebra from Coton End presents further evidence of the batrachian nature of the *Labyrinthodon*.   It has concave articular cavities at the extremities of the body,—a condition now known among existing reptiles only in the Geckos, and in the lower or perennibranchiate division of Batrachians. It is a common structure in extinct Saurians, but the depth of the vertebral articular cavities in the *Labyrinthodon* exceeds that in the amphicœlian Crocodiles and in most Plesiosaurs. The body of the vertebra is elongate and sub-compressed, with a smooth but not regularly curved lateral surface, terminating below in a slightly-produced, longitudinal, median ridge ; and it exhibits the same exceptional condition in the reptilian class as do the vertebræ of existing Batrachians, in having the superior arch or neurapophysis anchylosed with the centrum. From each side of the base of the neural arch a thick and strong transverse process extends obliquely outwards and upwards.

A symmetrical bone, resembling the episternum of the *Ichthyosaurus* was associated with the preceding remains. It consists of a stem or middle, which gradually thickens to the upper end, where cross pieces are given off at right angles to the stem, and support on each a pretty deep and wide groove indicating strongly the presence of clavicles, and thus pointing out another distinction from crocodiles, in which clavicles are wanting. Most Batrachians possess these bones.

The modifications of the jaws, and more especially those of the bony palate of the *Labyrinthodon leptognathus*, prove the fossil to have been essentially Batrachian, but with affinities to the higher Sauria, leading, in the form of skull and the sculpturing of the cranial bones, to the crocodilian group, in the collocation of the larger fangs at the anterior extremities of the jaws to the *Plesiosaurus*, and in one part of the dental structure, in the form of the episternum, and the bi-concave vertebræ, to the *Ichthyosaurus*. Another marked peculiarity in this fossil is the anchylosis of the base of the teeth to distinct and shallow sockets, by which it is made to resemble the Sphyræna and certain other fishes. From the absence of any trace of excavation at the inner side of the base of the functional teeth, or of alveoli of reserve for the successional teeth, it may be concluded that the teeth were reproduced, as in the lower Batrachians and in many fishes, in the soft mucous membrane which covered the alveolar margin, and that they subsequently became fixed to the bone by anchylosis, as in the pike and Lophius.

Labyrinthodon pachygnathus.—The remains of this species, which have been obtained, consist of portions of the lower and upper jaws, an anterior frontal bone, a fractured humerus, an ilium with a great part of the acetabulum, the head of a femur, and two ungual phalanges. A portion, nine and a half inches long, of a right ramus of a lower jaw, in addition to the characters common to it and the fragment of the lower jaw of the

L. leptognathus, in the structure of the angular and dentary pieces, shows that the outer wall of the alveolar process is not higher than the inner, as in frogs and toads, the salamanders and menopome, in all of which the base of the teeth is anchy-losed to the inner side of an external alveolar plate. The smaller serial teeth are about forty in number, and gradually diminish in size as they approach both ends, but chiefly so towards the anterior part of the jaw. The sockets are close together, and the alternate ones are empty. The great laniary teeth were apparently three in each symphysis, and the length of the largest was one inch and a half. The base of each tooth is anchylosed to the bottom of its socket, as in scomberoid and sauroid fishes ; but the *Labyrinthodon* possesses a still more ichthyic character in the continuation of a row of small teeth anterior and external to the two or three larger tusks. The premaxillary bone presents the same peculiar modification as in the higher organized Batrachia, the palatal process of the premaxillary extending beyond the outer plate both externally and, though in a less degree, internally, where it forms part of the boundary of the anterior palatal foramen, whence the outer plate rises in the form of a compressed process from a longitudinal tract in the upper part of the palatal process ; it is here broken off near its margin, and the fractured surface gives the breadth of the base of the outer plate, stamping the fossil with a Batrachian character conspicuous above all the saurian modifications by which the essential nature of the fossil appears at first sight to be masked.

In the pre-frontal bone there are indications of crocodilian structure. Its superior surface is slightly convex and pitted with irregular impressions ; and from its posterior and outer part it sends downwards a broad and slightly concave process, which appears to be the anterior boundary of the orbit. This process presents near its upper margin a deep pit, from which a groove is continued forwards ; and in the corresponding orbital plate of the crocodile there is a similar but smaller foramen.

From these remains of the cranium of the *L. pachygnathus* it is evident that the facial or maxillary part of the skull was formed in the main after the crocodilian type, but with well-marked batrachian modifications in the premaxillary and inferior maxillary bones. The most important fact which they show is, that this sauroid Batrachian had subterminal nostrils, leading to a wide and shallow nasal cavity, separated by a broad and almost continuous palatal flooring from the cavity of the mouth; indicating, with their horizontal position, that their posterior apertures were placed behind the anterior or external nostrils, whereas in the air-breathing Batrachia the nasal meatus is short and vertical, and the internal apertures pierce the anterior part of the palate, suitable to their mode of breathing by deglutition. It may be inferred, therefore, that the apparatus for breathing by inspiration must have been present in the *Labyrinthodon* as in the crocodile; and that the skeleton of the *Labyrinthodon* will be found to be provided with well-developed costal ribs, and not, as in most of the existing Batrachians, with merely rudimentary styles. Since the essential condition of this defective state of the ribs of Batrachians is well known to be their fish-like mode of generation and necessary distension of the abdomen, it is probable that the generative economy of the Labyrinthodonts, in which the more complete ribs would prevent the excessive enlargement of the ovaria and oviducts, may have been similar to that of saurian reptiles.

Of the few bones of the extremities which have come under the writer's inspection, one presents all the characteristics of the corresponding part of the humerus of a toad or frog, viz., the convex, somewhat transversely extended articular end, the internal longitudinal depression, and the well-developed deltoid ridge. The length of the fragment is two inches, and the breadth is thirteen lines. The ridges are moderately thick and compact, with a central medullary cavity. In its structure, as well as in its general form, the present bone agrees with the batrachian, and differs from the crocodilian type.

In the right ilium, about six inches in length, and in the acetabulum, there is a combination of crocodilian and batrachian characters. The acetabular cavity is bounded on its upper part by a produced and sharp ridge, as in the frog, and not emarginate at its anterior part, as in the crocodile.

As the fragment of the ilium was discovered in the same block as the two fragments of the cranium and the portion of the lower jaws, it is probable that they may have belonged to the same animal ; and if so, as the portions of the head correspond in size with those of the head of a crocodile six or seven feet in length, but the acetabular cavity with that of a crocodile twenty-five feet in length, then the hinder extremities of the *Labyrinthodon* must have been of disproportionate magnitude compared with those of existing Saurians, but of approximate magnitude with some of the living anourous Batrachians. That such a reptile, of a size equal to that of the species whose remains have just been described, existed at the period of the formation of the new red sandstone, is abundantly manifested by the remains of those singular impressions to which the term *Cheirotherium* has been applied. Other impressions, as those of the *Cheirotherium Hercules*, correspond in size with the remains of the *Labyrinthodon salamandroïdes*, which have been discovered at Guy's Cliff. The head of a femur from the same quarry in which the ilium was found is shown to correspond in size with the articulate cavity of the acetabulum. The two toe-bones, or terminal phalanges, resemble those of Batrachians in presenting no trace of a nail, and from their size they may be referred to the hind feet of the *L. pachygnathus.*

An entire skull of the largest species discovered in the new red sandstones of Wurtemberg ; a lower jaw of the same

species found in the same formation in Warwickshire ; some
vertebræ, and a few fragments of bones of the limbs, have
served, with the indications of size and shape of the trunk of
the animal yielded by the series of consecutive footprints, as
the basis of the restoration of the *Labyrinthodon salamandroïdes*, at the Crystal Palace. It is to be understood, however, that, with the exception of the head, the form of the
animal is necessarily more or less conjectural.

Labyrinthodon scutulatus.—The remains to which this
specific designation has been applied compose a closely and
irregularly aggregated group of bones imbedded in sandstone,
and manifestly belonging to the same skeleton ; they consist
of four vertebræ, portions of ribs, a humerus, a femur, two
tibiæ, one end of a large flat bone, and several small osseous
externally sculptured dermal scutes, which show that the
crocodilian nature of a fossil reptile is not determinable by
scutes only. The mass was discovered in the new red sandstone at Leamington, and was transmitted to the writer in the
summer of 1840.

The vertebræ present bi-concave articular surfaces similar
to those of the other species. In two of them the surfaces
slope in a parallel direction obliquely from the axis of the
vertebræ, as in the dorsal vertebræ of the frog, indicating a
habitual inflexion of the spine, analogous to that in the humped
back of the frog. The neurapophyses are anchylosed to the
vertebral body. The spinous process rises from the whole
length of the middle line of the neurapophysial arch, and its
chief peculiarity is the expansion of its elongated summit into
a horizontally-flattened plate, sculptured irregularly on the
upper surface. A similar flattening of the summit of the
elongated spine is exhibited in the large atlas of the toad.
The body of the vertebra agrees with that of the *L. leptognathus*. The humerus is an inch long, regularly convex at the
proximal extremity, and expanded at both extremities, but

contracted in the middle. A portion of a somewhat shorter
and flatter bone is bent at a subacute angle with the distal
extremity, and resembles most nearly the anchylosed radius
and ulna of the Batrachia.

The femur wants both the extremities ; its shaft is subtrihedral and slightly bent, and its walls are thin and compact,
including a large medullary cavity. The tibiæ exhibit that
remarkable compression of their distal portion which characterizes the corresponding bone in the Batrachia ; they likewise
have the longitudinal impression along the middle of the
flattened surface. Were more of the skeleton of the above-
defined species of *Labyrinthodon* known, they might present
differences of subgeneric value. Such differences in the forms
and proportions of the skull, and in the form and relative
position of the orbits, of specimens that have been discovered
subsequently in the triassic sandstones of Germany, have been
so interpreted.

In the *Labyrinthodon* (*Mastodonsaurus*) *Jaegeri* — the
largest of the species—the skull is triangular, the two condyles
projecting from the middle of the base ; the sides are straight,
and converge to the obtuse apex. The orbits are oval, narrowest anteriorly, and are situated nearly midway between the
fore and back part of the skull. The nostrils are very small,
and are as wide apart as the orbits.

Labyrinthodon (*Trematosaurus*) *Braunii*, Von Meyer.—
The name *Trematosaurus* was given by Braun to a labyrinthodont reptile, in reference to the parietal foramen, at that
time deemed to be peculiar to it, but now known to be common to all the family. The genus was founded on an unusually
perfect skull discovered in the richly fossiliferous bunter-
sandstein of Bernburg. It is about one foot long, and, relatively to its basal breadth, it is longer and narrower than in
L. Jaegeri, the sides converging at a more acute angle. The
orbits are elliptical, situated in the middle of the skull, and

wider apart than in *L. Jaegeri;* the nostrils are relatively nearer together, their interspace being only half that in the *L. Jaegeri.*

Labyrinthodon (Metopias *) *diagnosticus,* H. von M.—In this species the skull is broader in proportion to its length than in the foregoing ; the sides are convex as they converge to the obtuse muzzle. The orbits are small, of a wide elliptical form, situated in the anterior third of the skull; they are twice as wide apart as are the nostrils. The parietal foramen is near the occipital ridge. The remains of this species are from the upper beds of the keuper sandstone in Wirtemberg.

The *Labyrinthodon (Capitosaurus) arenaceus,* Münster, is distinguished by a much broader and almost truncate muzzle. The orbits are elliptic, and situated almost wholly in the hinder third of the cranium ; their interspace is the same as that between the nostrils, which are relatively as large as in *L. Braunii.*

The name *Zygosaurus* appears to have been applied with better grounds, by Eichwald, to a labyrinthodont reptile from the Permian cupriferous beds at Orenburg. It has the parabolic skull of *L. Jaegeri* and *L. diagnosticus;* the orbits large, and divided by an interval less than their own diameter. The temporal fossæ are relatively larger, and bounded by stronger zygomatic arches, and seem not to have been roofed over by bone. The dentition is strictly labyrinthodont.

Odontosaurus Voltzii is a genus and species founded by Von Meyer on a portion of a lower jaw, containing fifty teeth lodged in rather a deep groove, but apparently presenting the labyrinthic structure. The specimen is from the bunter sandstone of Soultz-les-Bains.

Xestorrhytias Perrini.—By this name M. von Meyer would indicate certain flat cranial bones, sculptured like those of

---

* This generic term has been applied to another fossil by Eichwald.

Labyrinthodon, but with a peculiarly polished ganoid-like surface, from the muschelkalk of Lunéville.

In all the foregoing forms of Labyrinthodonts, represented by complete crania, with the exception perhaps of *Zygosaurus*, the supplemental osseous plates roofing over the temporal fossæ are present, as in *Archegosaurus*, viz., the "post-orbital" and the "super-squamosal" bones. In all of them the occipital condyles are distinct, forming a pair ; and in all the vomer is divided and bears teeth. The structure and disposition of the entire dental system is strictly labyrinthodont.

The relation of these remarkable reptiles to the saurian order has been advocated to be one of close and true affinity, chiefly on the character of the extent of ossification of the skull, and of the outward sculpturing of the cranial bones. But the true nature of some of these bones appears to have been overlooked, and the glance of research for analogous structures has been too exclusively upward. If directed downward from the Labyrinthodonts to the *Archegosauri* and certain ganoid fishes, it suggests other conclusions.

The conformity of pattern in the dermal, semidermal, or neurodermal bones of the outwardly well-ossified skull of *Polypterus, Lepidosteus, Sturio*, and other salamandroid-ganoid fishes, with well-developed lung-like air-bladders, and of the same skull-bones in the *Archegosaurus* and the Labyrinthodonts ; the persistence of the notochord (*chorda dorsalis*) in *Archegosaurus*, as in *Sturio ;* the persistence of the notochord and branchial arches in *Archegosaurus*, as in *Lepidosiren ;* the absence of occipital condyle or condyles in *Archegosaurus*, as in *Lepidosiren ;* the presence of labyrinthic teeth in *Archegosaurus*, as in *Lepidosteus* and *Labyrinthodon ;* the large median and lateral throat-plates in *Archegosaurus*, as in *Megalichthys*, and in the modern *Arapaima* and *Lepidosteus ;*—all these characters point to one great natural group, peculiar for the extensive gradations of development, linking and blending together fishes and reptiles within the limits of such group. The salamandroid (or so-called "sauroid") Ganoids—*Lepidosteus* and *Polypterus*—are the most ichthyoid, the true Labyrinthodonts are the most sauroid, of the group. The *Lepidosiren* and *Archegosaurus* are intermediate gradations, one having more of the piscine, the other more of the reptilian characters. The *Archegosaurus* conducts the march of development from the fish proper to the labyrinthodont type; the *Lepidosiren* conducts it to the perennibranchiate batrachian type. Both illustrate the artificiality of the supposed class distinction between fishes and reptiles, and the naturality of the "Hæmatocrya," or cold-blooded Vertebrata, as the one unbroken progressive series. There is nothing in the known structure of the so-named *Archegosaurus* or *Mastodonsaurus* that truly indicates a belonging to the saurian or crocodilian order of reptiles. The exterior ossifications of the skull and the canine-shaped labyrinthic teeth are both examples of the salamandroid modification of the ganoid type of fishes.

The small proportion of the fore limb of the *Mystriosaurus* in nowise illustrates this alleged saurian affinity; for though it be as short as in *Archegosaurus*, it is as perfectly constructed as in the crocodile, whereas the short fore limb of *Archegosaurus* is constructed after the simple type of that of the *Proteus* and *Siren*. But the futility of this argument of the sauroid affinities is made manifest by the proportions of the hind limb of *Archegosaurus*. As in *Proteus* and *Amphiuma*, it is as stunted as the fore limb; whereas in *Mystriosaurus*, as in other Teleosaurians, the hind limbs are relatively larger and stronger than in the existing crocodiles.

Order 3.—ICHTHYOPTERYGIA.*

The bones of the head still include the supplementary "post-orbitals" and "supra-temporals," but there are small

* Ιχθύς, *a fish;* πτέρυξ, *a fin.*

temporal and other vacuities between the cranial bones :
a "foramen parietale," a single convex occipital condyle,
and one vomer which is edentulous. Two antorbital
nostrils. Vertebral centra, ossified, biconcave. Pleurapophyses of the trunk long and bent, the anterior ones
with bifurcate heads. Teeth with converging folds of
cement at their base ; implanted in a common alveolar
groove, and confined to the maxillary, premaxillary, and
premandibular bones. Premaxillaries much exceeding
the maxillaries in size. Orbit very large ; a circle of
sclerotic plates. Limbs natatory ; with more than five
multi-articulate digits ; no sacrum. With the retention
of characters which indicate, as in the preceding orders,
an affinity to the higher Ganoid fishes, the present exclusively marine Reptilia more directly exemplify the
ichthyic type in the proportions of the premaxillary
and maxillary bones, in the shortness and great number
of the biconcave vertebræ, in the length of the pleurapophyses of the vertebræ near the head, in the large
proportional size of the eyeball and its well-ossified
sclerotic coat, and especially in the structure of the
pectoral and ventral fins.

It has been usual to unite the present with the following
order in the same group, called *Enaliosauria* or sea-lizards.
They were adapted for marine life, but breathed the air like
the *Cetacea :* they were, however, "cold-blooded," or of a low
temperature, like crocodiles and other reptiles. The proof
that the Enaliosaurs respired atmospheric air immediately,
and did not breathe water by means of gills like fishes, is
afforded by the absence of the bony framework of the gill
apparatus, and by the presence, position, and structure of the
air-passages leading from the nostrils to the mouth, and also
by the bony mechanism of the capacious chest or thoracic-

abdominal cavity ; all of which characters have been demonstrated by their fossil skeletons. With these characters the
sea-lizards combined the presence of two pairs of limbs shaped
like fins, and adapted for swimming.

The group of reptiles so termed includes all those which
have any part of the thoracic-abdominal cavity encompassed
by moveable ribs. The first character distinguishes them
from the *Batrachia* and *Chelonia* with fin-shaped limbs.

The *Enaliosauria,* however, do not form a strictly natural
group ; for this is based upon a single character relating to
merely the medium of life and locomotion. Some of the
labyrinthodont reptiles may have had their limbs in structure
and shape as paddles ; but more important modifications of
structure would keep them apart, like the lower *Batrachia*
and the *Chelonia,* from the more lizard-like reptiles called
Enaliosauria.

In this group there are two divisions,—one characterized
by having five digits in the fin, the other by having more
than that typical number. The pentadactyle division may
be sub-divided into those in which the ilio-pubic arch is
attached to a sacrum and those on which it is freely suspended or not so attached. The polydactyle division presents
a general type of structure more conformable with that of
which the Archegosaurs and Labyrinthodonts manifest two
phases of development, and in which the ascent from the
gano-salamandroid fishes reaches its culminating point in
Ichthyosaurus.

Genus ICHTHYOSAURUS.—The name (from the Greek *ichthys,*
a fish, and *sauros,* a lizard) was devised to indicate the closer
affinity of the Ichthyosaur, as compared with the Plesiosaur,
to the class of fishes. The Ichthyosaur (fig. 68) is remarkable for the shortness of the neck and the equality of the
width of the back of the head with the front of the chest,
impressing the observer of the fossil skeleton with a conviction

that the ancient animal must have resembled the whale tribe
and the fishes, in the absence
of any intervening constriction
or neck.

This close approximation in
the Ichthyosaurs to the form of
the most strictly aquatic vertebrate animals of the existing
creation, is accompanied by an
important modification of the
surfaces forming the joints of
the back-bone, each of which
surfaces are hollow, leading to
the inference that they were
originally connected together
by an elastic bag or "capsule"
filled with fluid—a structure
which prevails in the class of
fishes, in the Labyrinthodonts
and a few extinct aquatic reptiles, in the existing perennibranchiate Batrachia, but not
in any of the whale or porpoise
tribe.

With the above modifica-
tions of the head, trunk, and
limbs, in relation to swimming,
there co-exist corresponding
modifications of the tail. The
bones of this part are much
more numerous than in the
Plesiosaurs, and the entire tail
is consequently longer ; but it
does not show any of those

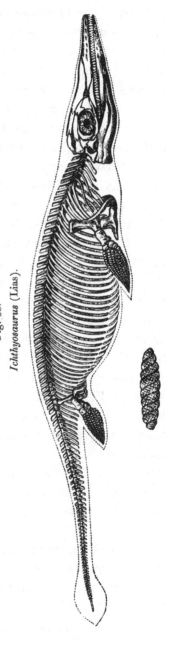

Fig. 68. *Ichthyosaurus* (Lias).

modifications that characterize the bony support of the tail fin in fishes. The numerous caudal vertebræ of the *Ichthyosaurus*, gradually decrease in size to the end of the tail, where they assume a compressed form, or are flattened from side to side, and thus the tail, instead of being short and broad as in fishes, is lengthened out as in crocodiles.

The very frequent occurrence of a fracture of the tail, about one-fourth of the way from its extremity, in well preserved and entire fossil skeletons, is owing to that proportion of the end of the tail having supported a cutaneous and perishable caudal fin.[*] The only evidence which the fossil skeleton of a whale would yield of the powerful horizontal tail-fin characteristic of the living animal, is the depressed or horizontally-flattened form of the bones supporting such fin. It is inferred, therefore, from the corresponding bones of the *Ichthyosaurus* being flattened in the vertical direction, or from side to side, that it possessed a tegumentary tail fin expanded in the vertical direction. The shape of a fin composed of such perishable material is of course conjectural, as is the outline in fig. 68. Thus, in the construction of the principal swimming organ of the *Ichthyosaurus* we may trace, as in other parts of its structure, a combination of mammalian (beast-like), saurian (lizard-like), and piscine (fish-like) peculiarities. In the great length and gradual diminution of the tail we perceive its saurian character; in the tegumentary nature of the fin, unsustained by bony fin-rays, its affinity to the same part in the mammalian whales and porpoises is shown; whilst its vertical position makes it closely resemble the tail fin of the fish.

The horizontality of the tail fin of the whale tribe is essentially connected with their necessities as warm-blooded animals breathing atmospheric air; without this means of displacing a mass of water in the vertical direction, the head of the whale

---

[*] Trans. Geol. Soc., 2d series, vol. v., p. 511.

could not be brought with the required rapidity to the surface to respire ; but the Ichthyosaurs, not being warm-blooded or quick breathers, would not need to bring their head to the surface so frequently or so rapidly as the whale ; and moreover, a compensation for the want of horizontality of their tail fin was provided by the addition of a pair of hind paddles, which are not present in the whale tribe. The vertical fin was a more efficient organ in the rapid cleaving of the liquid element, when the Ichthyosaurs were in pursuit of their prey, or escaping from an enemy.

The general form of the cranium of the *Ichthyosaurus* resembles that of the ordinary cetaceous dolphin (*Delphinus tursio*) ; but the *I. tenuirostris* rivals the *Delphinus gangeticus* in the length and slenderness of the jaws. The essential difference in the sea-reptile lies in the restricted size of the cerebral cavity, and the vast depth and breadth of the zygomatic arches, to which the seeming expanse of the cranium is due ; still more in the persistent individuality of the elements of those cranial bones which have been blended into single though compound bones in the sea-mammal. The *Ichthyosaurus* further differs in the great size of the premaxillary, and small size of the maxillary bones, in the lateral aspects of the nostrils, in the immense size of the orbits, and in the large and numerous sclerotic plates, which latter structures give to the skull of the *Ichthyosaurus* its most striking features.

The true affinities of the Ichthyosaur are, however, to be elucidated by a deeper and more detailed comparison of the structure of the skull ; and few collections now afford richer materials for pursuing and illustrating such comparisons than the palæontological series in the British Museum.* The two supplemental bones of the skull, which have no homologues

---

* The anatomical reader is referred to the writer's "Report on British Fossil Reptiles," Trans Brit. Assoc. 1839, and to the Annals and Magazine of Natural History, 1858, p. 388.

in existing Crocodilians, are the post-orbital and super-squamosal; both, however, are developed in *Archegosaurus* and the Labyrinthodonts. The post-orbital is the homologue of the inferior division of the post-frontal in those Lacertians —*e. g.*, *Iguana*, *Tejus*, *Ophisaurus*, *Anguis*, in which that bone is said to be divided. But in *Ichthyosaurus* the post-orbital resembles most a dismemberment of the malar. Its thin obtuse scale-like lower end overlaps and joins by a squamous suture the hind end of the malar : the post-orbital expands as it ascends to the middle of the back of the orbit, then gradually contracts to a point as it curves upward and forward, articulating with the super-squamosal and post-frontal. The super-squamosal may be in like manner regarded as a dismemberment of the squamosal; were it confluent therewith, the resemblance which the bone would present to the zygomatic and squamosal parts of the mammalian temporal would be very close; only the squamosal part would be removed from the inner wall to the outer wall of the temporal fossa. The super-squamosal, in fact, occupies the position of the temporal fascia in *Mammalia*, and should be regarded as a supplemental sclero-dermal plate, closing the vacuity between the upper and lower elements of the zygomatic arch, peculiar to certain air-breathing *Ovipara*. In the *Ichthyosaurus* it is a broad, thin, flat, irregular-shaped plate, smooth and slightly convex externally, and wedged into the interspaces between the post-frontal, post-orbital, squamosal, tympanic, and mastoid.

The principal vacuities or apertures in the bony walls of the skull of the *Ichthyosaurus* are the following :—In the posterior region the "foramen magnum," the occipito-parietal vacuities, and the auditory passages ; on the upper surface the parietal foramen and the temporal fossæ ; on the lateral surfaces the orbits and nostrils, the plane of the aperture in both being vertical ; on the inferior surface the palato-nasal, the pterygo-sphenoid, and the pterygo-malar vacuities. The occipito-parietal vacuities are larger than in *Crocodilia*, smaller than in *Lacertilia;* they are bounded internally by the basi-, ex-, and super-occipitals, externally by the parietal and mastoid. The auditory apertures are bounded by the tympanic and squamosal. The tympanic takes a greater share in the formation of the "meatus auditorius" in many lizards; in crocodiles it is restricted to that which it takes in *Ichthyosaurus*.

The orbit is most remarkable in the *Ichthyosaurus*, amongst reptiles, both for its large proportional size and its posterior position; in the former character it resembles that in the lizards, in the latter that in the crocodiles. It is formed by the pre- and post-frontals above, by the lacrymal in front, by the post-orbital behind, and by the peculiar long and slender malar bar below. In crocodiles and in most lizards the frontal enters into the formation of the orbits, and in lizards the maxillary also. The nostril is a longish triangular aperture, with the narrow base behind; it is bounded by the lacrymal, nasal, maxillary, and premaxillary bones. It is proportionally larger than in the *Plesiosaurus*, and is distant from the orbit about half its own long diameter. Like the orbit, the plane of its outlet is vertical.

The pterygo-palatine vacuities are very long and narrow, broadest behind, where they are bounded, as in lizards, by the anterior concavities of the basi-sphenoid, and gradually narrowing to a point close to the palatine nostrils. These are smaller than in most lizards, and are circumscribed by the palatines, ecto-pterygoid, maxillary, and premaxillary. The pterygo-malar fissures are the lower outlets of the temporal fossæ; their sudden posterior breadth, due to the emargination of the pterygoid, relates to the passage of the muscles for attachment to the lower jaw. The parietal foramen is bounded by both parietals and frontals; its presence is a mark of labyrinthodont and lacertian affinities; its formation is like that in *Iguana* and *Rhynchocephalus*. The temporal fossæ are bounded above by the parietal internally, by the mastoid and post-frontal externally; they are of an oval form, with the great end forward. In their relative size and backward position they are more crocodilian than lacertian.

In the *Ichthyosaurus communis* there are seventeen sclerotic plates forming the fore part of the eyeball. In a well-preserved example, the pupillary or corneal vacuity, as bounded by those plates, is of a full oval form, $1\frac{1}{2}$ inch in long diameter, the length of the plates (or breadth of the frame) being from 8 to 10 lines. In the same skull the long diameter of the orbit is 4 inches. The deep position of the sclerotic circle in this cavity showed how they had sunk, by pressure of the external mud, as the eyeball became collapsed by escape of the humours in decomposition.

Whenever the antecedent forms of an extinct genus of any class are known, the characters of such genus should be compared with those of its predecessors in such class, rather than with its successors or with existing forms, in order to gain an insight into its true affinities.

We derive a truer conception of the affinities of the *Ichthyosaurus* by comparison with the Labyrinthodonts and other triassic reptiles, as we do of the *Plesiosaurus* by comparison with the muschelkalk *Sauropterygia*, than of either by comparison with modern Lacertians and Crocodilians. It is commonly said that the *Ichthyo-* or the *Plesio-saurus* resembles more the lizards in such and such characters, and in a less degree the crocodiles, as in such a character. The truer expression would be that the lizards, which are the predominating form of Saurians at the present day, have retained more of the osteological type of the triassic and oolitic reptiles, and that the crocodiles deviate further from them or exhibit a more modified or specialized structure. The posterior position of the nostrils, the small size and position of the palato-pterygoid foramen, are marks of affinity to *Plesiosaurus*, in common with which genus the cranial structure of the *Ichthyosaurus* exhibits a majority of lacertian characters.

In comparing the jaws of the *Ichthyosaurus tenuirostris* with those of the gangetic Gharrial, an equal degree of strength and of alveolar border for teeth result from two very different proportions in which the maxillary and premaxillary bones are combined together to form the upper jaw. The prolongation of the snout has evidently no relation to this difference; and we are accordingly led to look for some other explanation of the disproportionate development of the premaxillaries in the *Ichthyosaurus*. It appears to me to give additional proof of the collective tendency of the affinities of the *Ichthyosaurus* to the lacertian type of structure. The backward or antorbital position of the nostrils, like that in whales, is related to their marine existence. But in the Lacertians in which the nostrils extend to the fore part of the head, their anterior boundaries are formed by the premaxillaries: it appears, therefore, to be in conformity with the lacertian affinities of the Ichthyosaur that the premaxillaries should still enter into the same relation with the nostrils, although this involves an extent of anterior development proportionate to the length of the jaws, the forward production of which sharp-toothed instruments fitted them, as in the modern dolphins, for the prehension of agile fishes.

That the Ichthyosaurs occasionally sought the shores, crawled on the strand, and basked in the sunshine, may be inferred from the bony structure connected with their fore fins, which does not exist in any porpoise, dolphin, grampus, and whale; and for want of which, chiefly, those warm-blooded, air-breathing, marine animals are so helpless when left high and dry on the sands. The structure in question in the Ichthyosaur is a strong osseous arch, inverted and spanning across beneath the chest from one shoulder-joint to the other; and what is most remarkable in the structure of this "scapular"

arch is, that it closely resembles, in the number, shape, and
disposition of its bones, the same part in the singular aquatic
mammalian quadruped of Australia, called *Ornithorynchus*,
and *Platypus*, or duck-mole. The Ichthyosaur, when so
visiting the shore either for sleep or procreation, would lie or
crawl prostrate, or with its belly resting or dragging on the
ground.

The most extraordinary feature of the head was the enormous magnitude of the eye : and from the quantity of light
admitted by the expanded pupil, it must have possessed great
powers of vision, especially in the dusk.   It is not uncommon
to find in front of the orbit in fossil skulls, a circular series of
petrified thin bony plates, ranged round a central aperture,
where the pupil of the eye was placed.   The eyes of many
fishes are defended by a bony covering consisting of two
pieces ; but a compound circle of overlapping plates is now
found only in the eyes of turtles, tortoises, lizards, and birds.
This curious apparatus of bony plates would aid in protecting
the eye-ball from the waves of the sea when the *Ichthyosaurus*
rose to the surface, and from the pressure of the dense element
when it dived to great depths ; and they show, writes Dr.
Buckland (*Bridgewater Treatise*), "that the enormous eye of
which they formed the front, was an optical instrument of
varied and prodigious power, enabling the *Ichthyosaurus* to
descry its prey at great or little distances, in the obscurity of
night, and in the depths of the sea."

Of no extinct species are the materials for a complete and
exact restoration more abundant and satisfactory than of the
Ichthyosaurus ; they plainly show that its general external
figure must have been that of a huge predatory abdominal
fish, with a longer tail and a smaller tail fin ; scaleless, moreover, and covered by a smooth or finely wrinkled skin, analogous to that of the whale tribe.

The mouth was wide, and the jaws long, and armed with

numerous pointed teeth, indicative of a predatory and carnivorous nature in all the species; but these differed from one another in regard to the relative strength of the jaws, and the relative size and length of the teeth.

Masses of masticated bones and scales of extinct fishes, that lived in the same seas and at the same period as the *Ichthyosaurus*, have been found under the ribs of fossil specimens, in the situation where the stomach of the animal was placed; smaller, harder, and more digested masses, containing also fish-bones and scales, have been found, bearing the impression of the structure of the internal surface of the intestine of the great predatory sea-lizard. One of these "coprolites" is figured beneath the skeleton in fig. 68.

In tracing the evidences of creative power from the earlier to the later formations of the earth's crust, remains of the *Ichthyosaurus* are first found in the lower lias, and occur more or less abundantly through all the superincumbent secondary strata up to, and inclusive of, the chalk formations. They are most numerous in the lias and oolite, and the largest and most characteristic species have been found in these formations.

More than thirty species of *Ichthyosaurus* are known to the writer, many of which have been described or defined.

Order 4.—SAUROPTERYGIA.*

No post-orbital and supra-temporal bones : large temporal and other vacuities between certain cranial bones ; a foramen parietale ; two antorbital nostrils ; teeth simple, in distinct sockets of premaxillary, maxillary, and pre-mandibular bones, rarely on the palatine or pterygoid bones ; maxillaries larger than premaxillaries. Limbs natatory ; not more than five digits. A sacrum of one or two vertebræ for the attachment of the pelvic arch in some, numerous cervical vertebræ in most. Pleura-

* Σαύρος, *a lizard;* πτέρυξ, *a fin.*

P

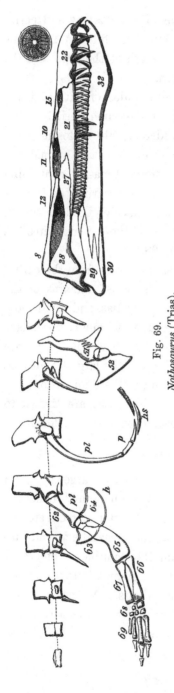

Fig. 69.

Nothosaurus (Trias).

pophyses with simple heads; those of the trunk long and bent.

Genus NOTHOSAURUS, Münster.

Sp. *Nothosaurus mirabilis,* Münster.—In fig. 69 is given an analysis of the chief characters as yet ascertained of the species which may be regarded as the type of its family; by comparing this diagram with that of the *Archegosaurus* (fig. 65), the advance in the organization of the aquatic reptiles will be readily traced and understood.

The skull is no longer defended by a continuous covering of sculptured plate-bones; the vacuities behind the orbits for the temporal muscles are large and widely open. These vacuities are fenced externally by two long and slender horizontal bony bars; the upper one is formed by the mastoid (fig. 69, 8), and the post-frontal (12); the lower one by the malar (27), and squamosal (28); the latter answering to the true zygomatic arch in Mammals. The squamosal abuts by its hinder expanded end against the almost vertical tympanic pedicle, which gives attachment to the lower jaw. This shows the reptilian compound structure : 29 marks the surangular element, 30 the angular one, 32 the dentary. In the side-view of the skull in fig. 69, 22 is the premaxillary, 21 the maxillary, 15 the nasal—the cavity below being the nostril, 10 is the prefrontal—between which and 21 is the lacrymal, 11 the frontal above the orbit. The premaxillary teeth and corresponding premandibular ones are unusually long, strong, and sharp ; there are two similar teeth in each maxillary ; the remaining serial teeth are smaller, but equally acute. There are no teeth on the palate.

The almost entire and undisturbed vertebral column, from the muschelkalk of Bayreuth, figured by Von Meyer in pl. 23 of his work on muschelkalk Saurians, and attributed by him to *Nothosaurus mirabilis*, gives the earliest indication of that modification of the trunk-bones which reaches its maximum in the *Plesiosaurus* (fig. 71), in which it was first detected by the sagacity of Conybeare.*

Twenty of the anterior vertebræ of this series, in *Nothosaurus*, which begins with the atlas, have the whole or part of the rib-pit situated on the centrum as in the first vertebra in fig. 69 ; the pit is wholly there on fourteen vertebræ ; it begins to ascend upon the neural arch in the fifteenth, as in the second vertebra, given in fig. 69, and is wholly placed there on the twenty-first vertebra.

According, therefore, to the characters by which the writer has proposed† to distinguish the cervical from the dorsal vertebræ, *Nothosaurus* has twenty of the former. In the specimen referred to, nineteen consecutive vertebræ show the rib-pit supported wholly on an outstanding diapophysis from the neural arch, as in the third vertebra in fig. 69 ; these are

---

* Trans. Geol. Soc., vol. vi., 1822, and vol. i., 2d series, p. 381, 1824.
† Report of British Fossil Reptiles, 1839, pp. 50, 58.

to be reckoned therefore as dorsal vertebræ. In the cervical vertebræ the rib-pit is large, vertically reniform, not divided by a groove ; its circumference slightly projects in *Notho-saurus*.

There is no clear evidence of any of the cervical ribs being terminally expanded and hatchet-shaped, as in *Plesiosaurus ;* those of the back are vertically longer than in *Plesiosaurus*, and more convex.

In the sacral vertebræ, fourth in fig. 69, the rib-pits again begin to sink upon the centrum.

There are two distinct sacral vertebræ in *Nothosaurus*. They are known by their long, straight, terminally-bent, and convergent pleurapophyses, the first of which overlaps a little the second. To the convergent ends of these riblets, the ilium (fig. 69, 62, *pl*) was doubtless ligamentously affixed. In the first caudal vertebra the par- and di-apophyses stand out much farther than in the sacrum ; but rapidly shorten in the second and third caudals. The compound process in each supports a short stiliform straight riblet, as in the fifth figured vertebra (fig. 69) ; the anterior and succeeding caudals support hæmal arches and spines, after the disappearance of the pleur-apophyses. The hæmal arch disappears in about the eighth vertebra from the end, and finally the neural arch. The terminal centrums are subelongate and subcompressed. Both *Nothosaurus* and *Pistosaurus* had abdominal ribs, of which the median piece (fig. 69, *hs*) was subsymmetrical, the two rays diverging at a very open angle, and terminating in a point or a fork ; the side-pieces (*p*) seem not to have been so numerous as in *Plesiosaurus*.

The scapula (fig. 69, 51) is a short and strong bone, its blade appearing as a short and narrow sub-compressed process extending from the subquadrate, thick and expanded end which affords the articular surfaces for the coracoid, clavicle, and humerus.

The clavicle, which is an exogenous process in *Plesiosaurus*, is here united by a strong oblique suture to the scapula. It expands into, or sends off from its outer part, a broad, flat, obtuse process, near the suture ; then contracts and bends inwards to the episternum, to which it is articulated also by suture.

The coracoid (fig. 69, 52) sends forward a broad and short flattened process, separated by a narrow notch from the scapular part of its head ; it then contracts and soon expands into a broad, flat, sub-triangular plate, the broad and straight border of which articulates with that of the opposite coracoid. A wide unossified interval separates the coracoid from the episternum. The ossification of the coracoid in the direction of this interval gives the peculiar longitudinal or fore-and-aft extent to those bones in the Plesiosaur, in which they unite with the episternum.

The pelvic arch presents a closer correspondence with that in the *Plesiosaurus* (fig. 71). The ischium (fig. 69, 63), contracting beyond its articular head, there expands into a flat subtriangular plate. The pubis (fig. 69, 64) is a subcircular flat bone, with a notch near the articular end.

The bones of the limbs, although evidently those of fins or paddle-shaped extremities, are better developed than in *Plesiosaurus*, and more resemble the corresponding bones in the turtle (*Chelones*). The tuberosities or processes for muscular attachment near the head of the humerus (omitted in the diagram) are better marked, especially that on the concave side of the shaft ; the distal end is thicker and less expanded. The whole bone is more curved than in any *Plesiosauri*. The femur (fig. 69, 65) is relatively longer and less expanded at its distal end. The bones of the fore arm, like those of the leg (ib. 66 and 67), are longer than in *Plesiosaurus*. The articular surfaces present the foramina with raised borders, which characterize those in *Plesiosauri*, and which indicate the fibrocartilaginous nature of the joints.

There is a ligamentous or unossified space at the back part
of both carpus and tarsus (fig. 69, 68). At present there is
evidence of but four digits in both the fore and hind paddles
of *Nothosaurus*; the metapodial and phalangeal bones are of
the elongate flattened simple form, characteristic of supports
of a tegumentary fin.

One species of *Nothosaurus* (*N. Schimperi*, Von. M.) is from
the lower division of the trias, called "grès bizarré" of Soulz-
les-Bains; the other representatives of the genus (*N. giganteus*,
N. venustus, *N. Münsteri*, *N. Andriani*, *N. angustifrons*, and
N. mirabilis), are from the muschelkalk of Bayreuth and
Luneville.

Genus PISTOSAURUS, Von Meyer.

Sp. *Pistosaurus longœvus.*—In this genus the facial part of
the skull contracts abruptly in front of the orbits; so that,
viewed from above, it resembles a long-necked bottle; the
orbits are situated in the posterior half of the skull, and the
nostrils are lateral. From the muschelkalk of Bayreuth.

Genus CONCHIOSAURUS, Von Meyer.

Sp. *Conchiosaurus clavatus.*—The facial part of the skull is
less prolonged than in *Pistosaurus*, and the nostrils are
terminal. The teeth are twelve in number on each side, are
subequal, with a pyriform crown, and are placed at widish
intervals. From the muschelkalk at Laineck, near Bayreuth.

Genus SIMOSAURUS,* Von Meyer.

Sp. *Simosaurus Gaillardoti.*—The fossils, chiefly cranial,
on which this genus is founded, occur in the dolomitic
muschelkalk near Ludwigsberg, and in the muschelkalk of
Luneville. The skull presents the large temporal fossæ, the
divided nostrils, and the general depressed form and composition of that of *Nothosaurus* and *Pistosaurus*. But its facial
part is much shorter; the muzzle is neither prolonged nor
terminally expanded, but forms the obtuse end of the short

* Σιμος, *snub-nosed, flat-nose.*

depressed face, of which the premaxillary part is the narrowest. The nostrils, consequently, although distant from the orbits by half the diameter of the latter, are yet nearer the fore end of the skull than in the above-cited Sauropterygian genera. The nostrils are relatively nearer to each other, the intervening bony tract being due to the premaxillaries, which, relatively to the breadth of the skull, are much narrower in *Simosaurus* than in *Notho-* or *Pisto-saurus*.

The profile of the skull rises from the internasal to the interorbital regions much more than in the Nothosaur, and the depth of the skull behind the orbit is greater in proportion to its length. The post-frontals are most clearly produced backwards, along the upper border of the zygoma to the mastoids. The malars are co-extended, and connected with the post-frontals, but terminate freely and obtusely a little beyond the co-prolonged hind part of the maxillary, without being met by or joining a squamosal.

Most complete and extensive is the ossification of the roof of the mouth in this genus. The pterygoids are expanded into one broad unbroken imperforate flat expanse of bone, from about one-third of the distance from the snout to the occipital condyle; they are united by a median suture, and underlap the whole of the sphenoid. The teeth, compared with *Nothosaurus*, are few and large, and are subequal, save one or two at the fore and hind extremity of the series. The crown expands a little above the fang, is conical, sub-bifurcate, and impressed by a few coarse longitudinal ridges; some are obtuse, others acute; but all are shorter and thicker than in *Notho-* or *Pisto-saurus*.

The vertebræ have flat or very slightly concave articular surfaces on the body; the neural arch articulates therewith by suture. In these characters, and in their general proportions, they resemble those of *Notho-* and *Plesio-saurus*. It is significant of some difference in respect of the arrangement of the vertebræ in the same column, that although specimens from the tail, and from different parts of the back, have been obtained, no cervical vertebra with any probability belonging to this genus has yet been found. The caudal centrum presents two well-defined, rather prominent, hypapophyses for the hæmal arch.

The coracoid in the contraction of the body reminded Cuvier of that of the *Ichthyosaurus*, but its expanded median part was differently shaped. The pubis, like that of the *Plesiosaurus*, resembles to a certain degree the pubis in *Chelonia*. The few bones of the limbs which have been found still more resemble, as do those of *Pistosaurus*, the corresponding bones of marine *Chelonia*. Accordingly, there have been entered in palæonto-logical catalogues an *Ichthyosaurus Lunevillensis* (De la Beche), a *Plesiosaurus Lunevillensis* (Münster), and a *Chelonia Lunevillensis* (Gray and Keferstein) ; but all these are parts of one and the same genus of Enaliosaurian, the "Saurien des environs de Lunéville" of Cuvier, the "Simosaurus" of H. von Meyer.

Genus PLACODUS.—The cranial structure in this genus of muschelkalk reptile is closely similar to that in *Simosaurus*, but its proportions are different ; it is as broad as long ; the greatest breadth being behind, whence the sides converge to an obtuse muzzle ; the entire figure viewed from above being that of a right-angled triangle, with the corners rounded off. The temporal fossæ are the widest, and zygomatic arches the strongest, in the whole class *Reptilia ;* the lower jaw presents a like excessive development of the coronoid processes. These developments, for great size and power of action of the biting and grinding muscles, relate to a most extraordinary form and size of the teeth, which resemble paving-stones, and were evidently adapted to crack and bruise shells and crusts of marine Invertebrata.

The teeth of the upper jaw consist of an external or maxillary series, and an internal or palatal series. The maxillary series are supported in a marginal row of alveoli by the premaxillary and maxillary bones; the palatal series are implanted in the palatine and pterygoid bones. The maxillo-premaxillary teeth are five in number on each side, two implanted in the premaxillary, and three in the maxillary. The premaxillary teeth are subequal, smaller than the maxillary teeth; their crowns are subhemispheric in *P. laticeps,* but in *P. Andriani* they present a bent, pointed, prehensile character. In *P. laticeps* the first maxillary tooth has a full oval crown, $4\frac{1}{2}$ lines by 4 in diameter; the second measures $5\frac{1}{2}$ lines by $4\frac{1}{2}$ lines in diameter; the third is subcircular, 8 lines in diameter, on the right side. The palatal series begins on the inner side of this tooth, and consists of two teeth on each side. The first tooth has a full elliptical crown, 10 lines by 8; the second tooth, developed in the broad pterygoid bone, presents a full oval shape, 1 inch 9 lines by 1 inch 3 lines in diameter. In *Placodus gigas* and *P. Andriani* the palatal teeth, three in number on each side, are all of large size, slightly increasing from before backwards; they are situated close together, forming on each side a series a little curved with the convexity outwards, and the interspace between the two series is very narrow. The first tooth is triangular, the second and third are quadrangular; each with the angles rounded, and the transverse diameter exceeding the fore and aft or longitudinal one. The maxillary teeth are much smaller than the palatal ones, have a rounded or subquadrate crown, are four in number, and of subequal dimensions. The premaxillary teeth, three in number on each side, are more remote and distinct from the maxillary teeth than in *Placodus rostratus* and *P. laticeps;* their crowns are more elongated and conical than in *P. laticeps;* the prehensile power of the prolonged premaxillary part of the jaw being obviously greater in *Placodus gigas* than in *P. laticeps* or *P. rostratus.* The size of the last tooth in *P. laticeps* surpasses that of any of the teeth in the previously discovered species. In proportion to the entire skull, it is the largest grinding tooth in the animal kingdom, the elephant itself not excepted.

All these teeth are implanted by short simple bases in distinct hollow sockets, subject to the same law of displacement and succession as in other reptiles. By some it may be deemed requisite to separate generically the *Placodi* with two teeth from those with three teeth in each palatal series ; but the *Placodus rostratus* offers a transitional condition in the small relative size of the first two palatal teeth, and in the rounded form of all the teeth, from the *P. Andriani* to the *P. laticeps*.

We cannot contemplate the extreme and peculiar modification of form of the teeth in the genus *Placodus* without a recognition of their adaptation to the pounding and crushing of hard substances, and a suspicion that the association of the fossils with shell-clad Mollusks in such multitudes as to have suggested special denominations to the strata containing *Placodus* (*e. g.*, muschelkalk, terebratulitenkalk, etc.), is indicative of the class whence the *Placodi* derived their chief subsistence.

No doubt the most numerous examples of similarly-shaped teeth for a like purpose are afforded by the class of fishes, as, *e. g.*, by the extinct Pycnodonts, and by the wolf-fish (*Anarrhichas lupus*) and the Cestracion of the existing seas. But the reptilian class is not without its instances at the present day of teeth shaped like paving-stones, of which certain Australian lizards exhibit this peculiarity in so marked a degree that the generic name *Cyclodus* has been invented to express that peculiarity. Amongst extinct reptiles, also, a species of lizard from the tertiary deposits of the Limagne in France presents round obtuse teeth, of which the last, in the lower jaw, is suddenly and considerably larger than the rest.

Nothosaurus, Simosaurus, and *Pistosaurus* present the same evidences of lacertian affinities in the division of the nostrils by the median extension of the premaxillary backwards to the nasals, the same thecodont dentition, and the same circumscription of the orbits and temporal fossæ as in *Placodus* : there is also a general family likeness in the upward aspect of these apertures, accompanying an extreme depression of the skull. The muzzle, though varying greatly in length in these genera, presents the same obtuseness ; and the alveolar border of the jaws the same smooth outward convexity which we observe in the *Placodus*. The peculiar confluence of the elements of the upper and lower zygomatic arches,—*i. e.*, of post-frontal and malar,—forming the broad wall of bone behind the orbit, is continued still farther backwards in the *Simosaurus*. In *Pistosaurus* the elongated post-frontal, malar, and squamosal are united together in one deep zygomatic arch, which has the mastoid and tympanic for its hinder abutment.

It is remarkable that hitherto no vertebræ or other bones of the trunk or limbs have been found so associated with the teeth of *Placodus*, as to have suggested their belonging to the same species. Usually, after the indication of a reptile by detached teeth, the next step in its reconstruction is based upon detached vertebræ. The twelve or more evidences of *Placodus*, afforded by bone as well as tooth, are all portions of the skull. It is possible that some of the singularly modified vertebræ from the muschelkalk, next to be described, may belong to the *Placodus ;* and the same surmise suggests itself in reference to some of the limb-bones from the muschelkalk that cannot be assigned to other known saurian genera.

The obvious adaptation of the dentition of *Placodus* to the crushing of very hard kinds of food, its close analogy to the dentition of certain fishes known to subsist by breaking the shells of whelks and other shell-clad Mollusks, and the cha-

racteristic abundance of fossil shells in the strata to which the
remains of *Placodus* are peculiar, concur in producing the
belief that the species of this genus were reptiles frequenting
the sea-shore, and probably good swimmers.   But as at present
we have got no further than the head and teeth in the reconstruction of this mezozoic form of molluscivorous reptile, the
present notice will conclude with a remark suggested by the
disposition and form of the teeth.   In all the species, under
the rather wide range of specific varieties of the dentition,
there are two rows of the crushing teeth in the upper jaw, and
only one row in the lower jaw, on each side of the mouth ;
and the lower row plays upon both upper rows, with its
strongest (middle) line of force directed against their interspace.   Thus the crushing force below presses upon a part
between the two planes or points of resistance above, on the
same principle on which we break a stick across the knee ;
only here the fulcrum is at the intermediate point, the moving
powers at the two parts grasped by the hands.   It is obvious
that a portion of shell pressed between two opposite flat surfaces might resist the strongest bite, but subjected to alternate
points of pressure its fracture would be facilitated.*

Genus TANYSTROPHÆUS.

Sp. *Tanystrophæus conspicuus*, H. von Meyer.—Certain long,
slender, hollow bones (fig. 70, A), from the German muschelkalk, were referred by Count Münster to the class *Reptilia*,
under the name of *Macroscelosaurus*, under the impression that
they were bones of the limbs.   H. von Meyer subsequently,
in more perfect specimens, observing that each slightly expanded extremity of the long bone was terminated by a symmetrical oval concave articular surface, surmounted by a pair
of symmetrical lateral incurved plates, resembling confluent

---

* Previous to the writer's Memoir on *Placodus* in the Philosophical Transactions (1858), all palæontologists had referred the genus to the pycnodont
order of fishes.

neurapophyses, with articular surfaces, and with their sometimes confluent bases arching over a neural canal (as in figure
B, in cut 70), recognized their vertebral character; and, adopt-

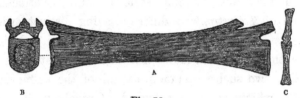

Fig. 70.

A, B, *Tanystrophœus* (Trias); C, *Ichthyosaurus*.

ing the determination of their reptilian nature, but repudiating
the idea of their being limb-bones, he discarded Münster's
name, and substituted for it that of *Tanystrophœus*,* indicative
of their peculiar proportions as vertebræ. Although the
articular ends are for the most part symmetrical, the long
intervening body is not so. It is subcompressed, usually
broader and flatter below than above; sometimes more flattened on one side than on the other, giving an irregular, vertically oval, or triangular cross section. A low median ridge is
not uncommon on the lower surface towards the ends of the
vertebra; and similar less regular ridges project from the
sides of the otherwise smooth outer surface. The centrum is
excavated by a canal, resembling a medullary one, but more
probably filled, in the recent state, as in the long caudal style
of the frog, with unossified cartilage. The walls of this cavity
are compact, and in thickness about one-sixth of the diameter
of the bone. The terminal neural arches support each a low
median ridge or rudimental spine, which soon subsides. The
trace of neural canal in like manner disappears, or is continued
by two distinct slender canals, which traverse for a certain
extent the substance of the thicker upper wall of the cavity
of the vertebral body. A single large vascular canal opens on
the wider surface midway between the two ends of the body.

* From τὰνύω, *to elongate*, στρεφω, *verto*.

There is no trace of transverse processes, rib-surfaces, or hæm-apophyses ; this, and the absence of the continuous neural canal, indicate these singular vertebræ to belong to the tail. From the long caudo-vertebral style of anourous Batrachia the vertebræ of *Tanystrophœus* differ in having distinct articular surfaces at both ends. The difference of shape and size in the few that have been found also indicates that there were more than two such vertebræ in the tail of the extraordinary animal to which they have belonged. Caudal vertebræ of the normal proportions and structure, from muschelkalk of the same localities with *Tanystrophœus* have been referred to *Nothosaurus*. It is possible, however, that one or other of the remarkable genera—*Simosaurus, Placodus, e.g.*—may have possessed the peculiar structure in the tail, or some part of it, which the tanystrophæan vertebræ indicate. The first four vertebræ of the neck or trunk of the *Fistularia tabaccaria* are those which most resemble in their proportions the vertebræ above described ; but none of the fistularian vertebræ have the articular concavity and the zygapophyses at both ends ; the first presents them at the fore end, and the last at the hind end, and the modifications of both these finished articular ends pretty closely correspond with those of *Tanystrophœus ;* but the second and third vertebræ of *Fistularia* are united with the first and fourth by sutural surfaces with deeply-inter-locking pointed processes.

Genus SPHENOSAURUS.

Sp. *Sphenosaurus Sternbergii*, Von. M.—The fossil vertebræ on which this genus is founded are imbedded in a sandstone, most like the bunter, from Bohemia or the south of Germany. Of the twenty-three vertebræ so preserved in nearly their natural position, and with their under surface exposed, five belong to the tail, the rest to the trunk. Of these, two are sacral, two lumbar, the rest are dorsal or thoracic, with long and slender ribs connected with them. The neural arch appears to have been suturally united to the centrum with large zygapophyses. The articular end of the centrum is vertical to its axis ; both are slightly concave. Between each centrum is a transversely oval, depressed ossicle, homologous with the cervical wedge-bones or hypapophyses in Enaliosaurs. This is the chief peculiarity in *Sphenosaurus*, and recalls a character in the vertebral column of *Archegosaurus*.

Genus PLESIOSAURUS.—The discovery of this genus forms one of the most important additions that geology has made to comparative anatomy. Baron Cuvier deemed the structure of the Plesiosaur "to have been the most singular, and its characters the most anomalous that had been discovered amid the ruins of a former world." "To the head of a lizard it united the teeth of a crocodile, a neck of enormous length, resembling the body of a serpent, a trunk and tail having the proportions of an ordinary quadruped, the ribs of a chameleon, and the paddles of a whale" (fig. 71). "Such," writes Dr. Buckland, "are the strange combinations of form and structure in the *Plesiosaurus*, a genus, the remains of which, after interment for thousands of years amidst the wreck of millions of extinct inhabitants of the ancient earth, are at length recalled to light by the researches of the geologist, and submitted to our examination, in nearly as perfect a state as the bones of species that are now existing upon the earth."

The first remains of this animal were discovered in the lias of Lyme Regis about the year 1822, and formed the subject of the paper by the Rev. Mr. Conybeare (afterwards dean of Llandaff), and Mr. (afterwards Sir Henry) De la Beche, in which the genus was established, and named *Plesiosaurus* ("approximate to the Saurians"), from the Greek words *plesios* and *sauros*, signifying "near" or "allied to," and "lizard," because the authors saw that it was more nearly allied to the lizard than was the *Ichthyosaurus* from the same formation.

The entire and undisturbed skeletons of several individuals,

of different species, have since been discovered, fully confirming the sagacious restorations by the original discoverers of
the *Plesiosaurus*.

Vertebral Column.—The vertebral bodies have their terminal articular surfaces either flat or slightly concave, or with
the middle of such cavity a little convex. In general the
bodies present two pits and holes at their under part. The
cervical vertebræ consist of centrum, neural arch, and pleurapophyses. The latter are wanting in the first vertebra ; but
both this and the second have the hypapophyses. The cervical
ribs are short, and expand at their free end, so as to have
suggested the term "hatchet-bones" to their first discoverers.
They articulate by a simple head to a shallow pit, which is
rarely supported on a process, from the side of the centrum ;
but is commonly bisected by a longitudinal groove, a rudimental indication of the upper and lower processes which
sustain the cervical ribs in *Crocodilia*.

The body of the atlas articulates with a large hypapophysis
below, with the neurapophysis above, with the body of the
axis behind, and with part of the occipital condyle in front ;
all the articulations save the last become, in *Plesiosaurus
pachyomus*, and probably with age in other species, obliterated
by anchylosis. The hypapophysis forms the lower two-thirds,
the neurapophysis contributes the upper and lateral parts, and
the centrum forms the middle or bottom of the cup for the
occipital condyle. The second hypapophysis is lodged in the
inferior interspace between the bodies of the atlas and axis ;
it becomes anchylosed to these and to the first hypapophysis.
The first pleurapophysis, or rudimental rib, is developed from
the centrum of the axis.

As the cervical vertebræ approach the dorsal, the lower
part of the costal pit becomes smaller, the upper part larger,
until it forms the whole surface, gradually rising from the
centrum to the neurapophysis (fig. 71).

The dorsal region is arbitrarily commenced by this vertebra, in which the costal surface begins to be supported on a diapophysis, which progressively increases in length in the second and third dorsal, continues as a transverse process to near the end of the trunk ; and on the vertebra above or between the iliac bones, it subsides to the level of the neurapophysis. In the caudal vertebra the costal surface gradually descends from the neurapophysis upon the side of the centrum ; it is never divided by the longitudinal groove which, in most *Plesiosauri*, indents that surface in the cervical vertebræ. The neural arches remain long unanchylosed with the centrum in all *Plesiosauri*, and appear to be always distinct in some species. The pleurapophyses gain in length, and lose in terminal breadth, in the hinder cervicals ; and become long and slender ribs in the dorsal region, curving outwards and downwards so as to encompass the upper two-thirds of the thoracic abdominal cavity. They decrease in length and curvature as they approach the tail, where they are reduced to short straight pieces, as in the neck, but are not terminally expanded ; they cease to be developed near the end of the tail. The hæmapophyses in the abdominal region are subdivided, and with the hæmal spine or median piece, form a kind of "plastron" of transversely-extended, slightly bent, median and lateral, overlapping, bony bars, occupying the subabdominal space between the coracoids and pubicals. In the tail the hæmapophyses are short and straight, and remain re-united both with the centrum above and with each other below. The hæmal spine is not developed in this region. This modification has been expressed by the statement that there were no chevron-bones in the Plesiosaur. The tail is much shorter in the *Plesio-* than in the *Ichthyo-saurus*.

The skull is depressed ; its length is rather more than thrice its breadth ; but the proportions somewhat vary in different species. The cranial part, or that behind the orbits, is quadrate; thence it contracts laterally to near the maxillo-premaxillary suture, where it continues either parallel or with a slight swelling before rounding into the obtuse anterior termination.

The orbits are at or near the middle of the skull: estimating the length of this by that of the lower jaw, they are in advance of the middle part in *Plesiosaurus Hawkinsii.* The orbits are rather subtriangular than round, being somewhat squared off behind, straight above, and contracted anteriorly. No trace of sclerotic plates has yet been discerned in any specimen. The temporal fossæ are large subquadrate apertures. The nostrils, which are a little in advance of the orbits, are scarcely larger than the parietal foramen. Beneath them, upon the palate, are two similar-sized apertures, probably the palatal nostrils.

The lower jaw presents an angular, surangular, splenial, and dentary element, in each ramus; the dentary elements being confluent at the expanded symphysis. There is no vacuity between the angular and surangular or any other element of the jaw. The coronoid process is developed, as in *Placodus,* from the surangular, but rises only a little higher than in crocodiles. The alveoli are distinct cavities, and there is a groove along their inner border in both jaws.

When the successional teeth first project in that groove, they give the appearance of a double row of teeth. All the teeth are sharp-pointed, long, and slender, circular in cross section, with fine longitudinal ridges on the enamel; the anterior teeth are the longest.

The scapula is a strong triradiate bone, the longest ray being formed by the acromial or clavicular process, which arches forward and inward to abut against the sternum or epicoracoid.

The proper body of the scapula* is short and straight,

* This is omitted in most of the published restorations of the *Plesiosaurus.*

somewhat flattened ; the thick articular end, which forms the
shortest ray, is subequally divided
by the articular surface for the
coracoid, and that for the head of
the humerus.

The coracoids are chiefly remarkable for their excessive expansion in the direction of the
axis of the trunk, extending from
the abdominal ribs forward, so as
to receive the entosternum, which
is wedged into their anterior interspace. The median borders meet
and unite for an extent determined by their degree of curvature
or convexity, which is always
slight. The coracoids unite anteriorly with the clavicles, as well
as with the episternum ; laterally
they articulate with the scapula,
combining to form the glenoid
cavity for the humerus.

The episternum has the same
general form as the median pieces
of the abdominal ribs, being, like
those pieces, a modified hæmal
spine, only more advanced in
position ; the lateral wings or
prolongations are broader and
flatter ; the median process is
short ; a longitudinal ridge projects from the middle of the internal surface. The humerus is
a moderately thick and long bone,

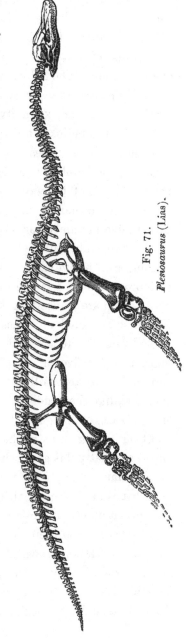

Fig. 71.

Plesiosaurus (Lias).

with a convex head, sub-cylindrical at its proximal end, becoming flattened and gradually expanded to its distal end, where it is divided into two indistinct surfaces for the radius and ulna. The shaft in most species is slightly curved backwards, or the hind border is concave, whilst the front one is straight. The radius and ulna are about half the length of the humerus ; the former is straight, the latter curved or reniform, with the concavity towards the radius ; both are flattened ; the radius is a little contracted towards its carpal end, and in some species is longer than the ulna. The carpus consists of a double row of flat rounded discs,—the largest at the radial side of the wrist ; the ulnar or hinder side appearing to have contained more unossified matter. The metacarpals, five in number, are elongate, slender, slightly expanded at the two ends, flattened, and sometimes a little bent. The phalanges of the five digits have a similar form, but are smaller, and progressively decrease in size ; the expansion of the two ends, which are truncate, makes the sides or margins concave. The first or radial digit has generally three phalanges, the second from five to seven, the third eight or nine, the fourth eight, the fifth five or six phalanges. All are flattened ; the terminal ones are nailless ; and the whole were obviously included, like the paddle of the porpoise and turtle, in a common sheath of integument. The pelvic arch consists of a short but strong and straight narrow moveable ilium, and of a broad and flat pubis and ischium ; the former subquadrate or subcircular, the latter triangular ; the fore-and-aft -expanse of both bones nearly equals that of the coracoids. All concur in the formation of the hip-joint. The ischium and pubis again unite together near their mesial borders, leaving a wide elliptic vacuity, or "foramen ovale," between this junction and their outer acetabular one. The pelvic paddle is usually of equal length with the pectoral one, but in *P. macrocephalus* it is longer. The bones closely correspond, in number, arrangement, and form, with those of the fore limb. The femur has the hind margin less concave, and so appears more straight. The fibula, in its reniform shape, agrees with its homotype the ulna. The tarsal bones are also smallest on the tibial side. Of existing reptiles, the lizards, and amongst these the old world Monitors (*Varanus*, Fitz.), by reason of the cranial vacuities in front of the orbits, most resemble the Plesiosaur in the structure of the skull. The division of the nostrils, the vacuities in the occipital region between the exoccipitals and tympanics, the parietal foramen, the zygomatic extension of the post-frontal, the palato-maxillary, and pterygo-sphenoid vacuities in the bony palate, are all lacertian characters, as contradistinguished from crocodilian ones.

But the antorbital vacuities between the nasal, pre-frontal, and maxillary bones are the sole external nostrils in the Plesiosaurs; the zygomatic arch abuts against the fore part of the tympanic, and fixes it. A much greater extent of the roof of the mouth is ossified than in lizards, and the palato-maxillary and pterygo-sphenoid fissures are reduced to small size. The teeth, finally, are implanted in distinct sockets. That the Plesiosaur had the "head of a lizard" is an emphatic mode of expressing the amount of resemblance in their cranial conformation. The crocodilian affinities, however, are not confined to the teeth, but extend to the structure of the skull itself.

In the simple mode of articulation of the ribs, the lacertian affinity is again strongly manifested; but to this vertebral character such affinity is limited; all the others exemplify the ordinal distinction of the Plesiosaurs from known existing reptiles. The shape of the joints of the centrums; the number of vertebræ between the head and tail, especially of those of the neck; the slight indication of the sacral vertebræ; the non-confluence of the caudal hæmapophyses with each other, are all "plesiosauroid." In the size and number of abdominal ribs and sternum may perhaps be discerned a first step in that series of development of the hæmapophyses of the trunk which reaches its maximum in the plastron of the Chelonia.

The connation of the clavicle with the scapula is common to the *Chelonia* with the *Plesiosauri;* the expansion of the coracoids—extreme in *Plesiosauri*—is greater in *Chelonia* than in *Crocodilia*, but is still greater in some *Lacertia.* The form and proportions of the pubis and ischium, as compared with the ilium, in the pelvic arch of the *Plesiosauri,* find the nearest approach in the pelvis of marine *Chelonia;* and no other existing reptile now offers so near, although it be so remote, a resemblance to the structure of the paddles of the Plesiosaur. Amongst the many figurative illustrations of the nature of the Plesiosaur in which popular writers have indulged, that which compares it to a snake threaded through the trunk of a turtle is the most striking ; but the number of vertebræ in the Plesiosaur is no true indication of affinity with the ophidian order of reptiles.

The reptilian skull from formations underlying the lias, to which that of *Plesiosaurus* has the nearest resemblance, is the skull of the *Pistosaurus;* in this genus the nostrils have a similar position and diminutive size, but are somewhat more in advance of the orbits, and the premaxillaries enter into the formation of their boundary : the premaxillary muzzle and the temporal fossæ are also somewhat longer and narrower. The post-frontals and mastoids more clearly combine with malars and squamosals in forming the zygomatic arch, which is of greater depth in *Pistosaurus;* the parietal foramen is larger ; there is no trace of a median parietal crest. On the palate, besides the vacuities between the pterygoids and pre-sphenoids, and the small foramina between the palatines, pre-maxillaries, and maxillaries, there is in *Pistosaurus* a single median foramen in advance of the latter foramina, between the pointed anterior ends of the pterygoids and the premaxillaries. In *Nothosaurus* the pterygoids extend back, under-lapping the basi-sphenoid, as far as the basi-occipital, the median suture uniting them being well marked to their ter-mination ; and there is no appearance of vacuities like the pterygo-sphenoid ones in *Plesio-* and *Pisto-saurus.* The tym-panics are relatively longer, and extend farther back in *Pisto-*than in *Plesio-saurus.* There is no trace of lacrymals in *Pistosaurus;* and its maxillaries are relatively larger than in *Plesiosaurus.* In *Pistosaurus* there are 18 teeth on each the upper jaw, including the 5 premaxillary teeth ; in *Plesiosaurus* there are from 30 to 40 teeth on each side. In *Pistosaurus* the teeth are relatively larger, and present a more oval trans-verse section: the anterior teeth are proportionally larger than the posterior ones than they are in *Plesiosaurus.* The dis-proportion is still greater in *Nothosaurus,* in some species of which the teeth behind the premaxillary and symphysial terminal expansions of the jaws suddenly become—*e. g.,* in *Nothosaurus mirabilis* (fig. 68)—very small, and form a straight, numerous, and close-set single series along the maxillary and corresponding part of the mandibular bone.

Both *Nothosaurus* and *Pistosaurus* had many neck-verte-bræ, and the transition from these to the dorsal series was effected, as in *Plesiosaurus,* by the ascent of the rib-surface from the centrum to the neurapophysis ; but the surface, when divided between the two elements, projected further outwards than in most *Plesiosauri.*

In both *Notho-* and *Pisto-saurus* the pelvic vertebra develops a combined process (par- and di-apophysis), but of relatively larger, vertically longer size, standing well out, and from near the fore part of the side of the vertebra. This process, with the coalesced riblet, indicates a stronger ilium, and a firmer base of attachment of the hind limb to the trunk, than in *Plesiosaurus.* Both this structure, and the greater length of the bones of the fore arm and leg show that the muschelkalk predecessors of the liassic *Plesiosauri* were better organized for occasional progression on dry land.

More than twenty species of *Plesiosaurus* have been described by, or are known to, the writer ; their remains occur in the oolitic, Wealden, and cretaceous formations, ranging from the lias upwards to the chalk, inclusive. A comparison of remains of various *Plesiosauri* has led to a conviction, that specific distinctions are accompanied with well-marked differences in the structure and proportions of answerable vertebræ, but are not shown in small differences of number in the cervical, dorsal, or caudal vertebræ.

When any region of the vertebral column presents an unusual excess of development in a genus, such region is more liable to variation, within certain limits, than in genera where its proportions are more normal. The differences of the number of cervical and dorsal vertebræ, ranging between 29 and 31 in the *Plesiosaurus Hawkinsii, e. g.*—as noted in the description of that species in the writer's Report on British Fossil Reptiles, 1839—indicate the range of variety observed in the only species of which, at that time, the vertebral column of different individuals could be compared.

Genus PLIOSAURUS, Ow.—M. von Meyer regards the number of cervical vertebræ and the length of neck as characters of prime importance in the classification of *Reptilia*, and founds thereon his order called *Macrotrachelen*, in which he includes *Simosaurus, Pistosaurus,* and *Nothosaurus,* with *Plesiosaurus.* No doubt the number of vertebræ in the same skeleton bears a certain relation to ordinal groups : the *Ophidia* find a common character therein ; yet it is not their essential character, for the snake-like form, dependent on multiplied vertebræ, characterizes equally certain Batrachians (*Cœcilia*) and fishes (*Murœna*). Certain regions of the vertebral column are the seats of great varieties in the same natural group of *Reptilia.* We have long-tailed and short-tailed lizards ; but do not therefore separate those with numerous caudal vertebræ, as "Macroura," from those with few or more. The extinct *Dolichosaurus* of the Kentish chalk, with its proccelian vertebræ, cannot be ordinarily separated, by reason of its more numerous cervical vertebræ, from other shorter-necked procœlian lizards. As little can we separate the short-necked and big-headed amphicœlian Pliosaur from the Macrotrachelians with which it has its most intimate and true affinities.

There is much reason, indeed, to suspect that some of the muschelkalk Saurians, which are as closely allied to *Notho-saurus* as *Pliosaurus* is to *Plesiosaurus*, may have presented analogous modifications in the number and proportions of the cervical vertebræ. It is hardly possible to contemplate the broad and short-snouted skull of the *Simosaurus*, with its proportionally large teeth, without inferring that such a head must have been supported by a shorter and more powerful neck than that which bore the long and slender head of the *Nothosaurus* or *Pistosaurus*. The like inference is more strongly impressed upon the mind by the skull of the *Placodus*, still shorter and broader than that of *Simosaurus*, and with vastly larger teeth, of a shape indicative of their adaptation to crushing molluscous or crustaceous shells.

Neither the proportions and armature of the skull of *Placodus*, nor the mode of obtaining the food indicated by its cranial and dental characters, permit the supposition that the head was supported by other than a comparatively short and strong neck. Yet the composition of the skull, its proportions, cavities, and other light-giving anatomical characters, all bespeak the close essential relationship of *Placodus* to *Simo-saurus* and other so-called "macrotrachelian" reptiles of the muschelkalk beds. I still, therefore, regard the fin-like modi-fication of the limbs as a better ordinal character than the number of vertebræ in any particular region of the spine. But by those who would retain the term *Enaliosauria* for the

large extinct natatory group of saurian reptiles, the essential
distinctness of the groups *Sauropterygii* and *Ichthyopterygii,*
typified by the *Ichthyosaurus* and *Plesiosaurus* respectively,
should be borne in mind.

Sp. *Pliosaurus brachydeirus,* Ow.—The generic characters
of *Pliosaurus* are given by the teeth and the cervical vertebræ.
As compared with those of *Plesiosaurus,* the teeth are thicker
in proportion to their length, are subtrihedral in transverse
section, with one side flattened, and bounded by lateral promi-

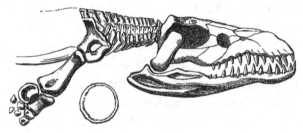

Fig. 72.
Pliosaurus (Kimmeridgian).

nent ridges from the more convex sides, which are rounded off
into each other, and alone show the longitudinal ridges of the
enamel ; these are there very well defined.   The vertebræ of
the neck, presenting a flat articular surface of the shape shown
in outline below the neck in fig. 72, are so compressed from
before backward as to resemble the vertebræ of the *Ichthyosaurus* (fig. 70, c), and as many as twelve may be compressed
within the short neck intervening between the skull and
scapular arch, as shown in fig. 72.   For the rest, save in the
more massive proportions of the jaws and paddle-bones, the
bony framework of *Pliosaurus* closely accords with that of
Plesiosaurus ; and, as the vertebræ of the trunk resume the
plesiosaurian proportions, they give little indication of the
genus of reptile to which they truly belong, when found detached
and apart.   Some individuals of *Pliosaurus* appear to have
attained a length of between 30 and 40 feet.   The remains of

this modified form of Enaliosaur are peculiar to the Oxfordian and Kimmeridgian divisions of the upper oolitic system. They have been discovered in these beds in Russia (*Pliosaurus Worinskii* and *Spondylosaurus* of Fischer), as well as in those counties of England where the Kimmeridge and Oxford clays have been deposited.

Genus POLYPTYCHODON.—A genus represented by species equalling in size those of *Pliosaurus.* The teeth have a strong conical crown with a sub-circular transverse section, and the longitudinal ridges of the enamel are set close all round the crown, whence the name of the genus, signifying " many-ridged tooth ;" they may be distinguished from the teeth of *Mosasaurus* or *Pliosaurus* by the absence of the smooth almost flattened facet of the crown, which, in those genera, is divided by two strong ridges from the rest of the crown. The teeth are implanted in distinct sockets, as in *Plesiosaurus.* The vertebræ found in the same strata, corresponding in size with the teeth, present the plesiosauroid type. Bones of a large paddle or natatory limb, from the chalk of Kent, may also belong to *Polyptychodon.* A portion of the cranium of *Polyptychodon interruptus,* from the chalk, shows the " foramen parietale," and a plesiosauroid type of temporal fossæ.

Remains of *Polyptychodon* have hitherto been met with only in the cretaceous formations : in the green-sand of Kent and Cambridge, also at Kursk, in Russia, and in the chalk of Kent and Sussex.

Order 5.—ANOMODONTIA.

Teeth wanting, or limited to a single maxillary pair, having the form or proportions of tusks : a " foramen parietale ;" two nostrils ; tympanic pedicle fixed ; vertebræ bicon-cave ; trunk-ribs long and curved, the anterior ones with a bifurcate head ; sacrum of more than two vertebræ. Limbs ambulatory.

FAM.—DICYNODONTIA.

A long ever-growing tusk in each maxillary bone ; pre-maxillaries connate, forming with the lower jaw a beak-
shaped mouth, probably sheathed with horn.

Genus DICYNODON, Ow.—In 1844 Mr. Andrew G. Bain,
who had been engaged in the construction of military roads
in the colony of the Cape of Good Hope, discovered, in the
tract of country extending northwards from the county of
Albany, about 450 miles east of Cape Town, several nodules
or lumps of a kind of sandstone, which, when broken, displayed
in most instances evidences of fossil bones, and usually of a
skull with two large projecting teeth.   Accordingly these evidences of ancient animal life in South Africa were first notified
to English geologists by Mr. Bain under the name of "Biden-
tals ;" and the specimens transmitted by him were submitted
to the writer for examination.   The results of the comparisons
thereupon instituted went to show that there had formerly
existed in South Africa, and from geological evidence, probably,
in a great lake or inland sea, since converted into dry land, a
race of reptilian animals presenting in the construction of their
skull (fig. 73) characters of the crocodile, the tortoise, and the
lizard, coupled with the presence of a pair of huge sharp-pointed
tusks, growing downwards, one from each side of the upper
jaw, like the tusks of the mammalian morse (*Trichecus*).   No
other kind of teeth were developed in these singular animals :
the lower jaw appears to have been armed, as in the tortoise,
by a trenchant sheath of horn.

The vertebræ, by the hollowness of the co-adapted articular
surfaces, indicate these reptiles to have been good swimmers,
and probably to have habitually existed in water ; but the
construction of the bony passages of the nostrils proves that
they must have come to the surface to breath air.   The pelvis

consists of a sacrum composed of 5 confluent vertebræ, with very broad iliac bones, and thick and strong ischial and pubic bones.

Some extinct plants allied to the Lepidodendron, with other fossils, render it probable that the sandstones containing the dicynodont reptiles were of the same geological age as those that have revealed the remains of the Rhynchosaurs and Labyrinthodonts in Europe.

The generic name *Dicynodon* is from the Greek words signifying "two tusks or canine teeth." Four species of this genus, having a rounded profile and less strongly ridged maxillaries, have been demonstrated from the fossils transmitted by Mr. Bain.

Sp. *Dicynodon lacerticeps*, Ow.[*]—This species is founded on a skull six inches in length, of which a reduced figure is given in cut 73, where *c* shows the canine tusks.

Sp. *Dicynodon testudiceps*, Ow.—In this species the skull, and the facial part more particularly, is shorter than in *D. lacerticeps*.

Sp. *Dicynodon strigiceps*, Ow.—The shortening of the jaws and blunting of the muzzle are carried to an extreme in this species, in which the nostrils are situated almost beneath the orbits.

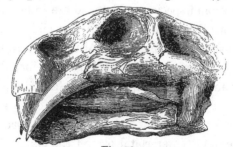

Fig. 73.
Skull and tusks of *Dicynodon lacerticeps*.

Sp. *Dicynodon tigriceps*, Ow.[†]—In this species the length of the skull is 20 inches, its breadth across the widest part of the zygomatic arches being 18 inches. It differs from the *D. lacerticeps* not only in size, but in the

[*] Trans. Geological Society, 2d series, vol. vii. (*dis*, two; *kunodos*, canine-tooth.)   [†] Trans. Geol. Soc., 2d series, vol. vii., p. 233.

relatively larger capacity of the temporal fossæ, and smaller size of the orbits. These cavities in *D. lacerticeps* occupy the middle third of the skull, but in *D. tigriceps* are wholly in the anterior half of the skull. The profile of the skull in *D. lacerticeps* begins to slope or curve down from a line parallel with the back part of the orbits, but in *D. tigriceps* it does not begin to bend down until in advance of the orbits.

Genus PTYCHOGNATHUS, Ow.—Three other species, showing a remarkable angular contour of the skull, with strongly ridged maxillary and upwardly produced mandibular bones, have been subgenerically separated under the name *Ptychognathus*. Their remains characterize the same formations as those of *Dicynodon*. No evidence of the Dicynodont family has yet been met with out of South Africa.

FAM.—CRYPTODONTIA.

Upper as well as lower jaws edentulous, or with inconspicuous teeth.

Genus OUDENODON, Bain.

Sp. *Oudenodon Bainii.*—The fossils on which the above genus and species are founded are from a bluish argillo-ferruginous limestone in South Africa, and form part of a collection transmitted to the British Museum by A. G. Bain, Esq.

One portion of the fossil skull includes all that part in advance of the temporal fossæ; the fore part of the temporal ridges, at the upper and back part of this fragment, curve as they diverge from each other to the back part of the orbit. The upper interorbital part of the cranium is nearly flat, with the orbital margins slightly raised, and terminating anteriorly in a low antorbital prominence; the least breadth of the interorbital space is one inch. A slight depression divides the antorbital from the supranasal tuberosities. The nasal bones form an almost flat rhomboid surface, from the contracted fore part of which the broad premaxillary part of the upper jaw inclines downward and forward at an open angle. This part is traversed by a low obtuse median ridge, and terminates below in a trenchant edentulous border.

The nostrils are small, oval, and separated from each other by the broad junction of the ascending branch of the premaxillary with them.

The maxillary bone presents the chief peculiarity, being traversed obliquely by a strong angular ridge, commencing a little anterior to the orbit, and terminating at the alveolar border, not far from the maxillo-premaxillary suture. The alveolar border gently curves to this termination, and shows no trace of a tooth or alveolus.

The compound structure of the lower jaw is shown at the fractured back part, where an upper (surangular?) element, thick and rounded above, is received into an outer and lower element, thin above, and thick and bent below, forming a groove for the reception of the upper element. On the outer side of the jaw, about the middle of the part preserved, there is a longitudinal depression or narrow vacuity, above which there is a low ridge. The symphysis is thick, long, and bent up in the form of a beak, terminating by an edentulous subtrenchant border; its fore and outer part is traversed by a low median ridge. The length of this portion of the skull is 6 inches; its breadth across the maxillary ridges is 2 inches 10 lines; the extent of the symphysis of the lower jaw is 2 inches 6 lines. *Oudenodon* is more closely allied to the African bidental reptiles of the following order than to the English *Rhynchosaurus*: so closely, in the construction of the skull, as to suggest the surmise that the absence of the two upper teeth may be a sexual character.

Genus RHYNCHOSAURUS, Ow.

Sp. *Rhynchosaurus articeps*, Ow.*—The fossils in which

---

* Transactions of the Cambridge Philosophical Society, vol. vii., part iii., 1842, p. 355, plates 5 and 6.

the above order, genus, and species of reptile have been based
are from the new red sandstone (trias) of Shropshire. They
occur at the Grinsill quarries, near Shrewsbury, in a fine-
grained sandstone, and also in a coarse burr-sandstone ; in the
latter the writer found imbedded some vertebræ, portions of
the lower jaw, a nearly entire skull, fragments of the pelvis
and of two femora : in the fine-grained sandstone, vertebræ,
ribs, and some bones of the scapular and pelvic arches are
imbedded. The bones present a very brittle and compact
texture ; the exposed surface is usually smooth, or very finely
striated, and of a light blue colour. The sandstones containing
these bones occasionally exhibit impressions of footsteps which
resemble those figured in the Memoir by Murchison and
Strickland (Geol. Trans., 2d series, vol. v., pl. xxviii. fig. 1) ; but
they differ in the more distinct marks of the claws, the less
distinct impression of a web, the more diminutive size of the
innermost toe, and an impression corresponding with the
hinder part of the foot, which reminds one of a hind toe pointing backwards, and which, like the hind toe of some birds,
only touched the ground with its point. The footprints are
likewise more equal in size, and likewise in their intervals,
than those figured in the above-cited Memoir : they measure
from the extremity of the outermost or fifth toe to that of
the innermost or first rudimental toe, about one inch and a
half. They are the only footprints that have as yet been
detected in the new red sandstone quarries at Grinsill.

As the fossil bones have always been found nearly in the
same bed as that impressed by the footsteps above described,
they probably belong to the same animal. In the vertebræ
both articular surfaces of the centrum are concave, and are
deeper than in the biconcave vertebræ of the extinct Crocodilians ; the texture of the centrum is compact throughout.
The neural arch is anchylosed with the centrum, without
trace of suture, as in most lizards ; it immediately expands

and sends outwards from each angle of its base a broad triangular process with a flat articular surface ; the two anterior
surfaces look directly upwards, the posterior ones downwards ;
the latter are continued backwards beyond the posterior extremity of the centrum ; the tubercle for the simple articulation of the rib is situated immediately beneath the anterior
oblique process. So far the vertebræ of the *Rhynchosaurus*,
always excepting their biconcave structure, resemble the
vertebræ of most recent lizards. In the modification next to
be noticed, they show one of the vertebral characters of the
Dinosauria. A broad obtuse ridge rises from the upper convex
surface of the posterior articular process and arches forwards
along the neural arch above the anterior articular process, and
gradually subsides anterior to its base : the upper part of this
arched angular ridge forms, with that of the opposite side, a
platform, from the middle line of which the spinous process is
developed. Nothing of this kind is present in existing lizards ;
the sides of the neural arch immediately converge from the
articular processes to the base of the spine, without the inter-
vention of an angular ridge formed by the sides of a raised
platform. The base of the spinous process is broadest behind,
and commences there by two roots or ridges, one from the
upper and back part of each posterior articular process. The
anterior margin of the spinous process is thin and trenchant ;
the height of the spine does not exceed the antero-posterior
diameter of its base ; it is obliquely rounded off. The spinal
canal sinks into the middle part of the centrum and rises to
the base of the spine, so that its vertical diameter is twice as
great at the middle as at the two extremities : this modification
resembles in a certain degree that of the vertebræ of the
Palæosaurus from the Bristol conglomerate.

The skull presents the form of a four-sided pyramid, compressed laterally, and with the upper facet arching down in a
graceful curve to the apex, which is formed by the termination

of the muzzle. The very narrow cranium, wide temporal fossæ on each side, bounded posteriorly by the parietal and the mastoid bones, and laterally by strong compressed zygomata; the long tympanic pedicle, descending freely and vertically from the point of union of the posterior transverse and zygomatic arches, and terminating in a convex pulley for the articular concavity of the lower jaw; the large and complete orbits, and the short, compressed, and bent down maxillæ, all combine to prove the fossil to belong to the lacertian division of the saurian order. The mode of articulation of the skull with the spine cannot be determined in the present specimen, but the lateral compression and the depth of the skull, the great vertical breadth of the superior maxillary bone, the small relative size of the temporal spaces, the vertical breadth of the lower jaw, prove that it does not belong to a reptile of the batrachian order. The shortness of the muzzle, and its compressed form, equally remove it from the Crocodilians. No Chelonian has the tympanic pedicle so long, so narrow, or so freely suspended to the posterior and lateral angles of the cranium.

The general aspect of the skull differs, however, from that of existing Lacertians, and resembles that of a bird or turtle, which resemblance is increased by the apparent absence of teeth. The dense structure of the produced ends of the pre-maxillaries indicates an analogy of function to the tusks of *Dicynodon ;* the premaxillaries are double, as in crocodiles and Chelonians ; but most of the essential characters of the skull are those of the lizard. The rami of the lower jaw are remarkable, as in *Bathygnathus,* for their great depth, but not the least trace of a tooth is discernible in the alveolar border of the dentary element.

The cranium, in my first described Rhynchosaur, was preserved with the mouth in the naturally closed state, and the upper and lower jaws in close contact. In this state we must suppose that they were originally buried in the sandy matrix which afterwards hardened around them ; and since lizards, owing to the unlimited reproduction of their teeth, do not become edentulous by age, we must conclude that the state in which the *Rhynchosaurus* was buried, with its lower jaw in undisturbed articulation with the head, accorded with its natural condition, while living, so far as the less perishable hard parts of its masticatory organs were concerned. Nevertheless, since a view of the inner side of the alveolar border of the jaws has not been obtained, we cannot be quite assured of the actual edentulous character of this very singular Saurian. The indications of a dental system are much more obscure in the *Rhynchosaurus* than in any existing Lacertian ; the dentations of the upper jaw are absolutely feebler than in the chameleon, and no trace of them can be detected in the lower jaw, where they are strongest in the chameleon. The absence of the coronoid process in the *Rhynchosaurus*, which is conspicuously developed in all existing lizards, corresponds with the unarmed condition of the jaw, and the resemblance of the *Rhynchosaurus* in this respect to the *Chelys ferox*, would seem to indicate that the correspondence extended to the toothless condition of the jaws. The resemblance of the mouth to the compressed beak of certain sea-birds, the bending down of the curved and elongated premaxillaries, so as to be opposed to the deep symphysial extremity of the lower jaw, are further indications that the ancient Rhynchosaur may have had its jaws encased by a bony sheath, as in birds and turtles, the dentinal ends of the premaxillaries projecting from, or forming, the deflected end of the upper mandible.

There are few genera of extinct reptiles of which it is more desirable to obtain the means of determining the precise modifications of the locomotive extremities than the *Rhynchosaurus*. The fortunate preservation of the skull has brought to light modifications of the lacertine structure leading towards

Chelonia and birds which before were unknown ; the vertebræ likewise exhibit very interesting deviations from the lacertian type. The entire reconstruction of the skeleton of the *Rhynchosaurus* may be ultimately accomplished, if due interest be taken in the collection and preservation of the fossils of the Grinsill quarries.

The cranium of a Rhynchosaurian reptile with small palatal teeth and obscure maxillary dentations, has been discovered in the problematical sandstones, containing the *Leptopleuron*, near Elgin ; and adds to the probability of their triassic age.

Order VI.—PTEROSAURIA.

Char.—Pectoral members, by the elongation of the anti-brachium and fifth digit, adapted for flight. Vertebræ proccœlian ; those of the neck very large, not exceeding eight in number ; those of the pelvis few and small. Most of the bones pneumatic. Head large ; jaws long, and armed with teeth.

The species of this order of reptiles are extinct, and peculiar to the mezozoic period. Although some members of the preceding order resembled birds in the shape or the edentulous state of the mouth, those of the present order make a closer approach to the feathered class in the texture and pneumatic character of most of the bones, and in the development of the pectoral limbs into organs of flight (fig. 74). This is due to an elongation of the antibrachial bones, and more especially to the still greater length of the metacarpal and phalangial bones of the fifth or outermost digit (fig. 74, 5), the last phalanx of which terminates in a point. The other fingers were of more ordinary length and size, and terminated by claws. The number of phalanges is progressive from the first (fig. 74, 1) to the fourth (4), which is a reptilian character. The whole osseous system is modified in accordance with the

possession of wings ; the bones are light, hollow, most of them
permeated by air-cells, with thin compact outer walls. The
scapula and coracoid are long and narrow, but strong. The
vertebræ of the neck are few, but large and strong, for the
support of a large head with long jaws, armed with sharp-

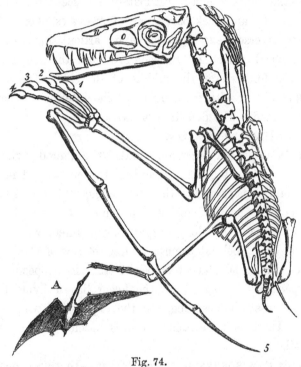

Fig. 74.

Fossil skeleton of *Pterodactylus crassirostris:* A, Sketch of living Pterodactyle.

pointed teeth. The skull was lightened by large vacuities, of
which one (*o*, fig. 74) is interposed between the nostril *n* and
the orbit *l.* The vertebræ of the back are small, and grow
less to the tail. Those of the sacrum are small, from three to
five in number : but the weak pelvis and hind limbs bespeak
a creature unable to stand and walk like a bird. The body
must have been dragged along the ground like that of a bat.
The *Pterosauria* may have been good swimmers as well as

flyers. The vertebral bodies unite by ball-and-socket joints, the cup being anterior, and in them we have the earliest manifestation of the "proccœlian" type of vertebra. The atlas consists of a discoid centrum, and of two slender neurapophyses ; the centrum of the axis is ten times longer than that of the atlas, with which it ultimately coalesces ; it sends off from its under and back part a pair of processes, above which is the transversely extended convexity articulating with the third cervical vertebra. In each vertebra there is a large pneumatic foramen at the middle of the side. The neural arch is confluent with the centrum. The anterior ribs have a bifurcate head. The dentition is thecodont.

Genus DIMORPHODON, Ow.

Sp. *Dimorphodon macronyx*, Bkd.—The Pterodactyles are distributed into sub-genera, according to well-marked modifications of the jaws and teeth. In the oldest known species, from the lias, the teeth are of two kinds ; a few at the fore part of the jaws are long, large, sharp-pointed, with a full elliptical base ; behind them is a close-set row of short, compressed, very small lancet-shaped teeth. In a specimen of *Dimorphodon macronyx*, from the lower lias of Lyme Regis, the skull was 8 inches long, and the expanse of wing about 4 feet. There is no evidence of this species having had a long tail.

Genus RAMPHORHYNCHUS, Von Meyer.—In this genus the fore part of each jaw is without teeth, and may have been encased by a horny beak, but behind the edentulous production there are four or five large and long teeth, followed by several smaller ones. The tail is long, stiff, and slender.

The *Ramphorhynchus longicaudus, R. Gemmingi*, and *R. Münsteri* belong to this genus. All are from the lithographic (middle oolitic) slates of Bavaria.

Genus PTERODACTYLUS, Cuv.—The jaws are provided with

teeth to their extremities ; all the teeth are long, slender'
sharp-pointed, set well apart. The tail is very short.

P. longirostris, Ok.—About 10 inches in length ; from
lithographic slate at Pappenheim. *P. crassirostris*, Goldf.—
About 1 foot long ; same locality (fig. 74). *P. Kochii*, Wagn.
—8 inches long ; from the lithographic slates of Kehlheim.
P. medius, Mnst.—10 inches long ; from the lithographic
slates at Meulenhard. *P. grandis*, Cuv.—14 inches long ;
from lithographic slates of Solenhofen. Two small and
probably immature Pterodactyles, showing the short jaws
characteristic of such immaturity, have been entered as species
under the names of *P. brevirostris* and *P. Meyeri*. The latter
shows the circle of sclerotic eye-plates.

The fragmentary remains of Pterodactyle from British
oolite—*e. g.*, Stonesfield slate, usually entered as *Pterodactylus
Bucklandi*—indicate a species about the size of a raven.

The evidences of Pterodactyles from the Wealden strata
indicate species about 16 inches in length of body. Those
(*P. Fittoni* and *P. Sedgwickii*, Ow.) from the greensand formation, near Cambridge, with neck-vertebræ 2 inches long, and
humeri measuring 3 inches across the proximal joint, had a
probable expanse of wing of from 18 to 20 feet. The *P.
Cuvieri*, Ow., and *P. compressirostris*, Ow., from the chalk of
Kent, attained dimensions very little inferior to those of the
greensand Pterodactyles.

More evidence is yet needed for the establishment of the
pterosaurian genus, on the alleged character of but two
phalanges in the wing-finger, and for which the term " Orni-
thopterus" has been proposed by Von Meyer.

With regard to the range of this remarkable order of flying
reptiles in geological time, the present evidence is as follows :
The oldest well-known Pterodactyle is the *Dimorphodon macronyx*, of the lower lias ; but bones of Pterodactyle have been
discovered in the coeval lias of Wirtemberg. The next in point

·of age is the *Dimorphodon Banthensis,* from the "Posidonomyen-
schiefer" of Banz in Bavaria, answering to the alum shale of
the Whitby lias; then follows the *P. Bucklandi* from the
Stonesfield oolite.   Above this come the first-defined and
numerous species of Pterodactyle from the lithographic slates
of the middle oolitic system in Germany, and from Cirin, on
the Rhone.   The Pterodactyles of the Wealden are as yet
known to us by only a few bones and bone fragments.   The
largest known species are those from the upper greensand
of Cambridgeshire.   Finally, the Pterodactyles of the middle
chalk of Kent, almost as remarkable for their great size, constitute the last forms of flying reptile known in the history of
the crust of this earth.

Order VII.—THECODONTIA.

Char.—Vertebral bodies biconcave : ribs of the trunk long
    and bent, the anterior ones with a bifurcate head :
    sacrum of three vertebræ : limbs ambulatory, femur
    with a third trochanter.   Teeth with the crown more
    or less compressed, pointed, with trenchant and finely
    serrate margins : implanted in distinct sockets.

Genus THECODONTOSAURUS.

Sp. *Thecodontosaurus antiquus.*—In 1836 certain reptilian
remains from the "dolomitic conglomerate" at Redland, near
Bristol, were described by Messrs. Riley and Stutchbury.*
The matrix has been referred to the Permian period ; it is
now thought by some good observers to be not older than the
triassic.

The teeth in these reptilian fossils are lodged in distinct
sockets ; they are arranged in a close-set series, slightly de-
creasing in size towards the posterior part of the jaw ; each
ramus of the lower jaw contained twenty-one teeth.   These

---

* Geological Transactions, 2d series, vol. v., p. 344.

are conical, rather slender, compressed and acutely pointed, with an anterior and posterior finely-serrated edge, the serratures being directed towards the apex of the tooth; the outer surface is more convex than the inner one; the apex is slightly recurved; the base of the crown contracts a little to form the fang, which is subcylindrical.

Genus PALÆOSAURUS, Riley and Stutchbury. In the same formation as contained the jaw and teeth of the *Thecodontosaurus* two other teeth were separately discovered, differing from the preceding and from each other ; the crown of one of these teeth measuring nine lines in length and five lines in breadth. It is compressed, pointed with opposite trenchant and serrated margins, but its breadth as compared with its length is so much greater than in the *Thecodontosaurus*, that upon it has been founded the genus *Palæosaurus*, and it is distinguished by the specific name of *platyodon*, from the second tooth, which is referred to the same genus under the name of *Palæosaurus cylindrodon*. The portion of the tooth of the *Palæosaurus cylindrodon* which has been preserved, shows that the crown is subcompressed and traversed by two opposite finely-serrated ridges, as in the *Thecodontosaurus ;* its length is five lines, its breadth at the base two lines.

The vertebræ associated with the two kinds of teeth above described are sub-biconcave, with the middle of the body more constricted, and terminal articular cavities rather deeper than in *Teleosaurus ;* but they are chiefly remarkable for the depth of the spinal canal at the middle of each vertebra, where it sinks into the substance of the centrum ; thus the canal is wider, vertically, at the middle than at the two ends of the vertebra : an analogous structure, but less marked, obtains in the dorsal vertebræ of the *Rhynchosaurus* from the new red sandstone of Shropshire.

Besides deviating from existing lizards in the thecodont dentition and biconcave vertebræ, the Saurians of the dolomitic

conglomerate also differ in having some of their ribs articulated
by a head and tubercle to two surfaces of the vertebra, as at
the anterior part of the chest in crocodiles and Dinosaurs.
The shaft of the rib was traversed, as in the Protorosaur and
Rhynchosaur, by a deep longitudinal groove. Some fragmentary bones indicate obscurely that the pectoral arch deviated
from the crocodilian, and approached the lacertian or enaliosaurian type, in the presence of a clavicle, and in the breadth
and complicated form of the coracoid. The sacrum includes at
least three vertebræ. The humerus appears to have been little
more than half the length of the femur, and to have been, like
that of the *Rhynchosaurus*, unusually expanded at the two
extremities. The femur is chiefly remarkable for a third
process or trochanter, just above the middle of the shaft,
which shows a medullary cavity. The distal condyles are
flattened, the outer one being the larger; there is a deep
depression between them posteriorly, and a very light one
anteriorly.

The tibia, fibula, and metatarsal bones manifest, like the
femur, the fitness of the Saurians for progression on land. The
ungual phalanges are sub-compressed, curved downwards,
pointed, and impressed on each side with the usual curved
canal.

The following conclusions may be drawn from the knowledge at present possessed of the osteology of the *Thecodontosaurus* and *Palæosaurus* : in their thecodont type of dentition,
biconcave vertebræ, double-jointed ribs, and proportionate size
of the bones of the extremities, they agree with the amphicœlian crocodiles ; but they combine a dinosaurian femur, a
lacertian form of tooth, and a lacertian structure of the pectoral and probably pelvic arch with these crocodilian characters ; and they have distinctive modifications, such as the
moniliform spinal canal, in which, however, the almost contemporary Rhynchosaur participates. It would be interesting

to ascertain whether the caudal vertebræ are characterized, as in the Thuringian Protorosaur, by double diverging spinous processes.

Genus BELODON, Von Meyer.

Sp. *Belodon Plieningeri.*—The reptile from the upper white keuper sandstone of Wirtemberg, described by Plieninger,[*] agrees in its essential characters so closely with the thecodont Saurians of the Bristol conglomerate as to add to the probability of both belonging to the same lower mezozoic period.

Three vertebræ are modified to afford adequate attachment to the iliac bones in *Belodon*, and this additional evidence of affinity to *Dinosauria* may have characterized also the English Thecodonts.

Genus CLADYODON, Ow.

Sp. *Cladyodon Lloydii.*—In the Memoir on the Triassic Red Sandstones of Warwick, by Murchison and Strickland, published in 1840, in the 2d series of the Geological Transactions, vol. v., a tooth, which is an extremely rare fossil in those English formations, was figured in pl. xxviii., fig. 6.

Having had the opportunity of studying the original specimen and fragments of some others of seemingly the same species from the new red sandstones of Warwick and Leamington, the writer recognized the affinity of the reptile possessing those teeth to the thecodont reptiles of the Bristol conglomerate, and indicated what appeared to be a generic modification of dental form by the term *Cladyodon.*[†] He subsequently received other specimens of the teeth characterizing this genus, which may be described as being two-edged, sub-compressed ; the sides more or less convex ; the edges more or less sharp, and

---

[*] Würtemb. naturf. Jahreshefte, viii., Jahrg. 1857, p. 389. Jaeger's *Phytosaurus* appears to have been founded on casts of the sockets of the teeth of *Belodon*.

[†] Reports of the British Association, "Brit. Fossil Reptiles," 1841, p. 155. (See fuller descriptions, with figures, in *Odontography*, pl. 62, A, fig. 4, *a, b.*)

frequently finely serrate ; the crown slightly bent sideways, the inner side towards the mouth-cavity. The teeth are sometimes lancet-shaped, through convergence of the edges towards point ; sometimes through one edge being convex and the other concave, the crown is slightly curved or sickle-shaped ; sometimes through use, the point is blunted. The enamel is very thin, smooth, showing under the lens a slight longitudinal striation, forming wrinkles. The dentine is disposed in concentric layers ; it is not labyrinthic ; the base of the tooth shows a conical pulp-cavity. These teeth indicate a Saurian about ten feet in length.

The writer cannot discern any generic, or even good specific distinctions, between the teeth from the Warwickshire keuper, on which in 1840 he founded the genus *Cladyodon,* and those from the Wirtemberg keuper, on which M. Von Meyer in 1844 founded the genus *Belodon.* Both are nearly allied to *Palæosaurus.*

The two following genera are referred provisionally and with doubt to the present order :—

Genus BATHYGNATHUS, Leidy.

Sp. *Bathygnathus borealis,* Leidy.—Allied to the *Cladyodon* and *Belodon* by the shape of the teeth is the Saurian from the new red sandstone of Prince Edward's Island, North America, the generic and specific characters of which have been deduced by Dr. Leidy [*] from a portion of lower jaw, containing seven teeth, but with interspaces from which others have been lost. The depth of the dentary bone is five inches ; a peculiarity which suggested the generic name (*bathus,* deep ; *gnathos,* jaw). The precise mode of implantation of the teeth is not described.

The fossil was discovered at a depth of 21 feet from the

---

[*] Journal of the Academy of Sciences, Philadelphia, vol. ii., p. 327, pl. xxxiii.

surface, in a red sandstone supposed to be of the same age as that of Connecticut, so remarkable for the various and singular foot-marks, referable, some to reptiles, and others to large birds.

Genus PROTOROSAURUS, Von. Meyer.

Sp. *Protorosaurus Speneri*, Von M.—The first fossil Saurian on record is that which marks the circumstance by its generic name, and honours its describer by the specific one. The slab of "copper-slate" from the Permian beds of Eisenach in Thuringia, displaying, either in fossils or impressions, the skull, vertebral column, and bones of the fore foot of the reptile in question, was figured and described by Spener, a physician at Berlin, in 1710.* The original specimen is now in the museum of the Royal College of Surgeons, London, where it forms part of the Hunterian series of fossils. It was obtained from a copper-mine near Eisenach, at a depth of 100 feet from the surface.

A second specimen, showing the two fore limbs, a hind limb, and part of the trunk, was described by Link in 1718.† Cuvier gives copies of portions of two other specimens in his *Ossemens Fossiles.*‡

The healthy, honest mind of Spener is shown by the conclusions which he formed from the state of preservation of his specimen ("omnia, enim, minutissima, etiam apophyses, spinæ," etc.), and from its association with equally well-preserved remains of fishes, and even of the delicate leaves of plants, against the notions of those fossils merely simulating, and never having been, the living organisms which they represented—notions which were then advocated under the sounding phrase of "plastic force," as they have lately been under that of "prochronism." Spener's only doubt was, whether the reptile had been a crocodile or a lizard ; but he

* Miscellanea Berolinensia, 4to, i., p. 99, figs. 24 and 25.
† Acta Eruditorum, 1718, p. 188, pl. ii.
‡ Ed. 8vo, 1836, pl. ccxxxvii., figs. 1 and 2.

inclined to the former view, on account of the proportions of the head to the trunk. He then enters upon speculations as to how a crocodile could have come into Germany ; and shows the usual effect of a mind biassed by a hypothetical diluvial catastrophe not demonstrated by observation and inductive research, and to the extent of such bias benumbed in the exercise of the faculty for the acquisition of natural truth.

The seven cervical vertebræ are proportionally larger than in any known recent or fossil terrestrial or aquatic Saurian ; they resemble in this respect the cervical vertebræ of Ptero-dactyles ; the tail is long, and its vertebræ differ from those of all other known reptiles, recent or fossil, in having the spinous processes bifurcate, diverging in the direction of the axis of the body.[*]

The muscular power of the neck is indicated by traces of bone-tendons. The dorsal vertebræ exceed eighteen in number, and have higher spines than in the modern Monitors; the dorsal ribs are long, and longitudinally impressed. The hind limb is much longer than the fore limb, and the leg is longer, in proportion to the thigh and foot, than in the Moni-tors. The teeth are sharp-pointed, slender ; there appear to be at least twenty in both upper and lower jaws in Spener's specimen.

It may be concluded, from the length and strength of the tail and the peculiar provision for muscular attachments in that part, and from the proportions of the hind limbs, that the *Protorosaurus* was of aquatic habits, and that the strength of its neck and head, and the sharpness of its teeth, enabled it to seize and overcome the struggles of the active fishes of the waters which deposited the old Thuringian copper-slates.

At Spynie and Cummingstone, in the neighbourhood of Elgin, N. B., in a stratum of a fine-grained whitish sandstone,

---

[*] A character first pointed out in the writer's "Report on British Fossil Reptiles," Trans. of Brit. Assoc., 1841, p. 155.

cemented by carbonate of lime, situated between "Old Red" and "Purbeck" formations, and resting conformably upon the former, evidences of Saurian (Crocodilian and Lacertian) reptiles, characteristic of triassic time, have been discovered. The remains of the large reptile, with pitted bony dermal scales, had been, on their first discovery, referred to a genus of fishes by Agassiz, under the name of *Staganolepis*, or "pitted-scale," probably from the belief that the formation belonged to the "Old Red System." I determined the crocodilian nature of the scales, and the affinity of the reptile to the Thecodonts, in the breadth of the coracoid or pubis as shown by the cast of the bone, at the meeting of the British Association at Leeds, in September 1858. I have since been favoured by Mr. Duff with a tooth, assoicated with scales of *Staganolepis*, which is "thecodont" in character, and like that of *Cladyodon*.

In the same sandstone, in the quarry at Cummingstone, near Elgin, a continuous series of thirty-four impressions have been observed. The impressions are in pairs, forming two parallel rows, the hind one being one inch in diameter.

I had some years before determined the true saurian nature of the impression of the skeleton of the trunk and part of the head of a small reptile discovered by Mr. Patrick Duff of Elgin, at Spynie, and noticed by him in the "Elgin Courant" of October 10th, 1851, as evidence of an air-breathing vertebrat in "Old Red Sandstone." The specimen was submitted by Mr. Duff to my examination, the result of which was given, Dec. 15, 1851, in the "Literary Gazette" of that week, as follows:—

"It is the impression, in two pieces, of a grey variety of the old red sandstone, of a long and slender four-footed vertebrate animal, four inches and a half in length, clearly belonging, by the form, proportions, and positions of the scapular and pelvic arches, and their appended limbs, to the reptilian class. The osseous substance has disappeared; the cavities in the sandstone which contained it remain, stained by a deposit of an ochreous tint. The impressions are so well defined, as clearly to show that there were twenty-six vertebræ between the skull and sacrum, two sacral vertebræ, and thirteen caudal vertebræ, before the tail disappears by dipping into an unexposed part of the matrix. Impressions of twenty-one pairs of ribs are preserved, all very slender, short where they commence near the head, but rapidly gaining length as they are placed further back. The cervical and anterior ribs are expanded, but not bifurcate, at their vertebral end ; all the ribs articulate close to the bodies of the vertebræ. In the crocodilian reptiles the anterior ribs are bifurcate, and the posterior ones, with a simple head, articulate with long diapophyses. The distinctive characters of the batrachian skeleton are the double occipital condyle ; ribs wanting, or very short and subequal ; a single sacral vertebra, and rib-shaped ilium. The first character cannot be determined, the occipital articulation not being preserved in the fossil. Instead of the second character, the fossil shows ribs of varied length, and most of them much longer than in the salamanders, newts, or any known Batrachian. With regard to the third character, the impression in the matrix clearly shows two sacral vertebræ and a short subquadrate pelvis.

"Both the humerus and the femur show the lacertian sigmoid shape, and near equality of length, which distinguish them alike from the crocodilian and batrachian orders ; they are likewise, as in lizards, relatively longer than in the newts and salamanders. Near the imperfect impression of the head may be seen the hollow bases of some large, slightly-compressed, conical teeth, which also tell for the saurian and against the batrachian nature of this ancient reptile. I propose to call it *Leptopleuron lacertinum.*[*] Many particulars of minor import, bearing upon the more immediate affinities of this most rare

---

[*] Λεπτός, *slender ;* πλευρόν, *rib ;* for this compound we have the authority of *Poikilopleuron,* already applied to an extinct genus of Saurians.

and interesting fossil, have been noted, and will be given, with
the figures, in my History of British Fossil Reptiles, for which
work Mr. Duff has kindly consented to place the specimen at
my disposal. In the meanwhile, I beg to offer the above *précis*
of the main characters of the fossil.—RICHARD OWEN."

Other palæontologists regarded the fossil as a batrachian
reptile; but no evidence, osteological or dental, has been
pointed out in support of this view.*

With regard to the geological age of the calcareous sandstone containing *Staganolepis* and *Leptopleuron,* the author has
remarked, in the article "Palæontology," when the belief of
some eminent geologists in the Devonian age of the stratum
is quoted—"As yet, however, no characteristic Devonian or
'Old Red' fossils of any class have been discovered associated
with the foregoing evidences of reptiles, which, according to
the determination of strata by characteristic fossils, would
belong to the secondary or mezozoic period."† It is, most
probably, of triassic age.

Order VIII.—DINOSAURIA.

Char.—Cervical and anterior dorsal vertebræ with par- and
di-apophyses, articulating with bifurcate ribs; dorsal
vertebræ with a neural platform, sacral vertebræ ex-

* The following notice of this determination of the fossil will be found in
the *Athenæum* of Dec. 13, in the title of a paper to be read at the ensuing
meeting of the Geological Society :—"Notice of the occurrence of Fossil Foot-
Tracks, and the Remains of a Batrachian Reptile, in the Old Red Sandstone of
Morayshire, by Capt. Brickenden and Dr. Mantell." The belief in the antiquity
of the stratum showing the impression of the skeleton—remains of the skeleton
there are none—appears to have weighed with these gentlemen in placing the
animal at the bottom of the air-breathing series of vertebrals, as well as in pro-
posing for it the name of *Telerpeton,* or "last of reptiles;" (τελος, the end or
issue of a thing, or τελειος, having reached its end, and ερπετον, a reptile), at least
as traced backwards in time. The term *Leptopleuron* has, however, the priority
of publication: being also the result of a truer exposition of the nature and
affinities of the fossil, and free from the signification of its appearance in time,
it will be, probably, preferred.

† Encyclopædia Britannica, vol. xxii., p. 130.

ceeding two in number ; body supported on four strong unguiculate limbs.

The well-ossified vertebræ, large and hollow limb-bones, and tritrochanterian femora of the thecodont reptiles of the Bristol conglomerate, together with the structure of the sacral vertebræ in the allied *Belodon*, indicate the beginning, at the triassic period, of an order of *Reptilia* which acquired its full development and typical characteristics in the oolitic period.

Genus SCELIDOSAURUS, Ow.—By this name is indicated a Saurian with large and hollow limb-bones, with a femur, having the third inner trochanter, and with metacarpal and phalangial bones, adapted for movement on land. The fossils occur in the lias at Charmouth, Dorsetshire.

Genus MEGALOSAURUS, Bkld.—The true dinosaurian characters of this reptile have been established by the discovery of the sacrum, which consists of five vertebræ, interlocked by the alternating position of neural arch and centrum. The articular surfaces of the free vertebræ are nearly flat ; the neural arch develops a platform which in the anterior dorsals supports very long and strong spines.

The compressed piercing and trenchant form of tooth which characterises the existing varanian lizards was manifested by the *Megalosaurus*. The specimen which is most illustrative of the dental peculiarities of this gigantic reptile is a portion of the lower jaw with a few teeth. The first character which attracts attention in this fossil is the inequality in the height of the outer and inner alveolar walls ; a similar inequality characterizes the jaws of almost all the existing lizards. But in these the oblique groove, so bounded, to which the bases of the developed teeth are anchylosed, is much more shallow, and is relatively wider ; and the teeth in all the stages of growth are completely exposed when the gum has been removed.

In the *Megalosaurus* the greater relative development of the inner alveolar wall, as compared with the dentigerous part of the jaw in existing Saurians, narrows the dental groove, and covers a greater proportion of the bases of the teeth, besides concealing more or less completely the germs of their successors. Moreover, instead of the mere shallow impressions upon the inner side of the outer alveolar plate to which the teeth are attached in modern lizards, there are distinct sockets formed by bony partitions connecting the outer with the inner alveolar wall in the jaw of the Megalosaurus. These partitions rise from the outer side of the inner alveolar wall in the form of triangular vertical plates of bone, and from the middle of the outer side of each plate a bony partition crosses to the outer parapet, completing the alveoli of the fully-formed or more advanced teeth ; the series of triangular plates forming a kind of zig-zag buttress along the inner side of those alveoli. The outer parapet rises an inch higher than the inner one.

Fig. 75 exhibits a portion of another jaw of the *Megalosaurus*, also from Stonesfield oolite, from which the inner wall has been removed to show the germ of a successional tooth *c*, about to succeed an old tooth *a*, which has been broken, and near to which is a newly-formed tooth, *b*, coming into place. These teeth will exemplify the shape of the crown of the tooth, which is subcompressed, slightly recurved, sharp edged, and sharp-pointed, the edges being minutely serrated ; the edge upon the convex or front border *b* becomes blunted as it descends about two-thirds of the way towards the base of the tooth ; that upon the concave hinder border *a* is continued to the base. The lower half of the crown is thicker towards the fore margin than towards the hind one ; so that a transverse section, like that above, *a*, in fig. 75, gives a narrow oval form pointed behind. At the upper half of the crown the sides slope more equally from the middle thickest part to both margins, and the section is a narrow pointed ellipse. The crown is covered by a smooth and polished enamel, which

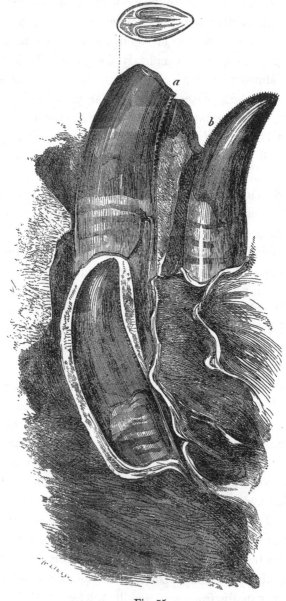

Fig. 75.
Section of jaw with teeth of the *Megalosaurus Bucklandi*, nat. size.

wholly forms the marginal serrations. The base of the tooth is coated with a smooth, lighter-coloured cement, forming a thin layer, and becoming a little thicker towards the implanted end of the tooth. The remains of the pulp are converted into osteo-dentine in the basal part of the completely formed tooth. Moderately magnified, the surface of the enamel presents a finely wrinkled appearance. The marginal serrations show, under a somewhat higher power, that the points are directed towards the apex of the tooth—a structure well adapted for dividing the tough tissues of the saurian integument.

The main body of the tooth consists of dentine, of that hard unvascular kind of which the same part of the teeth of existing crocodiles and most mammals is composed. No part of the dentine is pervaded by medullary canals, as in the Iguanodon.

A series of teeth from individual Megalosaurs, of different ages, are preserved in the British Museum and in the geological museum at Oxford; although differing in size, they preserve the characteristic form above described. In one specimen the point of the crown and the trenchant margins have been rubbed down to a smooth obtuse surface; it seems to have come from the hinder part of the dental series, where the teeth may have been smaller and less sharp, or more liable to be blunted by a greater share in the imperfect act of mastication, than the teeth in advance.

Successional teeth in different stages of growth are shown in the original portion of jaw of the Megalosaur in the Oxford museum. Some, more advanced, show their crowns projecting from alveoli already formed by the plates extending across from the triangular processes before described: vacant sockets, from which fully formed teeth have escaped, occur, generally in the intervals between these more advanced teeth. The summits of less developed teeth are seen protruding at the inner side of the basal interspaces of the triangular plate,

between them and the true internal alveolar parapet. There
can be no doubt that, in the course of the development of these
teeth, corresponding changes take place in the jaw itself, by
which new triangular plates and alveolar partitions are formed,
as the old ones become absorbed, analogous to those concomitant changes in the growth and form of the teeth, alveoli, and
jaws, which take place in so striking a degree in the elephant.
The peculiarity of the Megalosaur, as compared with the
crocodiles and lizards which have a like endless succession of
teeth, is the deeper position of the successional tooth (fig. 75,
c), in relation to the one (a) it is destined to replace, and the
great proportion of the tooth which is formed before it is protruded.   The anterior tooth a in this specimen shows at the
inner side of its base the commencing absorption stimulated by
the encroaching capsule of the successional tooth c below, the
crown of which is completed externally, though not consolidated.   On one of the fractured margins of this piece of jaw, a
part of the basal shell of an absorbed and shed tooth remains,
with part of the root of the successional tooth, which has risen
into place, but which shows its base full of matrix, the pulp
not having been calcified at that period of the tooth's growth.

In the proportion of the successional teeth which is formed
in the formative cavity in the substance of the jaw, the Megalosaur offers a closer resemblance to the mammalian class
than do any of the recent or extinct crocodilian or lacertian
reptiles.   But the evidence of uninterrupted and frequent
succession of the teeth in the Megalosaur is unequivocal;
and this part of the dental economy of the great carnivorous
reptile is strictly analogous to that which governs the same
system in the existing members of the class.   The different
forms of the teeth at different stages of protrusion did not fail
to attract the attention of the gifted discoverer of the great
predatory saurian, in whose words this notice of its dentition
may be fitly concluded :—

" In the structure of these teeth we find a combination of mechanical contrivances analogous to those which are adopted in the construction of the knife, the sabre, and the saw. When first protruded above the gum, the apex of each tooth presented a double cutting edge of serrated enamel. In this stage its position and line of action were nearly vertical ; and its form, like that of the two-edged point of a sabre, cutting equally on each side. As the tooth advanced in growth, it became curved backwards in the form of a pruning-knife, and the edge of serrated enamel was continued downwards to the base of the inner and cutting side of the tooth, whilst on the outer side a similar edge descended, but a short distance from the point ; and the convex portion of the tooth became blunt and thick, as the back of a knife is made thick for the purpose of producing strength. The strength of the tooth was further increased by the expansion of its side. Had the serrature continued along the whole of the blunt and convex portion of the tooth, it would in this position have possessed no useful cutting power ; it ceased precisely at the point beyond which it could no longer be effective. In a tooth thus formed for cutting along its concave edge, each movement of the jaw combined the power of the knife and saw ; whilst the apex, in making the first incision, acted like the two-edged point of a sabre. The backward curvature of the full-grown teeth enabled them to retain, like barbs, the prey which they had penetrated. In these adaptations we see contrivances which human ingenuity has also adopted in the preparation of various instruments of art."*

The oldest known beds from which any remains of *Megalosaurus* have been obtained are the lower oolites at Selsby Hill, and Chipping-Norton, Gloucestershire. Abundant and characteristic remains occur in the Stonesfield slate, Oxfordshire. Teeth of this genus have been found in the Cornbrash and Bath oolite ; both teeth and bones are common in the Wealden

* Buckland, "Bridgewater Treatise," p. 236.

strata and Purbeck limestone. Some of these fossils indicate
a reptile of at least 30 feet in length.

Genus HYLÆOSAURUS, Mtll.—Remains of the Dinosaurian
so called have hitherto been found only in Wealden strata, as
at Tilgate, Bolney, and Battle. The most instructive evidence
is that which was exposed by the quarreymen of the Wealden
stone at Tilgate, and was obtained and described by Mantell
in 1832. It consisted of a block of stone measuring 4½ feet
by 2½ feet (fig. 76), and included the following parts of the

Fig. 76.
Hylæosaurus (Wealden).

skeleton in almost natural juxtaposition :—10, anterior vertebræ, the first supporting part of the base of the skull ; several
ribs, 4, 4 ; some enormous dermal bony spines, 5, 6, 6, which
supported a strong defensive crest along the back ; two coracoids, 7, 7 ; scapulæ, 8, 8 ; and some detached vertebræ and
fragments of bones. In 1841 the writer showed that the
sacrum was dinosaurian, and included five vertebræ.

The teeth are relatively small, close-set, thecodont in implantation, with subcylindrical fang and a subcompressed
slightly expanded and incurved crown, with the borders
straight and converging to the blunt apex. They indicate
rather a mixed or vegetable diet than a carnivorous one. The

skin was defended by subcircular bony scales. The length of
the Hylæosaur may have been 25 feet.

Genus IGUANODON, Mtll.—Remains of the large herbivorous
reptiles of this genus have been found in Wealden and neocomian (greensand) strata. Femora, four feet in length, showing
the third inner trochanter, have been discovered. The sacrum
included five, and in old animals six, vertebræ ; the claw-bones
are broad, flat, and obtuse. There were only three well-
developed toes on the hind foot ; and singular large tridactyle
impressions, discovered by Beccles in the Wealden at Hastings,
have been conjectured to have been made by the Iguanodon.

With vertebræ, subconcave at both articular extremities,
having, in the dorsal region, lofty and expanded neural arches,
and doubly articulated ribs, and characterized in the sacral
region by their unusual number and complication of structure ;
with a Lacertian pectoral arch, and unusually large bones of
the hind limbs, excavated by large medullary cavities, and
adapted for terrestrial progression ;—the *Iguanodon* was distinguished by teeth, resembling in shape those of the Iguana, but
in structure differing from the teeth of that and every other
known reptile, and unequivocally indicating the former existence in the Dinosaurian order of a gigantic representative of
the small group of living lizards which subsist on vegetable
substances.

The important difference which the fossil teeth presented
in the form of their grinding surface was pointed out by
Cuvier,* of whose description Dr. Mantell adopted a condensed
view in his Illustrations of the Geology of Sussex, 4to, 1827,
p. 72. The combination of this dental distinction with the
vertebral and costal characters, which prove the *Iguanodon* not
to have belonged to the same group of Saurians as that which
includes the Iguana and other modern lizards, rendered it
highly desirable to ascertain by the improved modes of investi-

* Ossemens Fossiles, 1824, vol. v., pt. ii., p. 351.

gating dental structure, the actual amount of correspondence between the *Iguanodon* and Iguana in this respect. This has been done in the author's general description of the teeth of reptiles,[*] from which the following notice is abridged :—The teeth of the *Iguanodon* (fig. 77), though resembling most

closely those of the Iguana, do not present an exact magnified image of them, but differ in the greater relative thickness of the crown, its more complicated external surface, and, still more essentially, in a modification of the internal structure, by which the *Igua-nodon* equally deviates from every other known reptile.

Fig. 77.

Front and side views of a tooth of the lower jaw of the Iguanodon, nat. size.

As in the Iguana, the base of the tooth is elongated and contracted ; the crown expanded and smoothly convex on the inner side ; when first formed it is acuminated, compressed, its sloping sides serrated, and one surface, external in the upper jaw, internal in the lower jaw, is traversed by a median longitudinal ridge, and coated by a layer of enamel ; but beyond this point the description of the tooth of the *Igua-nodon* indicates characters peculiar to that genus. In most of the teeth that have hitherto been found, three longitudinal ridges traverse the ridged surface of the crown, one on each side of the median primitive ridge ; these are separated from each other and from the serrated margins of the crown by four wide and smooth longitudinal grooves. The relative

* Odontography, pt. ii., p. 249 ; Transactions of the British Association, 1838.

width of these grooves varies in different teeth; sometimes a fourth small longitudinal ridge is developed on the ridged side of the crown. The marginal serrations which, at first sight, appear to be simple notches, as in the Iguana, present under a low magnifying power (fig. 78), the form of transverse ridges, themselves notched, so as to resemble the mammilated margins of the unworn plates of the elephant's grinder; slight grooves lead from the interspaces of these notches upon the sides of the marginal ridges. These ridges or dentations do not extend beyond the expanded part of the crown; the longitudinal ridges are continued farther down, especially the median

Fig. 78.

Marginal ridges on the tooth of the Iguanodon, magn.

ones, which do not subside till the fang of the tooth begins to assume its subcylindrical form. The tooth at first increases both in breadth and thickness; then it diminishes in breadth, but its thickness goes on increasing; in the larger and fully formed teeth, the fang decreases in every diameter, and sometimes tapers almost to a point. The smooth unbroken surface of such fangs indicates that they did not adhere to the inner side of the maxillæ, as in the Iguana, but were placed in separate alveoli, as in the Crocodile and Megalosaur; such support would appear, indeed, to be indispensable to teeth so worn by mastication as those of the *Iguanodon*.

The apex of the tooth soon begins to be worn away, and it would appear, by many specimens, that the teeth were retained until nearly the whole of the crown had yielded to the daily abrasion. In these teeth, however, the deep excavation of the remaining fang plainly bespeaks the progress of the successional tooth prepared to supply the place of the worn-out grinder. At the earlier stages of abrasion a sharp edge is maintained at the ridged part of the tooth by means of the enamel which covers that surface of the crown; the prominent ridges upon that surface give a sinuous contour to the middle of the cutting edge, whilst its sides are jagged by the lateral serrations. The adaptation of this admirable dental instrument to the cropping and comminution of such tough vegetable food as the *Clathrariæ* and similar plants, which are found buried with the *Iguanodon*, is pointed out by Dr. Buckland, with his usual felicity of illustration, in his Bridgewater Treatise, vol. i., p. 246.

When the crown is worn away beyond the enamel, it presents a broad and nearly horizontal grinding surface (fig. 79),

and now another dental substance is brought into use, to give an inequality to that surface ; this is the ossified remnant of the pulp, which, being firmer than the surrounding dentine, forms a slight transverse ridge in the middle of the grinding surface ; the tooth in this stage has exchanged the functions of an incisor for that of a molar, and is prepared to give the final compression, or comminution, to the coarsely divided vegetable matters.

Fig. 79.

A worn tooth of the Iguanodon.

The marginal edge of the incisive condition of the tooth and the median ridge of the molar stage are more effectually established by the introduction of a modification into the texture of the dentine, by which it is rendered softer than in the existing Iguanæ and other reptiles, and more easily worn away. This is effected by an arrest of the calcifying process along certain cylindrical tracts of the pulp, which is thus continued, in the form of medullary canals, analogous to those in the soft dentine of the Megatherium's grinder, from the central cavity, at pretty regular intervals, parallel with the dentinal tubes, nearly to the surface of the tooth. The medullary canals radiate from the internal (upper jaw) or external (lower jaw) sides of the pulp-cavity, and are confined to the dentine forming the corresponding walls of the tooth. Their diameter is $\frac{1}{1 \cdot 250}$th of an inch. They are separated by pretty regular intervals, equal to from six to eight of their own diameters. They sometimes divide once in their course. Each medullary canal is surrounded by a clear space. Its cavity was occupied in the section described by a substance of a deeper yellow colour than the rest of the dentine.

The dentinal tubes present a diameter of $\frac{1}{1\cdot250}$th of an inch, with interspaces equal to about four of their diameters. At the first part of their course, near the pulp-cavity, they are bent in strong undulations, but afterwards proceed in slight and regular primary curves, or in nearly straight lines to the periphery of the tooth. The secondary undulations of each tooth are regular, and very minute. The branches, both primary and secondary, of the dentinal tubes are sent off from the concave side of the main inflections ; the minute secondary branches are remarkable at certain parts of the tooth for their flexuous ramifications, anastomoses, and dilatations into minute calcigerous cells, which take place along nearly parallel lines for a limited extent of the course of the main tubes. The appearance of interruption in the course of the dentinal tubes, occasioned by this modification of their secondary branches, is represented by the irregularly dotted tracts in the figure. This modification must contribute, with the medullary canals, though in a minor degree, in producing that inequality of texture and of density in the dentine, which renders the broad and thick tooth of the *Iguanodon* more efficient as a triturating instrument.

The enamel which invests the harder dentine, forming the ridged side of the tooth, presents the same peculiar dirty brown colour, when viewed by transmitted light, as in most other teeth. Very minute and scarcely perceptible undulating fibres, running vertically to the surface of the tooth, form the only discernible structure in it.

The remains of the pulp in the contracted cavity of the completely formed tooth are converted into a dense but true osseous substance, characterized by minute elliptical radiated cells, whose long axis is parallel with the plane of the concentric lamellæ, which surround the few and contracted medullary canals in this substance.

The microscopical examination of the structure of the *Iguanodon's* teeth thus contributes additional evidence of the perfection of their adaptation to the offices to which their more obvious characters had indicated them to have been destined.

To preserve a trenchant edge, a partial coating of enamel is applied; and, that the thick body of the tooth might be worn away in a more regularly oblique plane, the dentine is rendered softer as it recedes from the enamelled edge, by the simple contrivance of arresting the calcifying process along certain tracts of the opposite wall of the tooth. When attrition has at length exhausted the enamel, and the tooth is limited to its function as a grinder, a third substance has been prepared in the ossified remnant of the pulp to add to the efficiency of the dental instrument in its final capacity. And if the following reflections were natural and just, after a review of the external characters of the dental organs of the *Iguanodon*, their truth and beauty become still more manifest as our knowledge of their subject becomes more particular and exact :—

" In this curious piece of animal mechanism we find a varied adjustment of all parts and proportions of the tooth, to the exercise of peculiar functions, attended by compensations adapted to shifting conditions of the instrument during different stages of its consumption. And we must estimate the works of nature by a different standard from that which we apply to the productions of human art, if we can view such examples of mechanical contrivance, united with so much economy of expenditure, and with such anticipated adaptations to varying conditions in their application, without feeling a profound conviction that all this adjustment has resulted from design and high intelligence."

All trace of dinosaurian reptiles disappears in the lower cretaceous beds.

Order IX.—CROCODILIA.

Char.—Teeth in a single row, implanted in distinct sockets; external nostril single and terminal or sub-terminal. Anterior trunk vertebræ with par- and di-apophyses, and bifurcate ribs; sacral vertebræ two, each supporting its own neural arch. Skin protected by bony, usually pitted plates.

Sub-Order 1.—AMPHICŒLIA.*

Crocodiles closely resembling in general form the long and slender-jawed kind of the Ganges called "gavial" or "gharrial," existed from the time of the deposition of the lower lias.

Their teeth were similarly long, slender, and sharp, adapted for the prehension of fishes, and their skeleton was modified for more efficient progress in water by the vertebral surfaces being slightly concave, by the hind limbs being relatively larger and stronger, and by the orbits forming no prominent obstruction to progress through water. From the nature of the deposits containing the remains of the so-modified crocodiles, they were marine. The fossil crocodile from the Whitby lias, described and figured in the Philosophical Transactions, 1758, p. 688, is the type of these amphicœlian species. They have been grouped under the following generic heads :—*Teleosaurus, Steneosaurus, Mystriosaurus, Macrospondylus, Massospondylus,* to which must be added *Pœcilopleuron, Pelagosaurus, Æolodon, Suchosaurus, Goniopholis.*

Species of the above genera range from the lias to the chalk inclusive.

Suchosaurus of the Wealden is characterized by the compressed crown and trenchant margins of the teeth; *Goniopholis,* of the Purbeck beds, by some of the dermal scales having the same peg-and-pit interlocking as in the scales of the ganoid fish in fig. 52.

* *Amphi,* both; *koilos,* hollow; the vertebra being hollowed at both ends.

Sub-Order 2.—OPISTHOCŒLIA.*

The small group of Crocodilia so called is an artificial one, based upon more or less of the anterior trunk vertebræ being united by ball-and-socket joints, but having the ball in front, instead of, as in modern crocodiles, behind. Cuvier first pointed out this peculiarity† in a Crocodilian from the Oxfordian beds at Honfleur, and the Kimmeridgian at Havre. The writer has described similar opisthocœlian vertebræ from the great oolite at Chipping Norton, from the upper lias of Whitby, and, but of much larger size, from the Wealden formations of Sussex and the Isle of Wight. These specimens probably belong, as suggested by the writer in 1841,‡ to the fore part of the same vertebral column as the middle dorsal vertebræ, flat at the fore part, and slightly hollow behind, on which he founded the genus *Cetiosaurus.* The smaller opisthocœlian vertebræ described by Cuvier have been referred by Von Meyer to a genus called *Streptospondylus.*

In one species from the Wealden, dorsal vertebræ measuring 8 inches across are only 4 inches in length, and caudal vertebræ nearly 7 inches across are less than 4 inches in length. These characterize the species called *Cetiosaurus brevis.*§

Caudal vertebræ measuring 7 inches across and 5½ inches in length, from the lower oolite at Chipping Norton, and the great oolite at Enstone, represent the species called *Cetiosaurus medius.*

Caudal vertebræ from the Portland stone at Garsington, Oxfordshire, measuring 7 inches 9 lines across and 7 inches

---

* *Opisthos,* behind; *koilos,* hollow; vertebra concave behind, convex or flat in front.

† Annales du Muséum, tom. xii., p. 83, pl. x. xi.

‡ "Report on British Fossil Reptiles," Trans. Brit. Assoc. for 1841, p. 96.

§ They have since been referred to the dinosaurian order under the name of *Pelorosaurus,* but without any evidence of the true sacral characters of that order; the cavities of long bones are common to Crocodilians and Dinosaurs.

in length, are referred to the *Cetiosaurus longus.* The latter must have been the most gigantic whale-like of Crocodilians.

Sub-Order 3.—PROCŒLIA.*

The best and most readily recognizable characters by which the existing Crocodilians are grouped in appropriate genera are derived from modifications of the dental system.

In the caimans (genus *Alligator*) the teeth vary in number from $\frac{18-18}{18-18}$ to $\frac{22-22}{22-22}$; the fourth tooth of the lower jaw or canine, is *received into a cavity* of the palatal surface of the upper jaw, where it is concealed when the mouth is shut ; in old individuals the upper jaw is perforated by these large inferior canines, and the fossæ are converted into foramina.

In the true crocodiles (genus *Crocodilus*) the first tooth in the lower jaw perforates the palatal process of the intermaxillary bone when the mouth is closed ; the fourth tooth in the lower jaw is *received into a notch* excavated in the side of the alveolar border of the upper jaw, and is visible externally when the mouth is closed.

In the two preceding genera the alveolar borders of the jaws have an uneven or wavy contour, and the teeth are of unequal size.

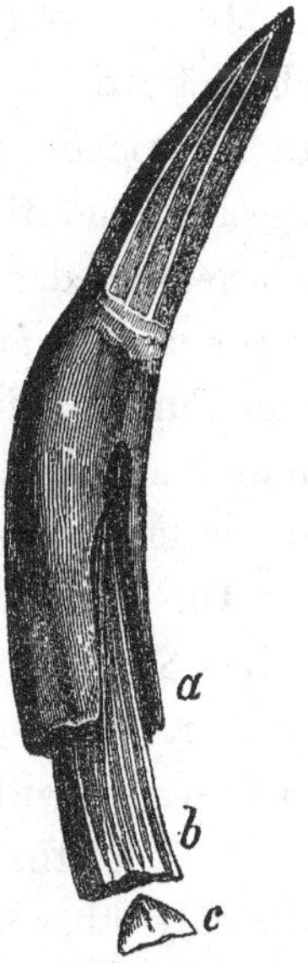

In the gavials (genus *Gavialis*) the teeth are nearly equal in size and similar in form

Fig. 80.
Teeth of the Gavial.

in both jaws, and the first as well as the fourth tooth in the lower jaw passes into a groove in the margin of the upper jaw, when the mouth is closed.

The number of teeth is always greater in the gavials than in the crocodiles or alligators. The first five pairs of teeth

---

* *Pros*, front ; *koilos*, hollow ; vertebra with the cup at the fore part and the ball behind.

T

above are supported by the premaxillary bones ; the first, second, and fourth of the lower jaw are the longest.

The eight or nine posterior teeth are nearly conical, the rest are sub-compressed antero-posteriorly, and present a trenchant edge on the right and left side, between which a few faint longitudinal ridges traverse the basal part of the enamelled crown (fig. 80).

The position of the opposite sharp ridges, and the direction of the flat sides of the crown, are reversed in the extinct crocodile (*Croc. cultridens*), which in other respects most nearly resembles the gavial in the form of the teeth.

In most of the extinct species of Crocodilians the teeth are characterized by more numerous and strongly developed longitudinal ridges upon the enamelled crown, than in the recent species ; and they are commonly longer, more slender, and sharp-pointed. But in one of the crocodiles with sub-biconcave vertebræ (*Goniopholis crassidens*), from the Wealden formation and Purbeck limestone, the teeth have crowns which are as round and as thick in proportion to their length as in the recent crocodiles or alligators.

The more ancient crocodiles, from the Oolite and Lias, called *Steneosauri* and *Teleosauri*, had jaws like those of the modern gavials, but sometimes longer and more attenuated, and armed with more numerous, equal, and slender teeth, adapted for the capture of fishes, which appear to have been the only other vertebrate animals existing at those periods in numbers sufficient to yield subsistence to carnivorous marine Saurians.

In all the *Teleosauri* the teeth are more slender, less compressed, and sharper pointed than in the gavial ; they are slightly recurved, and the enamelled crown is traversed by more numerous and better defined ridges—two of which, on opposite sides of the crown, are larger and more elevated than the rest. The fang is smooth, cylindrical, and always excavated at the base. The teeth of the *Steneosauri,* or extinct crocodiles with long and slender jaws, and with vertebræ sub-concave at both extremities, but with subterminal nostrils, differ from those of the *Teleosauri* in being somewhat thicker in proportion to their length, and larger in proportion to the jaws.

The teeth of both the existing and extinct crocodilian reptiles consist of a body of compact dentine, forming a crown covered by a coat of enamel, and a root invested by a moder-ately thick layer of cement. The root slightly enlarges or maintains the same breadth to its base (fig. 80, *a*), which is deeply excavated by a conical pulp-cavity extending into the crown, and is commonly either perforated or notched at its concave or inner side.

The tooth-germ *c* (figs. 80 and 81) is developed from the membrane covering the angle between the floor and the inner wall of the socket. It becomes, in this situation, completely enveloped by its capsule, and partially calcified, before the young tooth penetrates the interior of the pulp-cavity of its predecessor.

The matrix of the young growing tooth affects, by its pressure, the inner wall of the socket, as shown in figs. 80 and 81, and forms for itself a shallow recess; at the same time it

Fig. 81.

Section of jaw with teeth of the Alligator.

attacks the side of the base of the contained tooth : then, gaining a more extensive attachment by its basis and increased size, it penetrates the large pulp-cavity of the previously formed tooth either by a circular or semicircular perforation.

The size of the perforation in the tooth, and of the depression in the jaw, proves them to have been in great part caused by the soft matrix, which must have produced its effect by exciting absorbent action, and not by mere mechanical force. The resistance of the wall of the pulp-cavity having been thus overcome, the growing tooth and its matrix recede from the temporary alveolar depression, and sink into the substance of the pulp contained in the cavity of the fully-formed tooth.

As the new tooth grows, the pulp of the old one is removed ; the old tooth itself is next attacked, and the crown, being undermined by the absorption of the inner surface of its base, may be broken off by a slight external force, when the point of the new tooth is exposed.

The new tooth disembarrasses itself of the cylindrical base of its predecessor (fig. 81, a) with which it is sheathed, by maintaining the excitement of the absorbent process so long as the cement of the old fang retains any vital connection with the periosteum of the socket ; but the frail remains of the old cylinder, thus reduced, are sometimes lifted out of the socket upon the crown of the new tooth (as in fig. 81, a), when they are speedily removed by the action of the jaws. This is, however, the only part of the process which is immediately produced by violence ; an attentive observation of the more important previous stages of growth, teaches that the pressure of the growing tooth operates upon the one to be displaced only through the medium of the vital absorbent action which it has excited.

No sooner has the young tooth (fig. 80, b) penetrated the interior of the old one (fig. 80, a) than another germ c, begins to be developed from the angle between the base of the young tooth and the inner alveolar process ; or in the same relative position as that in which its predecessor began to rise, and the processes of succession and displacement are carried on uninterruptedly throughout the long life of these cold-blooded carnivorous reptiles.

From the period of exclusion from the egg, the teeth of the crocodile succeed each other in the vertical direction ; none are added from behind forwards like the true molars in Mammalia. It follows, therefore, that the number of the teeth of the crocodile is as great when it first sees the light as when it has acquired its full size ; and, owing to the rapidity of their succession, the cavity at the base of the fully-formed tooth is never consolidated.

The fossil jaws of the extinct Crocodilians demonstrate that the same law regulated the succession of the teeth at the ancient epochs when those highly-organised reptiles prevailed in greatest numbers, and under the most varied generic and specific modifications, as at the present period, when they are reduced to a single family composed of so few and slightly varied species as to have constituted in the system of Linnæus a small fraction of the genus *Lacerta.*

The large, thick, externally ridged or pitted scales, though common to the Crocodilian order, are not peculiar to them. The labyrinthodont *Anisopus,** the thecodont *Staganolepis,* the lacertian *Saurillus,* have left similar petrified scales.

Crocodilians with cup-and-ball vertebræ, like those of living species, first make their appearance in the greensand of North America (*Crocodilus basifissus* and *C. basitruncatus*). In Europe their remains are first found in the tertiary strata. Such remains from the plastic clay of Meudon have been referred to *C. isorhynchus, C. cœlorhynchus, C. Becquereli.* In the calcaire grossier of Argenton and Castelnaudry have been found the *C. Rallinati* and *C. Dodunii.* In the coeval eocene London clay at Sheppy Island the entire skull and characteristic parts of the skeleton of *C. toliapicus* and *C. Champ-*

---

* "*Labyrinthodon scutulatus,*" Trans. Geol. Soc , 2d series, vol. vi. p. 538, pl. 46.

soïdes occur. In the somewhat later eocene beds at Bracklesham occur the remains of the gavial-like *C. Dixoni.* In the Hordle beds have been found the *C. Hastingsiæ,* with short and broad jaws ; and also a true alligator (*C. Hantoniensis*). It is remarkable that forms of procœlian Crocodilia, now geographically restricted—the gavial to Asia, and the alligator to America—should have been associated with true crocodiles, and represented by species which lived, during nearly the same geological period, in rivers flowing over what now forms the south coast of England.

Many species of procœlian Crocodilia have been founded on fossils from miocene and pliocene tertiaries. One of these, of the gavial sub-genus (*C. crassidens*), from the Sewalik tertiary, was of gigantic dimensions.

Order X.—LACERTILIA.

(*Lizards, Monitors, Iguanæ.*)

Char.—Vertebræ procœlian, with a single transverse process
on each side, and with single-headed ribs ; sacral vertebræ not exceeding two : two external nostrils ; a foramen
parietale in most.

Small vertebræ of the lacertian type have been found in the Wealden of Sussex. They are more abundant, and are associated with other characteristic parts of the species, in the cretaceous strata. On such evidence have been based the *Raphiosaurus subulidens,* the *Coniasaurus crassidens,* and the *Dolichosaurus longicollis.*[*] The last-named species is remarkable for the length and slenderness of its trunk and neck, indicative of a tendency to the ophidian form. But the most remarkable and extreme modification of the lacertian type in the cretaceous period is that manifested by the huge species, of which a cranium five feet long was discovered in the upper

[*] Owen, "History of British Fossil Reptiles," 4to, pp. 173-183, pls. 2, 8, 9.

chalk of St. Peter's Mount, near Maestricht, in 1780. The
vertebræ are gently concave in front, and convex behind;
there are thirty-four between the head and the base of the
tail; a sacrum seems to have been wanting. The caudal
vertebræ have long neural and hæmal spines, both of which
arches coalesce with the centrum, and formed the basis of a
powerful swimming tail. The teeth are anchylosed to eminences along the alveolar border of the jaw, according to the
acrodont type. There is a row of small teeth on each pterygoid
bone. For this genus of huge marine lizard the name *Mosasaurus* has been proposed. Besides the *M. Hofmanni* of
Maestricht, there is a *M. Maximilliani*, from the cretaceous
beds of North America, and a smaller species, *M. gracilis*,
from the chalk of Sussex.* The *Leiodon anceps* of the Norfolk
chalk was a nearly allied marine Lacertian.†

Small pleurodont lizards, known at present only by jaws
and teeth, with associated pitted scutes, but which may have
had procœlian vertebræ, have been discovered in Purbeck
beds, and have been referred to the genera *Saurillus, Macel-
lodus,* etc.‡ Many small terrestrial Lacertians have left their
remains in European tertiary formations.

Order XI. — OPHIDIA.

(*Slow-worms, Serpents.*)

Char.—Vertebræ very numerous, procœlian, with a single
transverse process on each side; no sacrum; no visible
limbs.

The order *Ophidia,* as it is characterized in the system of
Cuvier, requires to be divided into two sections, according to
the nature of the food, and the consequent modification of the
jaws and teeth. Certain species, which subsist on worms,

---

* Op. cit., p. 185, pls. 1, 2, 9.   † Op. cit., p. 195, pl. 10.
‡ Quart. Journ. Geol. Soc., No. 40, 1854, p. 420.

insects, and other small invertebrate animals, have the tympanic pedicle of the lower jaw immediately and immoveably articulated to the walls of the cranium.   The lateral branches of the lower jaw are fixed together at the symphysis, and are opposed by the usual vertical movement to a similarly complete maxillary arch above ; these belong to the genera *Amphisbæna* and *Anguis* of Linnæus, the latter represented by our common slow-worm.   The rest of the Ophidians, including the ordinary serpents and constrictors, which form the typical members, and by far the greatest proportion, of the order, prey upon living animals of frequently much greater diameter than their own ; and the maxillary apparatus is conformably and peculiarly modified to permit of the requisite distension of the soft parts surrounding the mouth, and the transmission of their prey to the digestive cavity.

The earliest evidence of an ophidian reptile has been obtained from the eocene clay at Sheppy ; it consists of vertebræ indicating a serpent of 12 feet in length, the *Palæophis toliapicus*.   Still larger, more numerous, and better preserved vertebræ have been obtained from the eocene beds at Bracklesham, on which the *Palæophis typhæus* and *P. porcatus* have been founded.*   These remains indicate a boa-constrictor-like snake, of about 20 feet in length.   Ophidian vertebræ of much smaller size, from the newer eocene at Hordwell, support the species called *Paleryx rhombifer* and *P. depressus*.†   Fossil vertebræ from a tertiary formation near Salonica have been referred to a serpent, probably poisonous, under the name of *Laophis*.‡   A species of true viper has been discovered in the miocene deposits at Sansans, South of France.   Three fossil Ophidians from the Œningen slate have been referred to *Coluber arenatus*, *C. Kargii*, and *C. Owenii*.

* History of British Fossil Reptiles, pp. 139-149, pls. 2 and 3.
† Op. cit., p. 149, pl. 2, figs. 29-32.
‡ "Quarterly Journal of the Geological Society," vol. xiii , p. 196, pl. iv.

Order XII.—CHELONIA.

(*Tortoises and Turtles.*)

Char.—Trunk-ribs broad, flat, suturally united, forming with vertebræ and sternum an expanded thoracic abdominal case, into which, as into a portable chamber, the head, tail, and limbs can, usually, be withdrawn. No teeth : external nostril single.

Reference has already been made to the impressions in sandstones of triassic age in Dumfriesshire, referred by Dr. Duncan to tortoises. These impressions have been finely illustrated in the great work by Sir William Jardine on the footprints at Corncockle Muir. The earliest proof of chelonian life which the writer has obtained has been afforded by the skull of the *Chelone planiceps*, from the Portland stone ; and by the carapace and plastron of the extinct and singularly-modified emydian genera *Tretosternon* and *Pleurosternon** (fig. 82). In the first genus the plastron retains the central vacuity ; in the second genus an additional pair of bones is interposed between the hyosternals (*hs*) and hyposternals (*ps*). In the specimen figured (fig. 82), the plastron, and the under surface of the marginal pieces (2 to 12) of the carapace, of *Pleurosternon emarginatum* are shown. This fine Chelonite is now in the British-Museum.

True marine turtles (*Chelone Camperi, C. obovata, C. pulchriceps*) have left their remains in cretaceous beds.† The emydian *Protemys* is from the greensand near Maidstone.‡ The eocene tertiary deposits of Britain yield rich evidences of marine, estuary, and fresh-water tortoises. More species of true turtle have left their remains in the London clay at the mouth of the Thames than are now known to exist in the

---

* Monograph of the Fossil Chelonian Reptiles of the Wealden and Purbeck Limestones, 4to, 1853, Palæontographical Society.

† Owen, "Hist. Brit. Fossil Reptiles," pp. 155-168, pls. 41-46.

‡ Op. cit., p. 169, pl. 47.

whole world; and all the eocene *Chelones* are extinct. One of them (*C. gigas*, Ow.) attained unusual dimensions; the skull, now in the British Museum, measures upwards of a foot across its back part.* The estuary genus *Trionyx* (soft

Fig. 82.
Pleurosternon emarginatum (Purbeck).

turtle) is represented by many beautiful species in the upper eocene at Hordwell;† the fresh-water genera *Emys* and *Platemys* by as many species, both at Sheppy and Hordwell. In the pliocene of Œningen remains of a species of *Chelydra* have been discovered; this generic form is now confined to America.

* The upper end of the femur from Sheppy, in t. xxix. of Monograph of Fossil Reptilia of the London Clay, Palæontographical Society, 1850, belongs to this species. See also "Hist. of Brit. Foss. Reptiles," pp. 10-40, pls. 1-22.

† Op. cit., pp. 50-60, pls. 26-33.

Remains of land-tortoises (*Testudo*, Brong.) indicate several extinct species in the miocene and pliocene formations of continental Europe. Strata of like age in the Sewalik Hills have revealed the carapace of a tortoise 20 feet in length ; it is called by its discoverers, Cautley and Falconer, *Colossochelys atlas*. The same locality has also afforded the interesting evidence of a species of *Emys* (*E. tectum*, Gray) having continued to exist from the (probably miocene) period of the *Sivatherium* to the present day.

Order XIII.—BATRACHIA.

(*Toads, Frogs, Newts.*)

Char.—Vertebræ biconcave (*Siren*), procœlian (*Rana*), or opisthocœlian (*Pipa*) : pleurapophyses short, straight. Two occipital condyles ; two vomerine bones, in most dentigerous : no scales or scutes. Larvæ with gills, in most deciduous.

It is only in tertiary and post-tertiary strata that extinct species, referable to still existing genera or families of this order, have been found. The reptiles with amphibian or batrachian characters, of the carboniferous and triassic periods, combined those characters with others which gave them distinctions of ordinal value ; they illustrated, indeed, rather a retention of the more general cold-blooded vertebrate type, with concomitant piscine and saurian features, than any near affinity with the more specially modified naked or soft-skinned reptiles to which the name *Batrachia* is given in zoological catalogues of existing species.

Of the tailless or "anourous" Batrachia, toads of extinct species (*Palæophrynos Gessneri* and *P. dissimilis*) have been discovered in the Œningen beds ; and frogs, more abundantly, in both miocene and pliocene deposits of France and Germany. The batracholites from the tertiary lignites of the "Siebenge-

birge," near Bonn, show different stages of transformation of
the *Rana diluviana,* Gdf. Tertiary shales from Bombay have
made known to the author the small fossil *Rana pusilla.*

Of the salamander family, the most noted fossil is that
which was referred, when first discovered at Œningen in 1726,
to the human species, as *Homo diluvii testis.* Cuvier demonstrated its near affinities to the water-salamander (*Menopoma*)
of the United States : more recently a living species of salamander has been discovered in Japan which equals in size the
fossil in question—*Andrias Scheuchzeri.*

A retrospect of the foregoing outline of the palæontology
of the class of reptiles shows that, unlike that of fishes, it is
now on the wane ; and that the period when *Reptilia* flourished
under the greatest diversity of forms, with the highest grade
of structure, and of the most colossal size, is the mezozoic.
The progress of air-breathing vertebrates, graduating by close
transitional steps from the water-breathing class, has been
checked, as if it had been unequal to the exigencies and life-
capacities of the present state of the planet. Reptiles have
been superseded by air-breathers of higher types, which cannot
be directly derived from the class of fishes. A more generalized vertebrate structure is illustrated, in the extinct reptiles,
by the affinities to ganoid fishes shown by the *Ganocephala,*
Labyrinthodontia, and *Ichthyopterygia ;* by the affinities of the
Pterosauria to birds, and by the approximation of the *Dinosauria* to Mammals. It is manifested by the combination of
modern crocodilian, chelonian, and lacertian characters in the
Cryptodontia and *Dicynodontia ;* and by the combined crocodilian and lacertian characters in the *Thecodontia* and *Sauropterygia.* Even the *Chelonia* of the Purbeck period illustrate
the same principle, by the more typical number of modified
hæmapophyses, or abdominal ribs, entering into the composition
of their plastron.

The diagram (fig. 83) gives a concise view of the geological relations, or distribution in time, of the principal groups of the class *Reptilia*. In the column opposite the right hand, the dark mark shows that the ganocephalous group represented by the *Archegosaurus* began, culminated, and ended in the carboniferous period. The Labyrinthodonts, culminating in the trias, disappear at the base of the oolitic system. Of the true *Batrachia*, those retaining the tail appear to have been at their maximum during the upper tertiary period, and to have begun to decline after

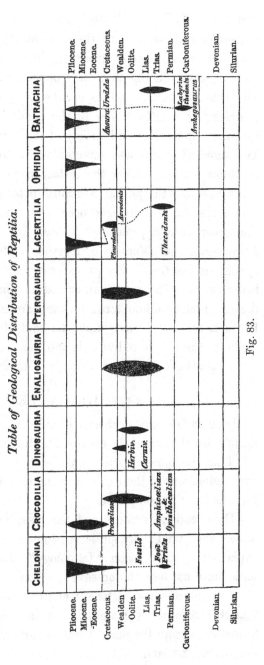

Table of Geological Distribution of Reptilia.

Fig. 83.

that time ; whilst the tailless genera and species are most numerous and various at the present day. The *Ophidia* resemble the *Anoura*, commencing in the older tertiary, and showing their maximum of development at the present day. The true procœlian, and especially the pleurodont, lizards, commencing a little earlier in the chalk, have also gone on increasing in number and variety of forms to the present day. The acrodont group was represented by *Mosasaurus*, with a maximum of size, during the cretaceous period. The Theco- donts have but the partial relationship to modern *Lacertilia* which the Labyrinthodonts bear to the modern *Batrachia*. The great ordinal groups of *Ichthyo-* and *Sauro-pterygia*, of *Pterosauria*, and *Dinosauria*, together with the amphi- and opistho-cœlian *Crocodilia*, passed away ere the tertiary time had dawned. The procœlian crocodiles which culminated in the lower and middle tertiary times, are now on the wane. Perhaps, also, the same might be said of the *Chelonia*, in regard to the size of individuals and the number of species of certain genera (*e. g. Chelone, Trionyx, Chelydra*).

CLASS III.—AVES.

Long before any evidence of birds from actual or recog- nizable fossil remains is obtained in tracing the progress of life from the oldest fossiliferous deposits upwards, we meet with indications of their existence impressed in sandstones of the triassic or liassic period.

These earliest evidences of the class are by footprints in some former tidal shore, preserved in one or other of the ways explained in the section " Ichnology." The fossil bones of birds have not been found save in strata of much later date than the impressed sandstones ; and they are much more rare than the remains of mammals, reptiles, and fishes, in any formations except the most recent in certain limited localities, —*e.g.*, New Zealand.

Sir C. Lyell has well remarked, that "the powers of flight possessed by most birds would insure them against perishing by numerous casualties to which quadrupeds are exposed during floods." The same writer further argues, that "if they chance to be drowned, or to die when swimming on water, it will scarcely ever happen that they will be submerged so as to become preserved in sedimentary deposits."* It is true that the carcase of a floating bird may not sink where it has died, but be carried far along the stream : ultimately, however, if not devoured, its bones will subside when the soft parts have rotted, and both the compactness of the osseous tissue, and the facts made known by the ornitholites of the greensand near Cambridge, of the London clay at Sheppy, and of the Montmartre eocene quarry-stone, show that they can be preserved in the fossil state. The length of time during which the carcase of a bird may float, doubtless exposes it the more to be devoured, and so tends to make more scarce the fossil remains of birds in sedimentary strata. Certain it is that the major part of the remains of extinct birds that have as yet been found are those of birds that were deprived of the power of flight, and were organized to live on land.

The existence of birds at the triassic period in geology, or at the time of the formation of sandstones which are certainly intermediate between the lias and the coal, is indicated by abundant evidences of footprints impressed upon those sandstones which extend through a great part of the valley of the Connecticut River, in Connecticut and Massachusetts, North America.

The footprints of birds are peculiar, and more readily distinguishable than those of most other animals. Birds tread on the toes only ; these are articulated to a single metatarsal bone at right angles equally to it, and they diverge more from each other, and are less connected with each other, than in

* Principles of Geology, ed. 1847, p. 721.

other animals, except as regards the web-footed order of
birds.

Not more than three toes are directed forward;* the fourth
when it exists, is directed backward, is shorter, usually rises
higher from the metatarsal, and takes less share in sustaining
the superincumbent weight. No two toes of the same foot in
any bird have the same number of joints. There is a constant
numerical progression in the number of phalanges (toe-joints),
from the innermost to the outermost toe. When the back toe
exists, it is the innermost of the four toes, and it has two
phalanges, the next has three, the third or middle of the front
toes has four, and the outermost has five phalanges. When
the back toe is wanting, as in some waders, and most wingless
birds, the toes have three, four, and five phalanges respectively.
When the number of toes is reduced to two, as in the ostrich,
their phalanges are respectively four and five in number ; thus
showing those toes to answer to the two outermost toes in
tridactyle and tetradactyle birds.

The same numerical progression characterizes the two
phalanges in most lizards from the innermost to the fourth ;
but a fifth toe exists in them which has one phalanx less than
the fourth toe. It is the fifth toe which is wanting in every
bird. In some *Gallinacea*, one or two (*Pavo bicalcaratus*) spurs
are superadded to the metatarsus ; but this peculiar weapon
is not the stunted homologue of a toe. Dr. Deane of Greenfield, United States, noticed, in 1835, impressions resembling
the feet of birds in some slabs of sandstone from Connecticut
River, and first, in a letter to Dr. Hitchcock, dated March 7,
1835, recorded his belief that they were the footsteps of a bird.
He prepared casts of the impressions, some of which he transmitted, with his opinion of their nature, to Professor Silliman,
Editor of the " American Journal of Science," in April 1835.
Dr. Hitchcock, President of Amherst College, United States, first

* Save in the Swift.

made public the fact, based on scientific comparison, and submitted to the geological world his interpretations of those
impressions as having been produced by the feet of living
birds, and he gave them the name of *Ornithichnites.*

It was a startling announcement, and a conclusion that
must have had strong evidence to support it, since one of the
kinds of the tracks had been made by a pair of feet, each
leaving a print 20 inches in length. Under the term *Ornithichnites giganteus,* however, Dr. Hitchcock did not shrink
from proclaiming the fact of the existence, during the period
of the deposition of the red sandstones of the valley of the
Connecticut, of a bird which must have been at least four
times larger than the ostrich.* The impressions succeeded
each other at regular intervals ; they were of two kinds, but
differing only as a right and left foot, and alternating with
each other, the left foot a little to the left, and the right foot
a little to the right, of the mid-line between a series of tracks.
Each footprint (fig. 84, *b* and *r*) exhibits three toes, diverging
as they extend forwards. The distance between the tips of
the inside and outside toes of the same foot was 12 inches.
Each toe was terminated by a short strong claw projecting
from the mid toe a little on the inner side of its axis, from
the other two toes a little on the outer side of theirs. The
end of the metatarsal bone to which those toes were articulated
rested on a two-lobed cushion which sloped upwards behind.
The inner toe (*r*) showed distinctly two phalangeal divisions,
the middle toe three, the outer toe (*b*) four. And since, in
living birds, the penultimate and ungual phalanges usually
leave only a single impression, the inference was just, that
the toes of this large foot had been characterized by the same
progressively-increasing number of phalanges, from the inner
to the outer one, as in birds. And, as in birds also, the toe
with the greatest number of joints was not the longest ; it

* American Journal of Science for 1836, vol. xxix., pl. i.

U

measured, *e. g.*, 12½ inches, the middle toe from the same base-
line measured 16 inches, the outer toe 12 inches. Some of
the impressions of this huge tridactylous footstep were so well
preserved as to demonstrate the papillose and striated character
of the integument covering the cushions on the under side of
the foot. Such a structure is very similar to that in the
ostrich. The average extent of stride, as shown by the distance
between the impressions, was between three and four feet;
the same limb was therefore carried out each step from six to
seven feet forward in the ordinary rate of progression.

These footprints, although the largest that have been observed on the Connecticut sandstones, are the most numerous.
The gigantic *Brontozoum*, as Principal Hitchcock proposes to
term the species, "must have been," he writes, "the giant rulers
of the valley. Their gregarious character appears from the
fact, that at some localities we find parallel rows of tracks a
few feet distance from one another."

The strata of red sandstone, with the above-described impressions, occupy an area more than 150 miles in length, and
from 5 to 10 miles in breadth. "Having examined this series
of rocks in many places, I feel satisfied that they were formed
in shallow water, and for the most part near the shore; and
that some of the beds were from time to time raised above the
level of the water and laid dry, while a newer series, composed
of similar sediment, was forming." "The tracks have been
found in more than twenty places, scattered through an extent
of nearly 80 miles from N. to S., and they are repeated through
a succession of beds attaining at some points a thickness of
more than 1000 feet, which may have been thousands of years
in forming."*

One of the evidences of birds from the Cambridge greensand, transmitted to the writer by their discoverer, Mr. Barret,
is the lower half of the trifid metatarsal, showing the outer toe-

* Lyell, Manual of Elementary Geology, 8vo, 1855, p. 343.

joint much higher than the other two, and projecting backwards. above the middle joint; it indicates a bird about the size of a woodcock.

In the conglomerate and plastic clay at the base of the eocene tertiary system at Mendon, near Paris, the leg and thigh bones (tibia and femur) of a bird (*Gastornis Parisiensis*) have been discovered : they indicate a genus now extinct. They belonged to a species as large as an ostrich, but more robust, and with affinities to wading and aquatic birds.*

In the eocene clay of Sheppy, fossil remains of birds have been found, indicating a small vulture (*Lithornis vulturinus*) ; also a bird, probably of the king-fisher family (*Halcyornis toliapicus*), and a species of the sea-gull family. In the same formation at Highgate, remains of a species of the heron family have been found.

The fossil bones of birds from the gypsum quarries at Montmartre were referred or approximated by Cuvier to eleven distinct species. Good ornitholites have been obtained from the Hordwell fresh-water deposits.

The most ancient example of a passerine bird is the *Protornis Glarisiensis*, founded on an almost entire skeleton discovered in the schistose rock of Glaris, referable to the older division of the eocene tertiary series. This skeleton is about the size of a lark, and in some respects similar to that bird.

Comparisons of the ornitholites of the eocene tertiaries show that the following ordinal modifications of the class of birds were at that period represented: the *raptorial*, or birds of prey, by species of the size of our ospreys, buzzards, and smaller falcons, and most probably also by an owl ; the *insessorial*, or tree-perching birds, by species seemingly allied to the nuthatch and the lark ; the *scansorials* or anisodactyles, by species as

---

* Hebert, "Comptes Rendus de l'Acad. des Sciences," 1855. Owen "On the Affinities of Gastornis Parisiensis," Quarterly Journal of Geological Society, vol. xii., 1856, p. 204.

large as the.cuckoo and king-fisher ; the *rasorials*, by a species
of small quail ; the *cursorials*, by a species as large as, but
with thicker legs than, an ostrich ; the *grallatorial*, by a curlew
of the size of the ibis, and by species allied to *Scolopax, Tringa*,
and *Pelidna*, of the size of our woodcocks, lapwings, and sand-
erlings ; and the *natatorial*, by species allied to the cormorant,
but one of them of larger size, though less than a pelican ; also
by a species akin to the divers *(Merganser)*.

The remains of birds become more abundant and varied as
we approach the present time ; especially in the miocene strata,
so richly developed in France, although wanting in Britain.
One of the most singularly-modified forms of beak is shown by
the flamingo. The fossil skull of a species of this genus
(Phœnicopterus) has been found in the miocene fresh-water
deposits of the plateau of Gergovia, near Clermonte-Ferrand ;
the entire metatarsal bone of a species of eagle *(Aquila)* or os-
prey *(Pandion)* in the same deposits at Chaptusal, Allier ; and
the humerus of a bird allied to and as large as the albatross,
in the *molasse coquillière marine* at Armagne. Remains of a
vulture, most probably a *Cathartes*, have been found in the
miocene lacustrine deposits of Cantal. Indications of all the
other orders of birds, save the great Cursores or *Struthionidæ*,
have also been discovered in miocene strata—those of wading
birds being the most numerous.

Fossil eggs of birds occur in miocene deposits in Auvergne ;
and impressions of feathers have been discovered in the pliocene
calcareous marls at Montebolca. In pliocene brick-earth deposits in Essex has been found a fossil metatarsal of a swan,
as large as, and not distinguishable from, the existing wild
swan ; in the pleistocene clay at Lawford a fossil humerus
like that of a wild goose. But most of the ornitholites of
this recent tertiary period have been discovered in ossiferous
caverns. They belong to birds closely resembling the falcon,
wood-pigeon, lark, thrush, teal, and a small wader. The writer

has received information of skeletons of birds found deeply imbedded in stratified clay at Aberdeen and Peterhead.

The most extraordinary additions to the present class have been obtained from the superficial deposits, turbaries, and caves in New Zealand.*

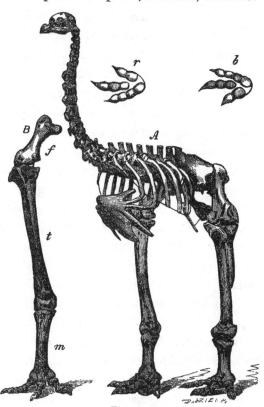

This island is remarkable for the absence of aboriginal species of land - mammals, and for the presence of a small bird with very rudimental wings, and the keel-less sternum and loose plumage of the Struthious order, but of a peculiar genus called *Apteryx*: the legs are very robust, and have three front toes and a very small back toe. Birds resembling the *Apteryx* in the

Fig. 84.

A. *Dinornis elephantopus*.
B. Leg-bones of *Dinornis giganteus*.
b, r. Impressions called Ornithichnites.

shape of the sternum and bony structure of the pelvis and hind limbs, some retaining also the small back toe, others apparently without it, formerly existed in New Zealand

* These remains are described in eight memoirs by the writer, published in the third and fourth volumes of the Transactions of the Zoological Society of London. The description of the first fragment of the bone, indicative of the *Dinornis*, is in vol. iii., p. 39, pl. 3.

under different specific forms ranging in height from 3 feet
to 10 feet. They have been referred by the writer to the
genera *Dinornis* and *Palapteryx*. The gigantic species are
interesting, as exhibiting birds equal to the formation of tridactyle impressions as large as those of the Connecticut sandstones,
called *Ornithichnites* (*Brontozoum*) *gigas* (fig. 84, *r*, *b*). In this
cut is given a figure of the leg-bones of *Dinornis giganteus* (*B*), in
which the tibia (*t*) measures upwards of a yard in length. In
the entire skeleton (*A*) of another species, the metatarsus is as
thick, but only half as long, as in the *D. giganteus ;* the framework of the leg is the most massive of any in the class of birds ;
the toe-bones almost rival those of the elephant ; whence the
name *Dinornis elephantopus*, given to this species. Several other
species of these extinct tridactyle wingless birds have been determined—*e. g., Dinornis ingens, D. struthioïdes, D. rheïdes, D.
dromïoides, D. casuarinus, D. robustus, D. crassus, D. geranoïdes,
D. curtus*. With these remains have been found bones of a bird
the size of a swan, but of an extinct genus (*Aptornis*) ; also
those of a large coot (*Notornis Mantelli*) which, founded originally on fossil remains, was afterwards discovered living in the
Middle Island of New Zealand. Two species of *Apteryx*, not
distinguishable from the existing kinds, were contemporaries
with the gigantic *Dinornis*, and the writer has received evidence that the *D. elephantopus* afforded food to the natives at
probably no very remote period. Some of the smaller kinds
of *Dinornis* may yet be found living on the Middle Island.

In Madagascar portions of metatarsal bones, indicating a
three-toed bird (*Epiornis*) as large as, but generically distinct
from, the *Dinornis giganteus*, have been discovered in alluvial
banks of streams ; and with them entire eggs, measuring from
13 to 14 inches in long diameter. The contents of one of these
eggs is computed to equal those of six ostrich eggs, or of one
hundred and forty-eight hen's eggs.

In the neighbouring island of Mauritius the dodo (*Didus*

ineptus) has been exterminated by man within the period of
two centuries ; and in the islands of Bourbon and Rodriguez
the " solitaire" (*Pezophaps*) has also become extinct. Both
these birds had wings too short for flight.

Class IV.—MAMMALIA.

(*Warm-blooded, Air-breathing, Viviparous Vertebrates.*)

Every calcified part of an animal, whether coral, shell, crust,
tooth, or bone, can preserve its form and structure when buried
in the earth during the changes there gradually operated in it,
until every original particle may have been removed and
replaced by some other mineral substance previously dissolved
in the water percolating the bed containing the fossil.   A bone,
or other part so altered, is said to be " petrified."   Not only are
all its outward characters preserved, but even the minutest
structure may be, and in most cases is, demonstrable in the
fine sections under the microscope.

Fossil bones and teeth have been discovered in every
intermediate stage of alteration, from their recent state to that
of complete petrifaction.   Recent bones consist of a soft—commonly called animal or organic—basis, hardened by earthy
salts, chiefly phosphate of lime.*   Fishes have the smallest
proportion, birds the largest proportion, of the earthy matter
in their bones.   The soft part is chiefly a gelatinous substance.

*Proportions of Hard and Soft Matter in the Bones of the
Vertebrate Animals.*

FISHES.

	Salmon.	Carp.	Cod.
Soft	60·62	40·40	34·30
Hard	39·38	59·60	65·70
	100·00	100·00	100·00

* That this combination of phosphorus and calcium has ever taken place in
nature, save under the influences of a living organism, remains to be proved.

REPTILES.

	Frog.	Snake.	Lizard.
Soft	35·50	31·04	46·67
Hard	64·50	69·96	53·33
	100·00	100·00	100·00

MAMMALS.

	Porpoise.	Ox.	Lion.	Man.
Soft	35·90	31·00	27·70	31·03
Hard	64·10	69·00	72·30	68·97
	100·00	100·00	100·00	100·00

BIRDS.

	Goose.	Turkey.	Hawk.
Soft	32·91	30·49	26·72
Hard	67·09	69·51	73·28
	100·00	100·00	100·00

The chemical nature of the hardening particles, and of the soft basis of bone, is exemplified in the subjoined table, including a species of each of the four classes of Vertebrata :—

Chemical Composition of Bones.

Ingredients.	Hawk.	Man.	Tortoise.	Cod.
Phosphate of lime, with trace of fluate of lime ...	64·39	59·63	52·66	57·29
Carbonate of lime	7·03	7·33	12·53	4·90
Phosphate of magnesia	0·94	1·32	0·82	2·40
Sulphate, carbonate, and chlorate of soda	0·92	0·69	0·90	1·10
Glutin and chondrin	27·73	29·70	31·75	32·31
Oil	0·99	1·33	1·34	2·00
	100.00	100·00	100·00	100·00

The most common change which bones first undergo is the loss of more or less of their original soft and soluble basis. This effect of long interment is readily tested by applying the specimen to the tongue, when the affinity of the pores of the earthy constituent, after having lost the gelatine for fluid, is so great, that the specimen adheres to the tongue like a piece of dry chalk. Bones and teeth in this state quickly absorb a solution of gelatine, and thus their original tenacity may be restored.* Petrified fossils need no such treatment; they are usually harder and more durable than the original bone itself.

The interpretation of such fossil remains requires a comparison of them with the corresponding parts of animals now living, or of previously determined extinct species. In the case of the vertebrate animals, such comparison is limited to the osseous and dental systems. The interpretation of a vertebrate fossil, therefore, presupposes a knowledge of the various modifications of the skeleton and teeth of the existing vertebrate animals; and the more extensive and precise such knowledge may be, the more successful will be the efforts, and the more exact the conclusions, of the interpreter.

The determination of the remains of quadrupeds is beset, as Cuvier truly remarks, with more difficulties than that of other organic fossils. Shells are usually found entire, and with all the characters by which they may be compared with their analogues in the museums, or with figures in the illustrated books, of naturalists. Fishes frequently present their skeleton or their scaly covering more or less entire, from which may be gathered the general form of their body, and frequently both the generic and specific characters which are derived from such internal or external hard parts. But the entire skeleton of a fossil quadruped is rarely found, and when it occurs, it gives little or no information as to the hair, the fur, or the colour of the species. Portions of the skeleton with the bones dislocated, or scattered pell-mell—detached bones and teeth, or their fragments merely—such are the conditions in which the petrified remains of the mammalian class most commonly present themselves in the strata in which they occur.

---

* The writer's experience of this effect led him to suggest the application of a similar process to the long-buried ivory ornaments from the ruins of Nineveh in the British Museum; it proved successful.

Prior to the time of Cuvier but little progress had been made in the interpretation of such fragmentary remains. The striking success which attended the application of the great comparative anatomist's science to this previously neglected field of study, was referred by Cuvier to principles in the organization of animal bodies, which he termed the " Correlation of Forms and Structures," and the " Subordination of Organs"—principles which his clear-thinking biographer, M. Flourens,* in common with most contemporary philosophers, has regarded as the most effective and successful instrument in the restoration of extinct animals. They will be exemplified in the course of the present section of this work.

A terminal phalanx modified to fit a hoof may give, as Cuvier declared, the modifications of all the bones of the fore limb that relate to the absence of a rotation of the fore leg, and all the modifications of the jaw and skull that relate to the mastication of food by broad-crowned complex molars.

But there are certain associated structures for the coincidence of which the physiological law is unknown. "I doubt," writes Cuvier, " whether I should have ever divined, if observation had not taught it me, that the ruminant hoofed beasts should all have the cloven foot, and be the only beasts with horns on the frontal bone."† We know as little why horns should be in one or two pairs on the frontal bone of those Ungulates only which have hoofs in one or two pairs ; whilst in the horned Ungulates with three hoofs, there should be either one horn, or two horns placed one behind the other in the middle line of the skull ; or why the Ungulates with one or three hoofs on the hind foot should have three trochanters on the femur, whilst those with two or four hoofs on the hind foot should have only two trochanters.

* Éloge Historique et l'Analyse Raisonnée des Travaux de G. Cuvier, 12mo, Paris, 1841, p. 42.

† Ossemens Fossiles, 8vo, ed. 1834, tom. i., p. 184.

"However," continues Cuvier, "since these relations are constant, they must have a sufficing cause; but as we are ignorant of it, we must supply the want of the theory by means of observation.* This, if adequately pursued, will serve to establish empirical laws almost as sure in their application as rational ones." "That there are secret reasons for all these relations, observation may convince us independently of general philosophy." "The constancy between such a form of such organ, and such another form of another organ, is not merely specific, but one of class, with a corresponding gradation in the development of the two organs."†

"For example, the dentary system of non-ruminant Ungulates is generally more perfect than that of the Bisulcates; inasmuch as the former have almost always both incisors and canines in the upper as well as the lower jaw; the structure of their feet is in general more complex, inasmuch as they have more digits, or hoofs less completely enveloping the phalanges, or more bones distinct in the metacarpus and metatarsus, or more numerous tarsal bones; or a more distinct and better developed fibula; or a concomitance of all these modifications. It is impossible to assign a reason for these relations; but, in proof that it is not an affair of chance, we find that whenever a bisulcate animal shows in its dentition any tendency to approach the non-ruminant Ungulates, it also manifests a similar tendency in the conformation of its feet. Thus the camels, which have canines and two or four incisors in the upper jaw, have an additional bone in the tarsus, resulting from

* "Puisque ces rapports sont constants, il faut bien qu'ils aient une cause suffisante; mais comme nous ne la connoissons pas, nous devons suppléer au défaut de la théorie par le moyen de l'observation." (Tom. cit., p. 184.)

† "En effet, quand on forme un tableau de ces rapports, on y remarque non-seulement une constance spécifique, si l'on peut s'exprimer ainsi, entre telle forme de tel organe, et telle autre forme d'un organe différent; mais l'on apercoit aussi une constance de classe et une gradation correspondante dans le développement de ces deux organes, qui montrent, presque aussi bien qu'un raisonnement effectif, leur influence mutuelle." (Tom. cit., p. 185.)

the scaphoid not being confluent with the cuboid ; and the small hoofs have correspondingly small phalanges. The musk-deer, which have long upper canines, have the fibula co-exten-sive with the tibia, whilst the other ruminants have a mere rudiment of fibula articulated to the lower end of the tibia." "There is then a constant harmony between two organs to all appearance quite strangers to each other, and the gradations of their forms correspond uninterruptedly even in the cases where one can render no reason for such relations." "But in thus availing ourselves of the method of observation as a supple-mentary instrument when theory abandons us, we arrive at astonishing details. The smallest articular surface (*facette*) of a bone, the smallest process, presents a determinate character relating to the class, to the order, to the genus, and to the species to which they belong ; so that whoever possesses merely the well-preserved extremity of a bone, may, with applica-tion, aided by a little tact (*adresse*) in discerning analogies, and by sufficient comparison, determine all these things as surely as if he possessed the entire animal."*

There have been, of course, instances, and will be, where for want of the "efficacious comparison," and the "tact in dis-cerning likeness," such results have not rewarded the endea-vours of the palæontologist ; and these shortcomings, and the mistakes sometimes made, even by Cuvier himself, have been cast in the teeth of his disciples, as arguments against the principles by which they believed themselves guided in their endeavours to complete the glorious edifice of which their master laid the foundations.

The writer has, therefore, quoted from the well-known "Preliminary Discourse" to Cuvier's great work on Fossil Remains, with a view to neutralize the efforts of statements reiterated in apparent ignorance of the clear and explicit man-ner in which Cuvier there defines the limits within which the

* Tom. cit., p. 187.

law of correlation of animal structures may be successfully applied, and indicates the instances in which—the physiological condition being unknown, and the coincident structures being understood empirically—careful observation and rigorous comparison must supply the place of the physiologically understood law.

Those who deny the existence of design in the construction of any part of an organized body, and who protest against the deduction of a purpose from the valves of the veins or the lens of the eye-ball, repudiate the reasoning which the palæontologist carries out from the hoof to the grinder, or from the carnassial molar to the retractile claw, through the guidance of the principle of a pre-ordained mutual adaptation of such parts ; but such minds are not, nor have been, those who have contributed to the real advancement of physiology or palæontology.

By reference to the "Table of Strata" (fig. 1), it will be seen that the earliest evidence of a vertebrate animal is of the cold-blooded water-breathing class in the upper Silurian period. Next follows that of a cold-blooded but air-breathing vertebrate, under the batrachian grade, in the carboniferous period. The warm-blooded air-breathing classes are first indicated, as birds, by footmarks in a sandstone of probably triassic but not older age ; and, as mammals, by fossil teeth from bone-beds of the upper triassic system in Wirtemberg, and of the same age near Frome, Somersetshire. Mammalian remains have also been found in a coal-field in North Carolina, which may be earlier, but cannot be later, than the lias formation.

Genus MICROLESTES.—The mammalian teeth from German and English trias indicate a very small insectivorous quadruped, to which the above generic name was given by Professor Plieninger. The German specimens were discovered in 1847 in a bone breccia at Diegerloch, about two miles from Stuttgardt, the geological relations of which are well determined as between lias and Keuper sandstone. The teeth of *Microlestes* from Frome, submitted to the writer by the discoverer, Mr. Charles Moore, F.G.S., in 1858, are four in number, two being molars of the upper jaw, each with four fangs ; one a molar with a narrower crown and two fangs from the lower jaw ; and the fourth a small, pointed, front tooth. The crowns of the molars are short vertically in proportion to their breadth ; the distinct enamel contrasts with the cement-covered fangs ; the grinding surface shows a wide and shallow depression, surrounded by small, low, obtuse cusps, three of equal size being on one side, a larger cusp near one end, and smaller and less regular cusps on the side opposite the three. One lower molar shows a similar type, but with the three marginal cusps less equal in size : a second smaller, and from a more anterior part of the series has three low cusps on one side, and but one cusp on the other side of the crown, the grinding surface of which presents an elongate triangular form. This tooth had two fangs. The crown of the largest of the upper molars does not exceed one line in its longest diameter. Amongst existing Mammals, some of the small molars of the marsupial and insectivorous *Myrmecobius* of Australia offer the nearest resemblance to these fossil teeth ; but a still closer one is presented by the small tubercular molars of the extinct oolitic Mammal called *Plagiaulax* (fig. 93, *m*, 1 and 2).

Genus DROMATHERIUM.—It would appear that the Mammal from the American triassic or liassic coal-bed (*Dromatherium sylvestre*, Emmons) also found its nearest living analogue in *Myrmecobius ;* for each ramus of the lower jaw contained 10 small molars in a continuous series, 1 canine, and 3 conical incisors, the latter being divided by short intervals.

Genus AMPHITHERIUM (*Thylacotherium*, Val.)*—This genus

---

* For the full description and demonstration of the mammalian nature of this much-discussed fossil, see Owen, History of British Fossil Mammals, 8vo, p. 29.

is founded upon a few specimens of lower jaw, one ramus of which (fig. 85) gave the entire dentition of its side,—viz., three small conical incisors (*i*), one rather larger canine (*c*), six pre-molars, unicuspid, with a small point at one or both sides of the base (*p*, 1-6), and six quinque-cuspid molars (*m*, 1·6) not departing very far from the type above described. The mo-lars, and most of the

Fig. 85.

Lower jaw and teeth of the *Amphitherium Prevostii* (twice nat. size.)

premolars, are implanted by two roots. The condyle of the jaw is convex, and is a little higher than the level of the teeth ; the coronoid process is broad and high ; the angle projects backward, with a feeble production inward. It is, again, to the marsupial *Myrmecobius*, amongst living forms, that the present genus is most nearly allied. The remains of *Amphitherium* are from the lower oolitic slates of Stonesfield (fig. 87, stratum 8).

Genus AMPHILESTES[*]—This genus is founded on a ramus of the lower jaw, from the Stonesfield oolitic slate, showing true molars of a compressed form, with a large middle cusp and a smaller, but well marked, one at the fore and back part of its base ; the "cingulum," or basal ridge, peculiar to mammalian teeth, traverses the inner ridge of the crown, where it develops three small cusps, one at the base of the large outer or princi-pal cusp, and the other two forming the anterior and posterior ends of the crown. This form of tooth is unknown in existing Mammalia, but is as well adapted for crushing the cases of coleopterous insects (elytra of which are found fossil in the same oolitic matrix) as are any of the multi-cuspid molars of small opossums, shrews, and bats. The *Amphilestes Broderipii* was somewhat larger than *Amphitherium Prevostii*.

[*] Owen, Hist. Brit. Foss. Mam., p. 58, fig. 19 (*Amphitherium Broderipii*).

Genus PHASCOLOTHERIUM.—Although the evidence of the very slight degree of inflection of the angular process of the lower jaw of *Amphitherium* may favour its affinity to the placental Insectivores, yet the range of variety to which that mandibular character is subject in the different genera of existing *Marsupialia* warns us against laying undue stress upon its feeble development in the extinct genus of the oolitic epoch, and incites us to look with redoubled interest at whatever other indications of a marsupial character may be present in the fossil remains of other genera and species of Mammalia that have been detected in the Stonesfield slate.

In the specimen of *Phascolotherium* (fig. 86) presented to the British Museum by William J. Broderip, Esq., F.R.S., its

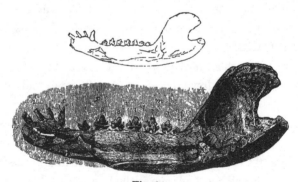

Fig. 86.
Lower jaw and teeth of the *Phascolotherium* (nat. size in outline),
Lower Oolite.

original describer,* which is as perfect in regard to the dentition as the jaw of the *Amphitherium* above described, the marsupial characters are more strongly manifested in the general form of the jaw, and in the extent and position of the inflected angle, while the agreement with the genus *Didelphys* in the number of the premolar and molar teeth is complete. The forms of the crowns of those teeth differ from those in *Didelphys*, and correspond so closely with those in the *Amphilestes Broderipii*,

* Zoological Journal, vol. iii., p. 408, pl. xl., 1828.

as to show the closer affinity of the Phascolothere with the latter oolitic Insectivora ; and, accordingly, whatever additional evidence of marsupiality is afforded by the *Phascolotherium*, may be regarded as strengthening the claims of both *Amphilestes* and *Amphitherium* to be admitted into the marsupial group. The general form and proportions of the coronoid process of the jaw of *Phascolotherium* resemble those in the zoophagous Marsupials ; and especially with that of the *Thylacynus* in regard to the depth and form of the entering notch between this process and the condyle.

The base of the inwardly-bent angle of the lower jaw progressively increases in *Didelphys, Dasyurus,* and *Thylacinus;* and judging from the fractured surface of the corresponding part of the fossil, it most nearly resembles the jaw of *Thylacinus*. The condyle of the jaw is nearer the plane of the inferior margin of the ramus in the Thylacine than in the Dasyures or opossums : and consequently, when the inflected angle is broken off, the curve of the line continued from the condyle along the lower margin of the jaw is least in the Thylacine. In this particular, again, the Phascolothere resembles that Australian Carnivore. In the position of the dental foramen, the Phascolothere, like the Amphithere, differs from the zoophagous Marsupials and placental Carnivora and Insectivora, and resembles the *Hypsiprymnus*, a marsupial Herbivore, that orifice being near the vertical line dropped from the last molar tooth. In the direction of the line of the symphysis, the Phascolothere resembles the Opossums more than the Dasyures or Thylacines. It is probable that the teeth at the fore part of the jaw showed the same correspondence. In the number of the molar series, the Phascolothere differs from *Amphitherium, Amphilestes,* and *Myrmecobius,* and resembles the Thylacine and Opossum, but without having the premolars $(p, 1, 2, 3)$ distinguished, as in them, from the true molars $(m, 1, 2, 3, 4)$, by smaller and more simple crowns. As, however, these two kinds of teeth can only be determined by their order of development and succession, the Phascolothere may well have had three premolars and four true molars.

The difference between these teeth in the lower jaw of *Didelphys* is shown by the addition, in the true molars, of a pointed tubercle on the inner side of the middle cone. In *Phascolotherium* a mere basal ridge or cingulum extends along the inner side of the middle cone. Such a ridge is present in the last molar of *Sarcophilus*, but not in the other molars.; but in these there are two small hind cusps on the same transverse line, whilst that cusp appears to be single in *Phascolotherium*. The cingulum, moreover, in the second to the penultimate of the molar series of this fossil, extends so far as to form a small talon at the fore and back part of the crown ; thus making five points, which are very distinct in the third to the penultimate tooth inclusive ; and by this character the dentition of *Phascolotherium* differs materially from any existing Marsupial, and repeats the type of molar which, as yet, would seem to be peculiar to the Insectivora of the oolitic epoch. There is a feeble indication of this structure in the antepenultimate and penultimate molars of *Thylacinus,* but the hinder division of the crown shows two small cusps on the same transverse line, besides the rudimental hindmost one ; and there is no cingulum. Upon the whole, it would seem that, though the affinity may not be close, *Phascolotherium* most resembles *Thylacinus* amongst existing Mammals ; but *Thylacinus* is now confined to Tasmania, and is there fast verging to extinction.

The resemblance shown by the lower jaw and its teeth of the Amphithere and Phascolothere to marsupial genera now confined to Australia and Tasmania, leads one to reflect on the interesting correspondence between other organic remains of the Oxfordshire oolite and other existing forms now confined to the Australian continent and surrounding sea. Here, for example, swims the Cestracion, or Port-Jackson shark, which has given the key to the nature of the "palates" from our oolites, now recognized as the teeth of congeneric larger forms of cartilaginous fishes.

Mr. Broderip, in his Memoir above quoted, observes, "that it may not be uninteresting to note that a recent species of *Trigonia* has very lately been discovered on the coast of Australia, that land of marsupial animals. Our specimen lies imbedded with a number of fossil shells of that genus." Not only *Trigoniæ* but *Terebratulæ* exist, and the latter abundantly, in the Australian seas, yielding food to the Cestracion, as their extinct analogues doubtless did to the allied Plagiostomes with crushing teeth, called *Acrodus, Psammodus*, etc. *Araucariæ*

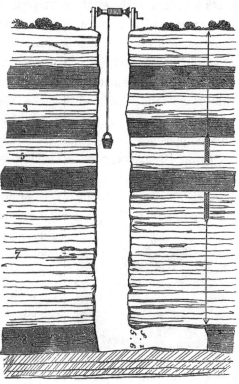

Fig. 87.
(*After Fitton.*)

1. Rubbly limestone (cornbrash).
2. Clay, with Terebratulites.
3. Limestone rock.
4. Blue clay.
5. Oolitic rock.
6. Stiff clay.
7. Oolitic rag, or limestone.
8. Sandy bed containing the Stonesfield slate.

and cycadeous plants, like those found fossil in oolitic beds, flourish on the Australian continent, where marsupial quadrupeds now abound; and thus appear to complete a picture of an ancient condition of the earth's surface, which has been superseded in our hemisphere by other strata and a higher type of mammalian organization.   Fig. 87 represents a section of the strata overlying the slates whence the fossil mammalian jaws, with associated Megalosaurs, Pterodactyles, and other oolitic organisms, have been obtained at Stonesfield in Oxfordshire.   The vertical thickness of the strata through which the shaft is sunk to the gallery is 62 feet; on the side opposite the right hand is marked the depth of the horizontal gallery, where the slate is dug which contains the fossils ; on the opposite side the strata are numbered in succession.

Genus STEREOGNATHUS.—The last evidence of a mammalian animal discovered in the Stonesfield slate is of peculiar interest, because it exhibits a type of grinding teeth quite distinct from any of the previously acquired jaws from that locality, and affords evidence of a small vegetable-feeding or omnivorous quadruped.   It consists of a portion of a lower jaw, imbedded in the characteristic matrix (fig. 88), about 9 lines in extent, and containing three molar teeth (*a, b, c*).   It is nearly straight ; the side exposed is convex vertically ; a slight bend

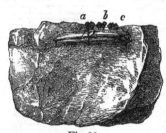

Fig 88.
Stereognathus; portion of jaw, imbedded in oolitic matrix (nat. size).

downwards, and decrease of vertical diameter towards the end, indicates it to be part of a left ramus.   This is unusually shallow, broad or thick below, the side passing by a strong convex curve into the lower part ; a very narrow longitudinal ridge, continued after its subsidence by a few fine lines, forms a tract which divides the lateral from the under surface ; elsewhere the bone is smooth, without conspicuous vascular perforations.   The depth or vertical diameter of the ramus is not more than two lines.   Of the three teeth remaining in this portion of jaw, the middle one is the least mutilated.   The

crown of this tooth (fig. 89, B) is of a quadrate form, 3 millimetres by $3\frac{1}{2}$ millimetres, of very little height, and supports
six subequal cusps in three
pairs, each pair being more
closely connected in the antero-posterior direction of the
tooth than transversely.

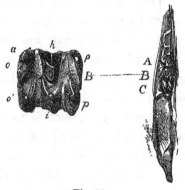

The outer side of the
crown (fig. 88, b), supported
by a narrow fang which contracts as it sinks into the
socket, shows two principal
cusps or cones, and a small

Fig. 89.

accessory basal cusp. The
hard and shining enamel
which covers these parts of

Stereognathus; upper view of portion
of jaw (nat. size), and magnified view
of the middle tooth, B (Stonesfield
Oolite).

the crown contrasts with the lighter cement that coats the root.
The two outer lobes or cones are subcompressed, and placed
obliquely on the crown, so that the hinder one (o', fig. 89) is a
little overlapped externally by the front one o, the fore part of
the base of the hinder one being prolonged inwards on the
inner side of the base of the front cone. The two middle cones
(h, i) are subcompressed laterally, with the fore part of their
base a little broader than the back part. The two inner cones
(p, p') have their inner surface convex, with their summits
slightly inclined forwards. The fore part of the base of the
hinder cone is prolonged obliquely towards the centre of the
crown, beyond the contiguous end of the base of the front cone,
so as to cause an arrangement like that of the two outer cones
(o, o'), the obliquity of the posterior cone of both the outer and
the inner pairs being such that they slightly converge as they
extend forwards.

This type of tooth differs from that of all other known recent
or extinct Mammals. The nearest approach to it is made by

the middle true molar of *Pliolophus vulpiceps*, a small extinct herbivorous Mammal from the London clay (fig. 96, *m*, ₂).

That the fragment in question is the jaw of a Mammal is inferred from the implantation of the tooth by two or more roots. Most Mammals are known to have certain teeth so implanted. Such complex mode of implantation in bone has not been observed in any other class of animals. Why two or more roots of a tooth should be peculiar to viviparous quadrupeds, giving suck, is not precisely known. That a tooth, whether it be designed for grinding hard or cutting soft substances, should do both the more effectually in the ratio of its firmer and more extended implantation, is intelligible. That a more perfect performance of a preliminary act of digestion should be a necessary correlation, or be in harmony, with a more complete conversion of the food into chyle and blood,— and that such more efficient type of the whole digestive machinery should be correlated, and necessarily so, with the hot blood, quick-beating heart and quick-breathing lungs, with the higher instincts, and more vigorous and varied acts of a Mammal, as contrasted with a cold-blooded reptile or fish,—is also conceivable. To the extent to which such and the like reasoning may be true, or in the direction of the secret cause of the constant relations of many-rooted teeth discovered by observation,—to that extent will such relations ascend from the empirical to the rational category of laws.

The interest which the above-described fossil from the Stonesfield oolitic slate excites is not exclusively due to its antiquity, its uniqueness, or its peculiarity ; much is attached to its relations as a test in palæontology of the actual value of a single tooth in the determination of other parts of the organization of the animal. According to our opinion of these unseen parts, we frame our expression of the nature and affinities, or of the place in the zoological system, of the extinct species. From the resemblance of the lower molars of *Stereognathus* to those of *Pliolophus*, which, though not close, is closer than to the teeth of any other known animal, it is probable that the *Stereognathus* was hoofed, and consequently herbivorous, or deriving the chief part of its subsistence from the vegetable kingdom. Cuvier has written,—" La première chose à faire dans l'étude d'un animal fossile est de réconnaitre la forme de ses dents molaires ; ou determine par la s'il est carnivore ou herbivore, et dans ce dernier cas, on peut s'assurer, jusqu'à un certain point de l'ordre d'herbivores auquel il appartient." * In the case in question the form of the molar teeth of one jaw is recognizable, but the herbivority of the fossil is not thereby determined. We can only infer it to be more probable that the fossil was a Herbivore than an Insectivore or a mixed-feeding Carnivore.

Admitting the herbivority of the fossil, it is not certain that it was hoofed ; there is nothing in the form and structure of the tooth to prove that. Both form and structure are compatible with the hoofless muticate type of herbivorous Mammal, as shown by the Manatee ; it is the small size of the *Stereognathus* which renders it less probable that it was a diminutive kind of Manatee, and more probable that it was a diminutive form of Ungulate. But seeing the manifold diversities of the multi-cuspid form of molar teeth in recent and extinct insectivorous unguiculate quadrupeds, it is not impossible but that the *Stereognathus* may have belonged to that order ; there is no known physiological law forbidding it.

The form of the cusps, and their regular symmetrical arrangement in the *Stereognathus*, as compared with the known modifications of multi-cuspid molars in certain small extinct forms of hoofed quadrupeds, constitute the grounds upon which an opinion is formed of its most probably belonging to the same section of *Ungulata*.

Then, is it not true, it may be asked, that by virtue of

* Ossemens Fossiles, 4to, tom. iii., 1822, p. 1.

certain established laws of correlated structures, an extinct animal may be re-constructed from a single tooth or from a fragment of bone ? Is the Cuvierian basis, or what has been so regarded, of palæontology unsound ? Not necessarily from aught that has been said or written on the subject of the *Stereognathus.* We do not know the comparative anatomy of the family of quadrupeds to which the *Stereognathus* belonged. What we do know of its teeth suggests that that family may have had modifications of the skeleton so far different from those of any, the modifications of which are known, as to have constituted a type of, perhaps, a marsupial family ; but a type as well marked, and as distinct, as the type of skeleton which Cuvier inductively studied in the feline *Carnivora* (fig. 128), and in the ruminant *Herbivora* (fig. 129), and by which preliminary study he was enabled to enunciate that beautiful law of the " correlation of forms and structures" to which allusion has been already made, and which will be illustrated by examples, and its mode of application pointed out, in another part of the present work.

In certain instances of constant coincidences of structure, as demonstrated by comparative anatomy, the sufficient—*i. e.* recognizable, intelligible, or physiological—cause of them is not yet known. But, as Cuvier in reference to such instances truly remarks, " Since these relations are constant, there certainly must be a sufficient cause for them." * In certain other cases Cuvier believed that he could assign that "sufficient cause," and he selects, as such, the correlated structures in a feline Carnivore, and in a hoofed Herbivore. The physiological knowledge displayed by him in his explanation of the condition of those correlations is most exact ; its application in the restoration of the *Anoplotherium* and *Palæotherium* most exemplary.

In the ratio of the knowledge of the reason of the coincidences of animal structures—in other words, as those coinci-

* Discours sur les Révolutions de la Surface du Globe, 4to, 1826, p. 50.

dences become "correlations"—is our faith in the soundness of the conclusions deduced from the application of such rational law of correlations ; and with the certainty of such application is associated a greater facility of its application. A knowledge of the physiological conditions governing the relations of the contents of the cavities of bones to the flight and other modes of locomotion in birds both enabled the writer to infer from one fragment of a skeleton that it belonged to a terrestrial bird deprived of the power of flight, and to predict that such a bird, but of less rapid course than the ostrich, would ultimately be found in New Zealand.*

This principle, however—those modes of thought—which Cuvier affirmed to have guided him in his interpretation of fossil remains, and which he believed to be a true clue in such researches, were repudiated or contested by some of his contemporaries.

Geoffroy St. Hilaire denied the existence of a design in the construction of any part of an organized body ; he protested against the deduction of a purpose from the contemplation of such structures as the valves of the veins or the converging lens of the eye.

Beyond the co-existence of such a form of flood-gate with such a course of the fluid, or of such a course of light with such a converging medium, Geoffroy affirmed that thought, at least his mode of thinking, could not sanely, or ought not to go.

The present is not the place for even the briefest summary of the arguments which have been adduced by teleologists and antiteleologists from Democritus and Plato down to Comte and Whewell. The writer would merely remark, that in the degree in which the reasoning faculty is developed on this planet and is exercised by our species, it appears to be a more healthy and normal condition of such faculty,—certainly one which has been productive of most accession to truths, as exemplified

* Transactions of the Zoological Society, vol. iii., p. 32, pl. 3.

in the mental workings of an Aristotle, a Galen, a Harvey, and a Cuvier,—to admit the instinctive impression of a design or purpose in such structures as the valves of the vascular system and the dioptric mechanism of the eye. In regard to the few intellects,—they have ever been a small and unfruitful minority,—who do not receive that impression and will not admit the validity or existence of final causes in physiology, the writer has elsewhere expressed his belief that such intellects are not the higher and more normal examples, but rather manifest some, perhaps congenital, defect of mind, allied or analogous to "colour-blindness" through defect of the optic nerve, or the inaudibleness of notes above a certain pitch through defect of the acoustic nerve.

The truth of a physiological knowledge of the condition of a correlated structure, and of the application of that knowledge to palæontology, is not affected by instances adduced from that much more extensive series of coincident structures of which the physiological condition is not yet known. Nor is the power of the application of the physiologically interpreted correlation the less certain because the merely empirically recognized coincidences have failed to restore, with the same certainty and to the same extent, an extinct form of animal.

Certain coincidences of form and structure in animal bodies are determined by observation. By the exercise of a higher faculty the reason, or a reason, of these coincidences is discovered, and they become correlations ; in other words, it is known not only that they do exist, but how they are related to each other. In the case of coincidences of the latter kind, or of "correlations" properly so called, the mind infers with greater certainty and confidence, in their application to a fossil, than in the case of coincidences which are held to be constant only because so many instances of them have been observed.

Because the application of the latter kind of coincidences is limited to the actual amount of observation at the period of such application, and because mistakes have been made through a miscalculation of the value of such amount, it has been argued that a rational law of the correlation of animal forms is inapplicable to the determination of a whole from a part ;* and it has not only been asserted that the results of such determination are unsound, but that the philosopher who believed himself guided by such law deceived himself and misconceived his own mental processes ! † But the true state of the case is, that the non-applicability of Cuvier's law in certain cases is not due to its non-existence, but to the limited extent to which it is understood.

The consciousness of that limitation led the enunciator of the law to call the attention of palæontologists expressly to the extent to which it could then be applied, as, for instance, to the determination of the class, but not the order ; or of the order, but not the family or genus, etc. ; and to caution them also as to the extent of the cases in which, the coincidences being only known empirically, he consequently enjoins the necessity of further observation, and of caution in their induction. Cuvier expresses, however, his belief that such coincidences must have a sufficient cause, and that cause once discovered, they then become correlations and enter into the category of the higher law. Future comparative anatomists will have that great consummation in view, and its result, doubtlessly, will be the vindication of the full value of the law in the interpretation of fossil remains as defined by the illustrious founder of palæontology.

Genus SPALACOTHERIUM, Ow.—The next stratum overlying the older oolites in which mammalian remains have been detected, is a member of the newest oolitic series at Purbeck, Dorsetshire, called the " marly " or " dirt-bed." In a series of

* De Blainville, Ostéographie, 4to, fasc. 1, 1839, p. 34.
† Prof. Huxley, "Lecture on Natural History," etc., Royal Institution of Great Britain, Feb. 15, 1856.

fossils discovered there by Mr. W. R. Brodie, and transmitted
for determination in 1854 to the writer, amongst the remains
of fishes and small Saurians, constituting the majority of the
specimens, were detected three unequivocal evidences of a
mammalian species, which were described under the name of

Fig. 90.

Spalacotherium tricuspidens (twice nat. size), Purbeck beds.

Spalacotherium* tricuspidens.
The specimen here selected
(fig. 90) to exemplify the
above extinct insectivorous
Mammal, is a right ramus of
the lower jaw. The posterior
half contains four teeth, and extends backward beyond the
dental series ; but instead of showing the compound structure
which that part of the jaw exhibits in the lizard tribe, it continues undivided ; the convex surface showing a smooth
depression for the insertion of the temporal muscle ; the lower
boundary answering to that going to the condyle and angle of
the jaw, and the upper one to that going to the coronoid process
in the ramus of the jaw of the mole and shrew. The crowns
of the teeth are long, narrow, and tricuspid, the inner part of
the crown being produced into a point both before and behind
the longer cusp which forms the chief outer division of the
crown. Each of these teeth is implanted by a fang divided
externally into two roots, in a distinct socket in the substance
of the jaw. The multicuspid crown, the divided root of the
tooth, its complex implantation, and the undivided or simple
structure of the ramus of the jaw, all concurred, therefore, to
prove the mammalian nature of this fossil.

The other specimens showed that the *Spalacotherium* had
ten molar teeth in each ramus of the lower jaw, preceded by a
small canine and incisors. The anterior molars are compressed,
increase in height and thickness to the sixth, and from the
seventh decrease in size to the hindmost, which seems to be

* From σπάλαξ, *a mole;* θηρίον, *a beast.*

the last of the series. The sharp multicuspid character of so much of the dental series as is here preserved, repeats the general condition of the molar teeth of the small insectivorous Mammalia in a striking degree : one sees the same perfect adaptation for piercing and crushing the tough chitinous cases and elytra of insects. The particular modification of the pointed cusps, as to number, proportion, and relative position, resembles in some degree that of the Cape mole (*Chrysochlora aurea*), but both in these respects and in the number of molars, the dentition accords more closely with that of the extinct *Amphitherium*. The chief interest in the discovery of the *Spalacotherium* is derived from its demonstration of the existence of Mammalia about midway between the older oolitic and the oldest tertiary periods.

Both the Oxford oolitic slate and the Purbeck marly shell-beds give evidence of insect life ; in the latter formation abundantly. The association of these delicate Invertebrata with remains of plants allied to *Zamia* and *Cycas*, is indicative of the same close interdependency between the insect class and the vegetable kingdom, of which our power of surveying the phenomena of life on the present surface of the earth enables us to recognize so many beautiful examples. Amongst the numerous enemies of the insect class ordained to maintain its due numerical relations, and organized to pursue and secure its countless and diversified members in the air, in the waters, on the earth and beneath its surface, bats, lizards, shrews, and moles now carry on their petty warfare simultaneously, and in warmer latitudes work together, or in the same localities, in their allotted task. No surprise need therefore be felt at the discovery that Mammals and Lizards co-operated simultaneously and in the same locality at the same task of restraining the undue increase of insect life during the period of the deposition of the Lower Purbeck beds.

Genus TRICONODON, Ow.

Sp. *Triconodon mordax.*—This name is proposed for a small
zoophagous Mammal, whose generic distinction is shown by
the shape of the crowns of the molar teeth of the lower jaw,
which consist of three nearly equal cones on the same longitudinal row, the middle one being very little larger than the
front and hind cone ; and these cones are not complicated by

any cingulum or accessory basal cusp.
The convex condyle is below the level
of the alveoli, and there is no angular
process projecting beneath it. The coronoid process is broad and high, with
its hinder point not extended so far

Fig. 91.

Jaw of *Triconodon mordax*
(nat. size), Purbeck.

back as the condyle ; the depression marking the insertion of
the temporal muscle extends nearly to the lower border of
the jaw.   There are the obscure remains of three broken
incisors, and the point of apparently a canine ; next come the
two stumps, or broken roots of a small premolar ; then the
crown of a second double rooted premolar, which show a
principal cone and a small anterior cusp ; the next tooth is
wanting ; then there is a larger premolar, with the two fangs
raised some way out of their socket : the crown of this tooth
shows a principal cone, with a small anterior and large posterior talon ; it rises, apparently from partial displacement,
higher than the succeeding molars ; these are three in number,
and present the characteristic three-coned structure already
described ; each cone is smooth, and convex externally.   The
three cones seem to answer to the three middle or principal
cones of the molars of *Amphilestes* and *Phascolotherium*, but
the front and hind cones are raised to near equality with the
middle cone in *Triconodon.*

The lower jaw of this species, in the relation of the condyle
to the lower border, resembles that of *Phascolotherium* more
than that of *Amphitherium*, but it differs from both ; there is
not the same gradual curve from the condyle to the symphysis

as in *Phascolotherium;* and the condyle, besides being on a
lower level, is divided by a less deep notch from the coronoid
process. This process is larger in proportion to the entire jaw ;
approaches more nearly to the quadrate or rhomboid form, the
upper border being less curved ; it affords a more extensive
surface of attachment to the principal biting muscles than in
most predatory extinct or recent quadrupeds. This character,
with the depth and strength of the jaw, suggested the specific
name. From the shape of the exposed part of the ramus, we
may conclude that the part answering to the angle is bent inwards, and that *Triconodon* was a genus of the marsupial order.
The specimen was discovered by Mr. Beccles in the same " dirt-
bed" at Purbeck as that in which *Spalacotherium* was found.

Genus PLAGIAULAX,* Fr.—The most remarkable of Mr.
Beccles' discoveries in the above formation are the mammalian
jaws indicative of the genus above named, of which two species
have been determined by Dr. Falconer.

Sp. *Plagiaulax Becclesii,* Fr.—Two specimens exemplified
the shape and pro-
portions of the entire
jaw of this species
(fig. 92). The fore-
most tooth (*i*) is a
very large one,
shaped like a canine,
but implanted by a
thick root in the

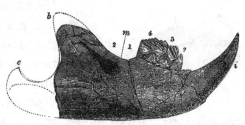

Fig. 92.

Plagiaulax Becclesii (twice nat. size), Purbeck.

fore part of the jaw, like the large lower incisor of a shrew or
wombat. The three anterior teeth in place have compressed
trenchant crowns, and rapidly augment in size from the first
(*2*) to the third (*4*). They are followed by sockets of two
much smaller teeth, shown in other specimens to have sub-

* An abbreviation for *Plagiaulacodon*, from πλάγιος, *oblique*, and αὐλάξ,
groove; having reference to the diagonal grooving of the premolar teeth.

tuberculate crowns resembling those of *Microlestes*. The large front tooth of *Plagiaulax* is formed to pierce, retain, and kill ; the succeeding teeth, like the carnassials of Carnivora, are, like the blades of shears, adapted to cut and divide soft substances, such as flesh. As in Carnivora, also, these sectorial teeth are succeeded by a few small tubercular ones. The jaw conforms to this character of the dentition. It is short in proportion to its depth, and consequently robust, sending up a broad and high coronoid process (*b*), for the adequate grasp of a large temporal muscle ; and the condyle (*c*) is placed below the level of the grinding teeth—a character unknown in any herbivorous or mixed-feeding Mammal ; whilst the lever of the coronoid process is made the stronger by the condyle being carried farther back from it than in any known carnivorous or herbivorous animal. The angle of the jaw makes no projection below the condyle, but is slightly bent inward, according to the marsupial type.

Sp. *Plagiaulax minor*, Fr.—In this species the first premolar

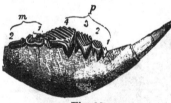

Fig. 93.

Plagiaulax minor (four times nat. size),
Purbeck (after Lyell).

(fig. 93, *p*, 1) is preserved ; the rest (*p*, 2, 3, and 4) show nearly the same shape and proportions as in *P. Becclesii*. The first molar (*m*, 1) has a broad depression on the grinding surface, surrounded by tubercles, of which three are on the outer border ; the marginal tubercles of the second smaller tooth are smaller and more numerous.

In the general shape and proportions of the large premolar (*p*, 4) and succeeding molars, *Plagiaulax* most resembles *Thylacoleo* (fig. 141, *p*, *m*, 1 and 2),—a much larger extinct predaceous Marsupial from tertiary beds in Australia. But the sectorial teeth in *Plagiaulax* are more deeply grooved ; whence its name. The single compressed premolar of the kangaroo-rat is also grooved ; but it is differently shaped, and is succeeded by four square-crowned double-ridged grinders adapted for vegetable food ; and the position of the condyle, the slenderness of the coronoid, and other characters of the lower jaw, are in conformity to that regimen. In *Thylacoleo* the lower canine or canine-shaped incisor projected from the fore part of the jaw close to the symphysis, and the corresponding tooth in *Plagiaulax* more closely resembles it in shape and direction than it does the procumbent incisor of *Hypsiprymnus*. From this genus *Plagiaulax* differs by the obliquity of the grooves on its premolars ; by having only two true molars in each ramus of the jaw, instead of four ; by the salient angle which the surfaces of the molar and premolar teeth form, instead of presenting a uniform level line ; by the broader, higher, and more vertical coronoid ; and by the very low position of the articular condyle.

The physiological deductions from the above-described characteristics of the lower jaw and teeth of *Plagiaulax* are, that it was a carnivorous Marsupial. It probably found its prey in the contemporary small insectivorous Mammals and Lizards, supposing no herbivorous form, like *Stereognathus*, to have co-existed during the upper oolitic period.

In the Woodwardian Museum at Cambridge is a specimen of anchylosed cervical vertebræ of a cetaceous animal as large as a grampus, but presenting specific distinctions from all known recent and fossil species. It is stated to have been found in the brown clay or "till" near Ely ; but in its petrified condition, colour, and specific gravity, it is so different from the true bones of the "till," and so closely like the fossils of the Kimmeridge clay, as to make it extremely probable that it has been washed out of that formation.

No evidence of the mammalian class has yet been met with in the chalk beds.

The examples of the Mammalia first met with in tertiary strata are the *Coryphodon* and *Palæocyon,* respectively representing the ungulate (herbivorous) and unguiculate (carnivorous) modifications of the class ; their remains have been found in the plastic clay and equivalent lignites in England and France.

Genus CORYPHODON, Ow.—Rarely has the writer felt more misgiving in regard to a conclusion based, in palæontology, on a single tooth or bone, than that to which he arrived after a study of the unique fragment of jaw with one tooth dredged up off the Essex coast, and on which he founded the genus *Coryphodon.**

The marked contraction of the part of the jaw near one end of the tooth seemed, at first view, clearly to show it to be the narrower fore part of the ramus ; in that case the tooth would have been a premolar, and of comparatively little value in the determination of a genus or species. But a closer inspection showed the line of abrasion of the summits of the two transverse ridges of the tooth to be on one side, and the general law of the relative apposition and reciprocal action of the upper and lower grinders in tapiroid Pachyderms determined that those oblique linear abrasions must be on the hinder side of the ridges. The smaller characters carried conviction against the showing of the larger and more catching ones. So, in determining the position of the nautilus in its pearly abode, when the animal without its shell was first brought to England in 1831, the reasons afforded by some small and inconspicuous parts in like manner outweighed the first impressions from more obvious appearances, as well as the bias from the general analogies of testaceous Univalves. Some contemporary naturalists asserted, and for a time it was believed, that the nautilus had been put upside down in its shell,† just as some

---

* History of British Fossil Mammals, 8vo, p. 299, figs. 103, 104.

† In plate i. of the writer's Memoir of the Nautilus, 4to, 1832.

contemporary anatomists surmised that the writer had mistaken the fore for the back part of the jaw of his *Coryphodon*, which, in that case, might only be the known *Lophiodon*. In both instances the conclusions founded on the less obvious characters have proved to be correct. And the writer would remark that, in the course of his experience, he has often found that the prominent appearances which first catch the eye, and indicate a conformable conclusion, are deceptive ; and that the less obtrusive phenomena which require searching out, more frequently, when their full significance is reasoned up to, guide to the right comprehension of the whole. It is as if truth were whispered rather than outspoken by Nature.

Truth, it is sometimes said, lies at the bottom of a well. The first additional glimpse that the writer obtained of the veritable nature of one of our most ancient tertiary Mammals was derived from the inspection of a fossil tooth brought up from a depth of 160 feet, out of the " plastic clay," during the operations of sinking a well in the neighbourhood of Camberwell, near London. It was a canine tooth,[*] belonging, from its size (near 3 inches in length), to a large quadruped, and, from the thickness and shortness of its conical crown, not to a carnivorous but to a hoofed Mammal, most resembling in shape, though not identical with, that of the crown of the canine tooth of some large extinct tapiroid Mammals, which Cuvier had referred to his genus *Lophiodon*, but which has proved to belong to *Coryphodon*.

The last lower molar of *Lophiodon* has three lobes ; the molar determined to be the ultimate one, in the fragment of lower jaw above referred to, resembles that of the tapir in the absence of a third or posterior lobe, but the posterior ridge or part of the cingulum is less developed than in the tapir. It presents two divisions in the form of transverse ridges or eminences, the front ridge being the largest, and with its edge·

[*] Hist. Brit. Foss. Mamm., p. 306, fig. 105.

most entire. From the outer end of each division a ridge is continued obliquely forward, inward, and downward : the anterior one extends to the antero-internal angle of the base of the crown ; the posterior one terminates at the middle of the interspace between the two chief divisions of the crown. The trenchant summit of the anterior ridge is slightly concave toward the fore part of the tooth, as in that of *Lophiodon ;* but its outer and inner ends rise higher, and appear as more distinct cones or points ; whence the generic name of *Coryphodon.* The posterior division is lower than the anterior one, and is bicuspid ; the trenchant margin connecting the outer and inner points does not extend across the crown parallel with the anterior ridge, as in *Tapirus* and *Lophiodon,* but bends back so as to form an angle, the apex of which rises into a third point.

Some lophiodontoid fossils from the lignites of Soissons and Laon, and from the plastic clay of Meudon in France, including the upper molar tooth figured by Cuvier in the chapter of the *Ossemens Fossiles* entitled " Animaux voisins de Tapirs," pl. vii., fig. 6, belong to the genus *Coryphodon.* Cuvier compares this tooth with one from Bastberg, which he figures in pl. vi., fig. 4, and which is certainly the last upper molar of a true *Lophiodon,* and points out truly that the Soissons tooth differs in the external border passing into the posterior one, so that, instead of being quadrangular, its crown is triangular ; but he explains this difference on the hypothesis that the Bastberg tooth was a penultimate molar. The reduction of the second or posterior ridge to a semi-circular one, developed at its middle and hindmost part into a prominent cone, so far agrees with the modification of the same part of the last molar of the lower jaw of the *Coryphodon* as to render it very probable that the last upper molar from Soissons, figured by Cuvier in pl. vii., fig. 6, above quoted, also belongs to the genus *Coryphodon.* Cuvier states that the entire skeleton was found, indicative of an animal as long and almost as large as a bull; but that the workmen employed in the sandpit (*sablonière*) preserved only that one tooth. Both the lower molar from Harwich, and the upper one from Soissons, indicate an animal of at least double the size of the American tapir.

Professor Hebert * has recently described a very instructive series of teeth and bones from the oldest eocene deposits in France, which he refers to the genus *Coryphodon*: the last molar is identical in form with the tooth from the plastic clay of Essex, on which the genus was originally founded.

Genus PLIOLOPHUS, Ow.—The most complete and instructive example of a Mammal from the next overlying division of the eocene tertiaries, viz., the "London clay," is that which the writer has described † under the name of *Pliolophus vulpiceps*. It is a hoofed Herbivore, but presents a dentition not exhibited by any later or existing species of Mammal.

The length of the skull (fig. 94) is 4 inches, its extreme

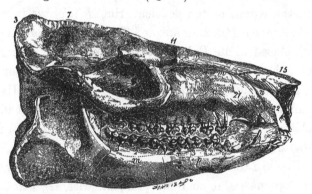

Fig. 94.
Skull of *Pliolophus vulpiceps* (half nat. size), London clay.

breadth 2 inches 2 lines, the height of the cranium opposite the first premolar tooth 9 lines. Its shape and characteristics

* Comptes Rendus de l'Acad. des Sciences, Paris, 26th January 1857 (*Coryphodon Oweni*, Hebert).

† Quarterly Journal of the Geological Society, vol. xiv., p. 54.

determine the hoofed nature of this species, and its affinities
to the Perissodactyla, or the order of Ungulata with toes in odd
number. The extent and well-defined boundary of the temporal fossæ by the occipital ($_3$), parietal.($_7$), and post-frontal
ridges, and their free communication with the orbits, give
almost a carnivorous character to this part of the cranium of
Pliolophus; but as in the hog, Hyrax, and Palæothere, the
greatest cerebral expansion is at the middle and toward the
fore part of the fossæ, with a contraction toward the occiput;
the brain-case not continuing to enlarge backward to beyond
the origin of the zygomata, as in the fox. The zygomatic
arches have a less outward span than in the *Carnivora.* In
this part of the cranial structure *Pliolophus* resembles *Palæotherium* more than it does any existing Mammal; but the
post-frontal processes are longer and more inclined backward.
The incompleteness of the orbit occurs in both *Anoplotherium*
and *Palæotherium,* as in *Rhinoceros, Tapirus,* and the hog tribe ;
but in the extent of the deficient rim, *Pliolophus* is intermediate between *Palæotherium* and *Tapirus.* The orbit is not
so low placed as in *Palæotherium, Tapirus,* and *Rhinoceros,*
nor so high as in *Hyrax* or *Sus.* The straight upper contour
of the skull ($_7$ to $_{15}$) is like that in the horse tribe and Hyrax,
and differs from the convex contour of the same part in the
Anoplothere and Palæothere. The size of the antorbital foramen (a) indicates no unusual development of the muzzle or
upper lip. In the conformation of the nasal aperture by four
bones (two nasals, $_{15}$, and two premaxillaries, $_{22}$), *Pliolophus*
resembles the horse, Hyrax, hog tribe, and Anoplothere, and
differs from the rhinoceros, tapir, and Palæothere, which have
the maxillaries, as well as the nasals and premaxillaries, entering into the formation of the external bony nostril.

The ungulate and herbivorous character of *Pliolophus* is
most distinctly marked by the modifications of the lower jaw,
especially by the relative dimensions of the parts of the ascend-

ing ramus which give the extent of attachment of the biting (temporal) and grinding (masseteric and pterygoid) muscles respectively. In the shape of the mandible *Pliolophus* most resembles *Tapirus* among existing Mammals, and the *Palæotherium* among the extinct ones in which that shape is known. As in almost every species of eocene quadruped yet discovered, the *Pliolophus* presents the type-dentition of the placental diphyodont series, viz.—

$$i\ \frac{3-3}{3-3},\ c\ \frac{1-1}{1-1},\ p\ \frac{4-4}{4-4},\ m\ \frac{3-3}{3-3} = 44.^{*}$$

The incisors are preserved in the lower jaw with marks of attrition on their crowns demonstrating corresponding teeth of the same number (six), and of similar size, in the upper jaw, from which the alveolar part of the premaxillaries had been broken away.

The canines are small in both jaws : they are separated by a vacant space from the outer incisors, and by a longer inter-

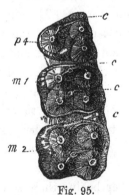

Fig. 95.
True molars, upper jaw (twice nat. size), *Pliolophus.*

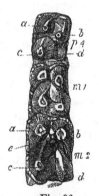

Fig. 96.
True molars, lower jaw (twice nat. size), *Pliolophus.*

val from the first premolars. These form a continuous series with the remaining teeth in the upper jaw, but are separated

* See Ency. Brit., art. ODONTOLOGY, vol. xvi., p. 478, for the "Homologies of the Teeth," and explanation of their symbols.

by a space of about half their breadth from the second pre-molar in the lower jaw. The succeeding teeth $(p\ 1, 2, 3, 4)$ increase in size to the penultimate molar in the upper, and to the last molar in the lower jaw $(p\ 4$ in figs. 95 and 96), which tooth has a third lobe.

In the last premolar upper jaw (fig. 95, $p\ 4$) the cingulum is uninterrupted along the outer side from its anterior well-developed talon (c) to the back part. The two outer cones resemble those of the true molars; but there is only one inner cone, and the crown of $p\ 4$ differs accordingly from that of m 1, in being triangular rather than square. A ridge is continued from the interspace between the anterior talon (c), and the outer anterior lobe obliquely inward and backward to the inner lobe, swelling into a small tubercle at the middle of its course.

The first molar $(m\ 1)$ presents four low thick cones, two internal and two external : each external cone is connected with its opposite internal one by a low ridge, swelling into a tubercle at the middle of its oblique course. The cingulum (cc) seems to be continued uninterruptedly round the crown of this tooth, thickest at the fore and back part, and at the interspace of the inner lobes ; and developing the small accessory antero-external tubercle. The second molar $(m\ 2)$ is similar to, but rather larger than, the first ; the tubercle on the oblique ridge connecting the two front lobes is less developed. The cingulum is obliterated on the inner side of the posterior lobe.

The last molar is rather narrower behind than $m\ 2$; the tubercle on the anterior of the oblique connecting ridges is smaller : that on the posterior ridge is almost obsolete.

In the last lower premolar (fig. 96, $p\ 4$) the division and development of the anterior lobe gives rise to a pair of cones, one external (a), the other internal (b), connected anteriorly by a basal ridge, in front of which is the fore part of the cingulum. The low posterior lobe (c) shows the rudiment of a second internal cone (d).

The first molar (fig. 96, m 1) has a pair of front lobes and a pair of hind lobes, with an oblique ridge continued from postero-internal lobe to the interspace between the front pair.

The second molar (m 2) shows an increase of size ; but its chief and most interesting modification is the development of a tubercle (e) between the two anterior lobes, making three cones on the same transverse line, and thus repeating the character of the molar tooth of *Stereognathus* (fig. 97, e). The oblique ridge from the outer and hinder lobe (c) abuts against the intermediate tubercle (e). The nearest approach to the above dentition is made by the extinct *Hyracotherium ;* also a fossil from the London clay.

Fig. 97.

True molar, lower jaw (magn.), *Stereognathus ooliticus.*

The third trochanter on the femur of *Pliolophus*, and the association of three metatarsals in one portion of the matrix, as if belonging to the same hind foot, confirm the essentially perissodactyle affinities of that genus as shown by the skull and teeth. *Pliolophus* and *Hyracotherium* form a well-marked section in the lophiodont family, which seems to have preceded the palæotherian family in the order of appearance, and to have retained more of the general ungulate type than that family. This is shown by the graduation of the tapiroid modification of the molar teeth into one more nearly resembling that of the *Anthracotheria* and *Chœropotami*, by the absence of the postero-internal cone on the ultimate premolar, by which all the premolars are, as in artiodactyles, less complex than the true molars, by the form and position of the nasal bones and by the structure of the external nostril.

Genus LOPHIODON, Cuv.—In the year 1800 Cuvier[*] first announced the discovery of the fossil remains of a quadruped

* Bulletin des Sciences, Paris, Nivose, an. viii., No. 34.

allied to and of the size of the tapir, in the lacustrine deposits of the Montagne Noir, near Issel, department of Aude in Languedoc. The outer incisor of the lower jaw was shortened to give room to the longer corresponding incisor above, as in the tapir ; the canines offered the same proportional develop- ment; but the three first molars (premolars) of the lower jaw presented a more simple structure, having the crown com- pressed, and forming two cones, the front one being the largest; —in short, a structure, the type of which is presented only by the first of the three premolars ($p\,2$) in the genus *Tapirus*.*

Years elapsed ere Cuvier obtained clear evidence of the structure of the upper molars of this new fossil Mammal. Such detached teeth as had been obtained from the fresh- water formations near Issel were referred, owing to the way in which they departed from the type of the upper molar teeth of the *Tapir*, to the genus *Rhinoceros*. This fact is indicative of the annectant affinities of the *Lophiodon* in the perissodactyle series.† Besides the character of form, the upper molar series of *Lophiodon* differs, like the lower one, from that in *Tapirus*, in the greater simplicity of the last two premolars ; these teeth have a single cone on the inner side in *Lophiodon ;* they have there two cones in *Tapirus*, forming the inner terminations of two transverse ridges, as in the true molars. These teeth in the *Lophiodon* differ from those in the *Tapirus* in the greater fore-and-aft expanse of the outer terminations of the transverse ridges, and the less depth of the cleft between them : a more complete coalescence of those parts causes a more entire outer wall of the crown, and com- pletes the transition to the Rhinoceros type, towards which the Palæotherium offers the next step.

Genus PALÆOTHERIUM, Cuv.—This extinct genus of quad- ruped was restored (fig. 98) by Cuvier through a series of admirably instructive steps, ultimately verified by a complete

* Ency. Brit., art. ODONTOLOGY, p. 471, fig. 136, $p\,2$. † Ibid., p. 470.

series of fossils, obtained chiefly from the upper eocene gypse-
ous formation at Montmartre and other parts of France.   The

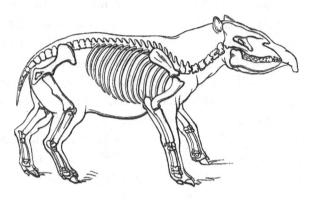

Fig. 98.
Restoration of the *Palæotherium* (Eocene Gyps).

molar teeth of *Palæotherium* (fig. 99) approach nearer to those
of the rhinoceros ; but in the number, kind, and general arrangement the entire dentition resembles that of *Pliolophus*.
The skull affords indications that the Palæothere possessed
a short proboscis.   It had three toes on each foot, each terminated by a hoof ; the middle one being the largest.   The

femur had a third trochanter,
and the dorso-lumbar vertebræ
were 21 in number.   Several
species of *Palæotherium* have
been determined, ranging from
the size of a sheep (*P. curtum*)
to that of a horse (*P. magnum*).
Fig. 99 gives the grinding
surface of an upper molar of
this species from the upper
eocene of the Bembridge beds,
Isle of Wight.   The crown is
divided into an anterior (*b, d*) and posterior (*a, c*) part by an

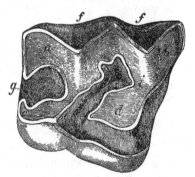

Fig. 99.
Upper molar, *Palæotherium magnum*
(Eocene).

oblique fissure (e), continued from near the middle of the inner surface of the crown obliquely across two-thirds of the tooth. Each division is subdivided partially into an outer (ab) and an inner (cd) lobes; the anterior division, by the terminal expansion (i) of the fissure (e), the posterior one by the fissure (g). The lobes (c and d) are bordered near their base by a ridge. This is the type of grinding surface, on which are superinduced the modifications of that surface in the upper molars of the rhinoceros and horse. The dental formula of *Palæotherium* is $i \frac{3-3}{3-3}$, $c \frac{1-1}{1-1}$, $p \frac{4-4}{4-4}$, $m \frac{3-3}{3-3} = 44$. The canines exceed in length the other teeth, and there are consequently vacancies in the dental series for the lodgment of the crowns of the canines when the mouth is shut.

Genus ANOPLOTHERIUM, Cuv.—With the same dental formula as in *Palæotherium*, the present genus, like *Dichodon* (fig. 102) has no interval in the series of teeth; neither the canine nor any other tooth rising above the general level. The grinding surface of the molar teeth somewhat resembles and prefigures the ruminant type; in the upper jaw the crown (fig. 100) is divided into a front (fc) and a back (fd) lobe by a valley (e) extending two thirds across. A second valley (gi) crosses its termination at right angles, forming a curved depression in each division, which it thus subdivides into two lobes, concave towards the outer

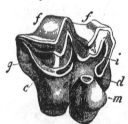

Fig. 100.

Upper molar, *Anoplotherium commune* (Eocene Gyps).

side of the tooth. There is a large tubercle (m) at the wide entry of the valley (e). The Anoplothere (fig. 101) was of a lighter and more elegant form than the Palæothere: its limbs terminated each in two digits, with the metapodial bones distinct, and the last phalanx hoofed. Some transitory characters of the embryo ruminant were retained throughout life by the Anoplothere. The species restored in fig. 101 was about

the size of a fallow-deer : it had a long and strong tail, and
was probably of aquatic habits. Smaller and more delicate
species of Anoplotherioids from upper eocene strata have been
referred to distinct genera by later palæontologists. The re-

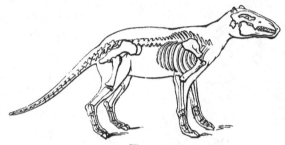

Fig. 101.
Restoration of the *Anoplotherium* (Eocene Gyps).

searches of Baron Cuvier, which resulted in the restoration of
the *Palæotherium* and *Anoplotherium*, are the most instructive
which the palæontologist can study. They form the third
volume of the 4to edition of the *Ossemens Fossiles*, 1822-5.

Genus DICHODON, Ow.—The upper eocene beds of Hampshire have yielded evidence of an extinct form of even-toed
(artiodactyle) hoofed quadruped, most interesting as a transitional form between the Anoplotherioids and the true Ruminants. Like the *Anoplotherium* the dental series is continuous, without break—a character which is only manifested by
mankind among existing Mammals—the crowns of the teeth,
in *Dichodon*, being all of nearly equal height, as they are in
man. On each side of both upper and lower jaws there are
in the *Dichodon* (fig. 102) three incisors (i 1, 2, 3), one canine
(c), four premolars (p 1, 2, 3, 4), and three true molars (m 1,
2, 3)—in all forty-four teeth, constituting the typical diphyodont* dentition which so many mammalian genera, on their
first appearance in the eocene strata, exhibit. It is formulized
as follows :—$i \frac{3\text{-}3}{3\text{-}3}, c \frac{1\text{-}1}{1\text{-}1}, p \frac{4\text{-}4}{4\text{-}4}, m \frac{3\text{-}3}{3\text{-}3} = 44$. From the first incisor

* See Ency. Brit., art. ODONTOLOGY, p. 439.

to the third premolar the teeth have a more or less trenchant crown. The back of the third premolar (p 3), and all the fourth premolar ($p.$ 4), show the crushing form of crown ; the pattern of which in the true molars, after the wearing down of the first sharp cusps, produces the double crescentic lines of enamel which are now peculiar to the Ruminants amongst hoofed quadrupeds. The first (p 1), second (p 2), and third (p 3) premolars, have their crown much extended from before backwards, with three progressively more developed and pointed compressed cusps on the same line : to which is added, in the upper jaw, an inner ridge, developed in the third premolar

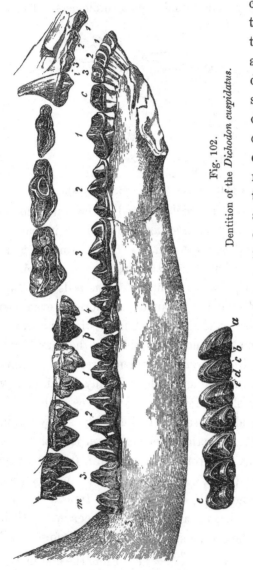

Fig. 102.

Dentition of the *Dichodon cuspidatus.*

(p 3) into an inner posterior cusp. The fourth premolar (p 4)

has a thicker and shorter crown with two pairs of cusps.
The upper true molars (m 1, 2, 3) have the two pairs of cusps
sharp and pointed, with a series of five low accessory points
developed from the outer part of the cingulum.   The lower
molars (m 1, 2, 3) have as complex crowns as the upper ones,
but with the accessory basal points (a, b, c, e) developed from
the inner, instead of the outer side of the crown, and with the
convex sides of the chief cusps turned in the opposite direction to those above.   At the upper part of
fig. 102 the outer side of the true molars, of
the last premolar, of the canine, and of the
incisors, is shown, together with the grinding
surface of the three anterior premolars in the
upper jaw.   Below these the inner surface
of the entire series of the lower teeth is shown, together with

Fig. 103.
Upper molar of
Dichodon.

the grinding surface of the three true molars, the last of
which (m 3) here supports a third pair of lobes ($e.$)   As
compared with the anoplotherian molar (fig. 100), the outer
lobes (a, b) of that of the *Dichodon* (fig. 103) are thicker and
sharper ; the inner ones (c, d)—especially the latter—are
developed to an equality with the outer ones, and more
distinctly separated from them.   The valley (m) extends
across the whole breadth of the tooth, and is crossed at right
angles by the fore-and-aft doubly-curved valley (g and i).
The extinct species showing the above characters, and on
which the genus was founded,[*] was nearly the size of a fallow-
deer : it is called *Dichodon cuspidatus,* in reference to the number of sharp points on the unworn molars.   The dentition
indicates that its food may have been of a peculiar character,
perhaps not exclusively of a vegetable nature.

In the same upper eocene formation of Hampshire have
been found instructive examples of some smaller members of
the extinct anoplotherioid family.

* Quarterly Journal of the Geological Society, tom. iv., 1847, p. 36, pl. 4.

Genus DICHOBUNE.—The genus *Dichobune* (from διχα, *bipar-titus ;* βουνὸς, *collis*) was proposed by Cuvier, in the second edition of his *Ossemens Fossiles,* 4to, tom. iii., 1822, p. 64, for the *Anoplotherium minus* of the original Memoir in the *Annales du Muséum,* tom. iii., 1803, and for the *A. leporinum* of the 4to edition, 1822, tom. i., pl. 2, fig. 3 ; and tom. iii., pp. 70 and 251. It is closely allied to the anoplotherioid genus *Xipho-don ;* the dental formula is the same, only there is a slight interval between the canine and the first premolar in both jaws ; the first three premolars are subcompressed, subtren-chant, but less elongated from behind forwards than in *Xipho-don.* Besides the two normally-developed and functional digits on each foot, there may be one, sometimes two, small supplemental digits.

The best illustration of the structure of the upper true molars is afforded by the figure of one of these teeth in the *Proceedings of the Geological Society,* May 20, 1846, published in the *Quarterly Journal,* vol. ii., p. 420. " The anoplotherian character of the tooth is shown by the large size of the lobe (*p, x,* fig. 1), and the subgeneric peculiarity by the continua-tion of its dentinal base with that of the inner and anterior lobe (*id*), at the early stage of attrition presented by the crown of the tooth in question. In the large and typical *Anoplo-theria,* the lobe (*p*) preserves its insular form and uninterrupted contour of enamel until the crown is much more worn down than in the present tooth (fig. 1). In this respect, as in the modifications of the lower molar teeth, the genus *Dichobune* shows its closer affinity to the true Ruminants ; but the little fold of enamel dividing the lobe *id* from *p* distinguishes the upper molar tooth in question from that of any Ruminant." (P. 421.)

A new and interesting species of this genus, called *Dicho-bune ovina,* has been founded upon an almost entire lower jaw with the permanent dental series, wanting only the four middle

incisors, which now forms part of the palæontological collection in the British Museum. The dental formula, as shown
by the mandibular teeth, and by the evidence on their crowns
of the presence of the teeth of the upper jaw, is the typical one
in diphyodont Mammalia, viz.—$i \frac{3-3}{3-3}$, $c \frac{1-1}{1-1}$, $p \frac{4-4}{4-4}$, $m \frac{3-3}{3-3} = 44$.
The canine, with a crown like that of the first premolar, and
not longer, is separated from it by an interval of half the
breadth of the crown, and by a narrower interval from the
outer incisor. The first premolar is divided by an interval of
scarce a line's breadth from the second. The rest of the molar
series are in contact. The total length of the lower jaw is
5 inches 11 lines $(0^{m\cdot}148)$; that of the molar series is 2 inches
11 lines $(0^{m\cdot}075)$; that of the three true molars is 1 inch $4\frac{1}{4}$
lines $(0^{m\cdot}035)$. The near equality in height of the crowns of
all the teeth, and their general character, show that the animal
belonged to the anoplotherioid family. The dentition of the
present species differs from that of *Dichodon* in the absence
of the accessory cusps on the inner side of the base of the true
molars ; and both from *Dichodon cuspidatus* and *Xiphodon
gracilis* in the minor antero-posterior extent of the premolars ;
it corresponds with *Dichobune* (as represented by the *D.
leporina*, Cuvier) in the proportions of the premolars and in
the separation of the canine from the adjoining teeth : to
this genus, therefore, the fossil is referable, provisionally,
in the absence of knowledge of the molars of the upper
jaw, which are the most characteristic : and the writer has
proposed to call the species, from the size of the animal represented by the fossil, *Dichobune ovina.* It is from Hampshire
eocene.

Genus XIPHODON.—The genus *Xiphodon* was indicated, and
its name proposed, by Cuvier, for a small and delicate, long
and slender-limbed, anoplotherian animal, which, in his first
Memoir (*Annales du Muséum*, tom. iii., p. 55, 1803), he had
called *Anoplotherium medium ;* but he altered the name, in

the second 4to edition of the *Ossemens Fossiles* (tom. iii., pp. 69 and 251, 1822), to that of *Anoplotherium gracile*.

The distinction indicated by Cuvier is now accepted by palæontologists as a generic one, and a second species (*Xiphodon Geylensis*) has been added by M. Gervais (*Paléontographie Française*, 4to, 1845, p. 90) to the type-species, *Xiphodon gracilis,* of which he figures an instructive portion of the dental series of both jaws, obtained from the lignites of Débruge near Apt. The dental formula of *Xiphodon* is the typical one, viz.—$i \frac{3-3}{3-3}$, $c \frac{1-1}{1-1}$, $p \frac{4-4}{4-4}$, $m \frac{3-3}{3-3} = 44$.

The teeth are arranged in a continuous series in both jaws. The canines and first three premolars have the crowns more extended antero-posteriorly, lower, thinner, transversely, and more trenchant, than in the type *Anoplotheria* (whence the name *Xiphodon*, or sword-tooth). The feet are didactyle, with metacarpals and metatarsals distinct. The tail is short. The lower true molars have two pairs of crescentic lobes with the convexity turned outwards. It was nearly allied to *Dichodon*.

Genus MICROTHERIUM.—Entire crania of *Microtherium,* from the lacustrine calcareous marls of the Puy-de-Dôme, are in the British Museum, and these show that the hinder division of the upper true molars was complicated by the additional (third) cusp.

With regard to *Microtherium*, the unusually perfect fossil skulls of that small Herbivore, which did not exceed in size the delicate chevrotains of Java and other Indo-Archipelagic islands—*e. g. Tragulus kanchil*—are of importance in regard to the question of their alleged affinity to the *Ruminantia*, on account of the demonstration they give of the persistent and functional upper incisor teeth. The little eocene even-toed Herbivores, like the larger Anoplotherioids, thus departed from the characters of the true Ruminants of the present day, in the same degree in which they adhered to the more general type of the artiodactyles. Had M. de Blainville, who believed them to be Ruminants, possessed no other evidence of the *Microtherium* than of the *Dichobune murina* and *Dichobune obliqua*, Cuv., he would have had the same grounds for referring the *Microtheria*, as the *Dichobunes*, to the genus *Tragulus* or *Moschus* (les Chevrotains); but the entire dentition of the upper jaw of the species *Anoplotherium murinum* and *A. obliquum*, referred by Cuvier to his genus *Dichobune*, must be known before the existence of Ruminants in the upper eocene gypsum of Paris can be inferred.

No doubt the affinity of these small Anoplotherioids to the Chevrotains was very close. Let the formative force be transferred from the small upper incisors to the contiguous canines, and the transition would be effected. We know that the ruminant stomach of the species of *Tragulus* is simplified by the suppression of the psalterium or third bag. The stomach of the small Anoplotherioids, whilst preserving a certain degree of complexity, might have been somewhat more simplified. The certain information which the gradations of dentition displayed by the above-cited extinct species impart, testifies to the artificial character of the order *Ruminantia* of the modern systems, and to the natural character of that wider group of even-toed hoofed animals for which has been proposed the term ARTIODACTYLA.*

Genus HYÆNODON, Laiz.—With the delicate and beautiful Herbivora of the upper eocene and lower miocene periods, there coexisted carnivorous quadrupeds, which, to judge by the character of their flesh-cutting teeth (carnassials), were more fell and deadly in their destructive task than modern wolves or tigers. Of these old extinct Carnivora a species of the remarkable genus *Hyænodon*, of about the size of a leopard, has left its remains in the upper eocene of Hordwell, Hampshire. Fig. 104 shows the dentition of the under jaw of another species of the same genus from miocene beds at

* Quarterly Journal of the Geological Society, vol. iv., 1847.

Débruge and Alais, France. The carnassial teeth (m, 1, 2, 3), instead of being one in number in each ramus of the jaw, as in modern Felines, were three in number, equally adapted by their trenchant shape, to work like scissor-blades on the teeth

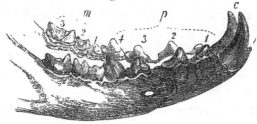

of the upper jaw, in the act of cutting flesh. After the small incisors came a pair of large piercing and prehensile canines (c), fol-

Fig. 104.

Dentition, lower jaw, of *Hyænodon*.

lowed by four compressed pointed and trenchant premolars (p, 1, 2, 3, 4) in each side of the jaw ; the whole of this carnivorous dentition conforming to the diphyodont type :—

$$i\frac{3\text{-}3}{3\text{-}3}, \ c\frac{1\text{-}1}{1\text{-}1}, \ p\frac{4\text{-}4}{4\text{-}4}, \ m\frac{3\text{-}3}{3\text{-}3}=44.$$

Genus AMPHICYON.—With the foregoing predecessor of the digitigrade Carnivora was associated a forerunner of the

plantigrade family, viz., a large extinct species having the molars tuberculated, after the pattern of those of the bears ; but retaining, like *Hyænodon*, the per-

Fig. 105.

Dentition, upper jaw, of *Amphicyon*.

fect type of diphyodont dentition. Fig. 105 shows the teeth of one side of the upper jaw of the *Amphicyon giganteus*. The first and second molars (m, 1 and 2) have each two tubercles on the outer side and one on the inner side ; the last tubercular molar (m, 3) is of very small size. Fossil remains of *Amphi-*

cyon have been found principally in the miocene deposits at Sansans, south of France. Those of a smaller species from the miocene at Eppelsheim, have been referred to the wolverine genus, as *Gulo diaphorus*, Kaup.

The proofs of the abundant mammalian inhabitants of the eocene continent were first obtained by Cuvier from the fossilized remains in the deposits that fill the enormous Parisian excavation of the chalk. But the forms which that great anatomist restored were all new and strange, specifically, and for the most part generically, distinct from all known existing quadrupeds. By these restorations the naturalist was first made acquainted with the aquatic cloven-hoofed Anoplothere, and with its light and graceful congeners, the Dichobunes and Xiphodon, with the great Palæotheres, which may be likened to hornless rhinoceroses, with the more tapiroid Lophiodon, with the large peccari-like Chœropotamus, and with about a score of other genera and species of placental Mammalia.

Almost the sole exception to the generic distinction of these eocene forms from modern ones was yielded by the opossum of Montmartre (*Didelphis Gypsorum*, fig. 106) ; and what made this discovery the more remarkable was the fact that all the known existing species of that marsupial genus are now confined to America, and the greater part to the southern division of that continent. An opossum appears to have been associated with the Hyracotherium in the eocene sand of Suffolk ; where likewise, a porcine beast with tusks like ordinary canines, and some remains of a monkey (*Eopithecus*), have been found. With respect to the *Didelphis Gypsorum*, its generic relations were deduced from characters of the lower jaw and teeth ; but these were associated with other parts of the skeleton in the same block of stone. When Cuvier expressed his convictions from the teeth and other parts first exposed and examined, his scientific associates were incredulous. He invited them, therefore, to witness a crucial test. The outline of the back part

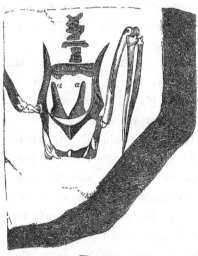

of the pelvis was exposed,
the fore part buried in the
matrix. By his delicate use
of the graving-tool, Cuvier
brought to light the fore-
part of the pelvis with the
two marsupial bones (fig.
106, *a*, *a*) in their natural
position. He thus demon-
strated that there had been
buried in the soft fresh-
water deposits, hardened in
after ages into the building-
stone of Paris, an animal

Fig. 106.

Pelvis and marsupial bones of *Didelphis*
Gypsorum (Eocene, Paris).

whose genus at the present
day is peculiar to America.

It is not uninteresting to remark that the Peccari, the nearest
existing ally to the old Chœropotamus, is, like the opossum,
now peculiar to America ; and that two species of tapir, the
nearest living allies to the Lophiodon and Palæothere, exist in
South America.

The marine deposits of the miocene epoch show the remains

of extinct genera of dolphins
(*Ziphius* and *Dioplodon*) and of
whales (*Balænodon*). Petrified
cetaceous teeth and ear-bones,
called "cetotolites" (fig. 107)

Fig. 107.

Cetotolite or fossil ear-bone of *Balæno-*
don gibbosus (Red Crag, Suffolk).

have been washed out of previous strata into the red crag

of Suffolk. These fossils belong to species distinct from any
known existing Cetacea, and which, probably, like some contemporary quadrupeds, retained fully-developed characters
which are embryonic and transitory in existing cognate

Mammals. The teeth of these Cetacea were determined in 1840, the ear-bones in 1843. The vast numbers of these fossils, and the proportion of phosphate of lime in them, led Professor Henslow* to call the attention of agricultural chemists to the red crag as a deposit of valuable manure. Since that period it has yielded a large supply, worth many thousand pounds annually, of the superphosphates. The red crag is found in patches from Walton-on-Naze, Essex, to Aldbro', Suffolk, extending from the shore to 5 or 15 miles and more inland. It averages in thickness 10 feet, but is in some places 40 feet. Broken-up septarian nodules form a rude flooring to the crag, left by the washing off of the London clay, and called "rough stone." The phosphatic fossils, or "cops" as they are now locally termed, occur in greatest abundance immediately above the "rough-stone." Thousands of cubic acres of earlier strata must have been broken up to furnish the cetacean nodules of the "red crag." This is a striking instance of the profitable results of a seemingly most unpromising discovery in pure science,—the determination of what in 1840 was regarded as a rare, unique, and most problematical British fossil.†

Our knowledge of the progression of mammalian life during the miocene period is derived chiefly from continental fossils. These teach us that one or two of the generic forms most frequent in the older tertiary strata still lingered on the earth, but that the rest of the eocene Mammalia had been superseded by new forms, some of which present characters intermediate between those of eocene and those of pliocene genera. The *Dinotherium* and narrow-toothed *Mastodon*, for example, diminish the interval between the *Lophiodon* and the elephant; the *Anthracotherium* and *Hippohyus*, that between *Chœropotamus* and *Hippopotamus*; the *Acerotherium* was a

* Proceedings, and Quart. Jour. Geol. Soc., 1843.

† History of British Fossil Mammals, 8vo, p. 536.

link connecting *Palæotherium* with *Rhinoceros;* the Hippo-
therium linked on *Paloplotherium* with *Equus.*

One of the most extraordinary of the extinct forms of the
cetaceous order has been restored from fossil remains discovered in formations of the miocene age in Europe and North
America. The teeth of this carnivorous whale, for which the
generic name *Zeuglodon* seems now to be generally accepted,
were first described and figured by the mediæval palæontologist Scilla, in his treatise entitled *De Corporibus Marinis*
(4to, 1747, tab. xii., fig. 1), and have since given rise to
various interpretations. The originals were obtained from
the miocene strata at Malta, and are now preserved in the
Woodwardian museum at Cambridge.

The remains of a gigantic species of the same genus, discovered by Dr. Harlan in miocene formations of Arkansas,
Mississippi, were described and figured by him as those of a
reptile, under the name of *Basilosaurus.** Teeth of a smaller

Fig. 108.
Deciduous and permanent teeth of the *Zeuglodon.*

species, discovered by M.
Grateloup, in miocene
beds of the Gironde and
Herault, were ascribed
by him also to a reptile,
under the name of *Squa-
lodon.*† In 1839 Dr. Har-
lan brought over his spe-
cimens of *Basilosaurus*
to London, and submit-
ted them to the writer's
inspection, by whom

they were determined to be mammalian and cetaceous. The
entire skeleton has since been obtained from miocene deposits
in Alabama, revealing a length of body of about 70 feet. The

* Medical and Physical Researches, p. 333.
† Act. Soc. Linn. de Bordeaux, 1840, p. 201.

skull is very long and narrow ; the nostril single, with an upward aspect, above and near the orbits. The jaws are armed with teeth of two kinds, set wide apart ; the anterior teeth have subcompressed, conical, slightly-recurved, sharp-pointed crowns, and are implanted by a single root ; the posterior teeth are larger, with more compressed and longitudinally extended crowns (fig. 108), conical, but with a more obtuse point, and with both front and hind borders strongly notched or serrated. The crown is contracted from side to side in the middle of its base, so as to give its transverse section an hour-glass form (fig. 109), and the opposite wide longitudinal grooves which produce this form become deeper as the crown approaches the socket, where they meet and divide the root into two fangs. The name *Zeu-*

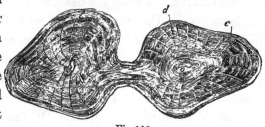

Fig 109.

Transverse section of a tooth of the *Zeuglodon*. Nat. size.

glodon (yoke-tooth) refers to this structure. The mode of succession of the teeth in this genus conforms to the general mammalian type more than does that of any of the existing carnivorous Cetaceans. In the figure given by Dr. Carus* of a portion of the jaw of *Zeuglodon cetoïdes,* a deciduous molar (fig. 108, *a*) is about to be displaced and succeeded, vertically, by a second larger molar. This mode of succession is not known in the *Platanista* or *Inia,* which among existing true Cetacea present teeth most like those of *Zeuglodon ;* but it is a mode of succession and displacement affecting certain teeth in the herbivorous Cetacea, or *Sirenia ;* and we thus seem to have in the *Zeuglodon* another of those numerous instances of a more generalized character of organization in older tertiary Mam-

* Nova Acta Cæs. Leop. Carol., vol. xxii., tab. xxxix. B, fig. 2, p. 340.

malia. In systematic characters, *Zeuglodon* typifies a distinct family or group, intermediate between *Cetacea* proper and *Sirenia*.

Of the latter family or order, however, represented at the present day by the Dugongs, Manatees, and Stellerians or Arctic Manatees (if the species still survives), there were abundant and more widely distributed representatives during the miocene period, having, upon the whole, the nearest affinity with the existing African Manatee (*Manatus Senegalensis*), but with associated characters of the Dugong (*Halicore*). There were, *e. g.*, two incisive tusks in the upper jaw, and four or five small incisors along the deflected part of each ramus of the lower jaw. The upper molars, with three roots, were thickly enamelled, like those of the Manatee, but with a pattern of grinding surface which led Cuvier to attribute detached specimens to a small species of *Hippopotamus*. The lower molars had two roots. All the bones have the dense or solid structure of those of the *Sirenia*. On the remains of this remarkable amphibious Mammal, discovered by Kaup in 1838, in the miocene beds at Eppelsheim, he founded the genus *Halitherium*. Other remains have been discovered in Piedmont, Asté, and many parts of France, from the " calcaire grossier" of the Gironde, containing Lophiodont fossils, up to the pliocene near Montpellier ; at which period the *Halitherium* seems to have become extinct.

Genus MACROTHERIUM, Lartet.—The edentate order, which is so abundantly and variously represented in South America, which has its Orycteropes and Pangolins in Africa, and its Manises in tropical Asia, has no living representative in Europe. Perhaps the most unexpected form of · Mammal to be revealed by fossil remains from European tertiary deposits, after a Marsupial, was a member of the edentate order. Cuvier, by whom the evidence of this extinct animal was first made known, prefaces his description of the single mutilated phalangeal bone, on which that evidence was founded, by the remark, that "nothing proves better the importance of the laws of comparative osteology than all the consequences which one may legitimately draw from a single fragment." One willingly admits the proof so afforded of the former existence of animals now unknown ; but one may demur to the conclusion that their extinction was due to some sudden catastrophe.

The single mutilated ungual phalanx on which Cuvier based his conclusions in regard to the species in question was discovered, associated with remains of *Rhinoceros*, *Mastodon*, *Dinotherium*, and *Tapir*, in a formation near Eppelsheim, Hesse-Darmstadt, which is now determined to belong to the miocene division of the tertiary series. This phalanx shows two distinctive characters of the edentate order :—1*st*, Its posterior surface for articulation with the antepenultimate phalanx is a double pulley, hollowed out on each side, with a salient crest between, constituting the firm kind of ginglymoid joint peculiar to certain *Edentata ; 2d*, The concave arch formed by that pulley curves furthest backward at its upper part, which would prevent the claw being retracted upward, as in the cat tribe, and constrain the flexion downward— " ainsi c'est necessairement un onguéal d'edenté."* To the foregoing characters are joined two others which Cuvier believed to determine " as necessarily" the genus. The species of *Myrmecophaga* have on the upper part of the pointed end of the claw-phalanx a groove, indicative of a disposition to bifurcate ; in the species of *Manis* the bifurcation is complete, the cleft extending as far as the middle of the claw-bone : so likewise in this fossil. The Pangolins (*Manis*) have not those bony sheaths which, in the sloths, some ant-eaters and armadillos, rise from the base and cover the root of the claw ; there was a like absence of any claw-sheath in the fossil. Thus the fossil claw-bone has no homologue in existing nature save those

* Ossemens Fossiles, 4to, t. v., pt. i., p. 193.

of the Manis ; and, " according to all the laws of co-existence, it is impossible to doubt that the most marked relations of the animal that bore it should have been with that genus of quadrupeds."* But what must have been its size ? The phalanx was not one of the largest on the foot—for it had not those slight raised borders which one sees in the large claw-bones of the Pangolins. This question, which Cuvier answered by the proportions of the short-tailed Manis, at 24 French feet, has had a more reasonable reply given to it by certain other bones of the skeleton subsequently discovered in the miocene tertiaries of France. These discoveries have likewise rectified and moderated the absolute application of the correlative law to the necessary determination of the genus as well as of the order. The relations of the double-jointed and cleft phalanx to the Edentata is beautifully confirmed ; but the additional fossils, and especially some evidences of teeth, have shown that it belonged to a peculiar and now extinct genus intermediate between the *Manis* and the *Orycteropus*. And these relations are deeply interesting, on account of the geographical position of both those edentate genera, on tracts of land, viz., which are now most contiguous to the continent containing the remains of the extinct osculant genus.

The locality in France is near the village of Sansan, near Auch, department of Gers, Haute-Pyrenées. The formation is a lacustrine deposit of the miocene period.

Portions of two molar teeth have been found, 1 inch 8 lines in greatest transverse diameter ; the tooth preserving the same size and shape through the whole length of the portion—viz., 1½ inch. They resemble in shape those of the Orycterope, but are less regular and have not the same tubular tissue. Their microscopic texture appears not to have been analyzed ; it would be important to determine whether it resembled that of the teeth of the sloths or armadillos. The

---

* Ossemens Fossiles, 4to, t. v., pt. i., p. 194.

humerus differs from that of the ant-eaters and armadillos by its greater length in proportion to its breadth, and by the peculiar flattening from before backwards of its lower half, and especially at the condyles, above which it is expanded transversely by both external and internal supra-condyloid ridges. It is not perforated above the inner condyle, as the same bone is in both the *Manis* and *Orycteropus*. In the degree to which it departs from the type of the ant-eaters it approaches that of the Megatherioids and sloths—viz., in its relative length, flattening at the distal end, and the imperforate character of that end. The radius also presents a sloth-like character in its greater proportionate length, which exceeds that of the humerus ; and in the compression of its lower slightly-expanded end. In both the Pangolin and Orycterope, the radius is shorter than the humerus. The ulna differs likewise from both that of the Pangolin and Orycterope, and still more from that in the Armadillos by the much smaller development of the olecranon, whereby, again, it more resembles that of the sloths. The femur is relatively longer and more slender than that of the terrestrial and fossorial Edentata ; it has not the third trochanter which characterizes it in the Orycterope, nor so marked a development of the great and small trochanters as in the Pangolin. In the flattened form of the shaft of the femur, and the position of the rotular surface near one side of the distal end, it resembles the femur of the Megatherium and Mylodon. It is shorter than the humerus ; whereas, in both the Pangolin and Orycterope the femur is longer : in this respect the femur of the Macrothere resembles that of the sloths. The great width of the popliteal space dividing the condyles is an edentate and more especially a megatherioid character. The internal condyle is much broader than the external one, as it likewise is in the Megatherioids ; it is certainly with the femur of the latter family of the Edentata, rather than with that of the

Proboscidians or Pachyderms, that one should compare the femur of the Macrothere: it is not so long or so slender relatively as in the sloths. The tibia is much shorter than the femur, and in the expansion of its proximal end and its relative length to the femur it resembles that of the Megatheroids more than that in the Pangolin or Orycterope; it was not anchylosed to the tibia as in the Armadillos, Glyptodons, and Megatherium, but remained a distinct bone, as in the Mylodon and sloths.

Genus PLIOPITHECUS, Gerv.—In the same miocene deposits of the south of France as those which contained the *Macrotherium*, fossil remains of two kinds of Quadrumana, resembling a small and large species of *Hylobates*, have been discovered. The smaller of these extinct apes, called *Pliopithecus antiquus* by Gervais, is based upon the lower jaw and dentition. The teeth occupy an extent of 1½ inch; the two incisors are narrower, the canine less, and the last molar is larger than in the siamang (*H. syndactyla*). As in this species the first premolar is uni-cuspid, and the hind talon of the second is more produced than in the chimpanzee and gorilla, and to the degree in which the fore-and-aft diameter of the tooth exceeds the transverse one, it departs farther from the human type; in the degree of the development of the talon or third lobe of the last lower molar, the *Pliopithecus* resembles the tailed monkeys (*Semnopithecus* and *Innus*).

Genus DRYOPITHECUS, Lart.—In the larger miocene ape (*Dryopithecus Fontani*, Lart.) the canine is relatively larger than in the *Hylobates*, and the incisors, to judge by their alveoli, are relatively narrower than in the chimpanzee and human subject. The first premolar has the outer cusp pointed, and raised to double the height of that of the second premolar, and its inner lobe is more rudimental than in the chimpanzee,*

---

* Compare Comptes Rendus de l'Acad. des Sciences, tom. xliii. (July 28, 1856, plate, fig. 7), with Trans. Zool. Soc., vol. iv., plate 32, fig. 3, p. 3.

and departs proportionally from the human type. The posterior lobe or talon of the second premolar is more developed, and the fore-and-aft extent of the tooth greater, than in the chimpanzee, thereby more resembling the second premolar of the siamang, and less resembling that of the human subject. The last (third) molar is undeveloped in the fossil jaw of the *Dryopithecus*, and its amount of departure from the human type, and approach to that of *Innus*, cannot be determined. The canine is more vertical in position than in *Troglodytes* or *Pithecus*, but this character is offered by some of the small South American apes, and cannot be cited as a mark of real affinity. From the portion of humerus associated with the jaw of *Dryopithecus*, the arm would seem to have been proportionally longer and more slender than in the chimpanzee and gorilla, with a cylindrical shaft, more like that in the long-armed apes (*Hylobates*), and less like the arm of the human subject.

The characters of the nasal bones, orbits, mastoid processes, relative length of upper limb to trunk, relative length of arm to fore arm, relative length and size of thumb, relative length of lower limb ; and, above all, the size of the hallux and shape of the astragalus and calcaneum, must be known before any opinion can be trusted as to the proximity of *Dryopithecus* to the human subject.

Genus MESOPITHECUS, Wagn.—In tertiary formations of Greece, at the base of Pentelicon, remains of a Quadrumane have been found, which Professor Wagner* regards as transitional between *Hylobates* and *Semnopithecus :* the third lobe of the last molar is, however, as well developed as in the latter genus.

In the pliocene deposits of Montpellier remains of a monkey occur, referred by Christol to a *Cercopithecus ;* and in pliocene brick-earth in Essex the writer has determined part of the fossil jaw and teeth of a *Macacus*.

* Abhanglungen der k. bayer Akademie, bd. ii., 1854, Munchen.

Genus DINOTHERIUM, Cuv. and Kp.—This name was given by Kaup, after the discovery of the singular shape and armature of the lower jaw, to the huge bilophodont Mammal, first made known by Cuvier under the name of "Tapir gigantesque." The length of the skull, from *f* to *d*, in fig. 110, is 3 feet 8 inches.

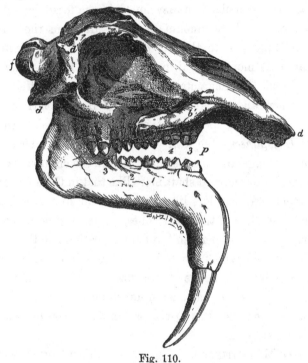

Fig. 110.
Skull of *Dinotherium giganteum* (Miocene, Eppelsheim).

The teeth in this skull, in addition to the two large deflected tusks of the lower jaw, are five in number on each side of both jaws. A study of the changes of dentition in fossils of young *Dinotheria* show that the first two teeth answer to the third and fourth premolars, as signified by the symbols *p*, 3 and 4; and that the rest are true molars (*m*, 1, 2, 3). Of these, the first tooth (*p*, 3), is rather trenchant than triturant; the third tooth (1) has three transverse ridges. The other grinders have two transverse ridges. This "bilophodont" or two-ridged type is shown by the molars of the *Tapir*, (fig. 111), *Lophiodon*, *Megatherium*, *Diprotodon*, *Nototherium*, *Kangaroo*, and *Manatee*. In the general shape of the skull and aspect of the nostrils *Dinotherium* most resembles *Manatus*. Bones of limbs have not

Fig. 111.
Molar series, lower jaw, Tapir.

yet been found so associated with teeth as to determine the ordinal affinities of *Dinotherium*. Yet cranial and dental evidences of the genus have been discovered in miocene deposits of Germany, France, Switzerland, and Perim Island, Gulf of Cambay.

Genus MASTODON, Cuv.—The earliest appearance of this genus of proboscidian or elephantoid Mammal is in tertiary strata of miocene age, and by a species in which the fore part of the lower jaw was produced into a pair of deep sockets containing tusks ; but these are only slightly deflected from the line of the grinding teeth (fig. 112, *C*). This species of *Mastodon*, discovered in the miocene of Eppelsheim, was called *longirostris* by Kaup ; but he afterwards recognized it as the same with a species which had been previously called *Mastodon arvernensis* (Croizet and Jobert).* Both belong to that section of *Mastodon* in which the first and second true molars have each four transverse ridges,† and for which Dr. Falconer proposes the name *Tetralophodon*. In the newer tertiary deposits of North America remains of a Mastodon (*M. Ohioticus*) have been discovered, in which the transverse ridges of the grinders are in shape more like those of the Dinothere than in any other Mastodon ; the first and second, moreover, are bilophodont, the third trilophodont ; but this is followed by two three-ridged molars and a last larger molar with four

* Beitrage zur Naeheren Kenntniss der Urweltlichen Saeugethiere, 4to, 1857, p. 19. The name *angustidens* was first applied by Cuvier to teeth of this type or species.

† First demonstrated by Kaup, Ossemens Fossiles de Darmstadt, 4to, 1835.

or five ridges.*　For the Mastodons with penultimate and
antepenultimate grinders with three ridges, Dr. Falconer pro-

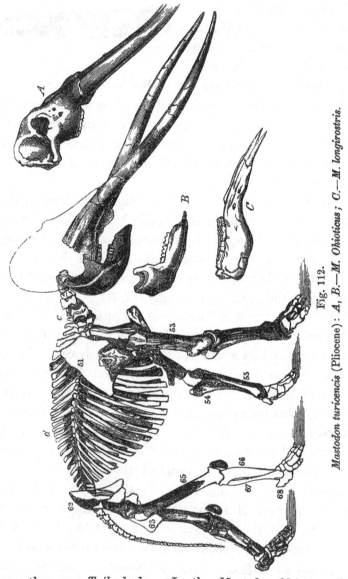

Fig. 112.

Mastodon turicensis (Pliocene) : A, B.—M. Ohioticus ; C.—M. longirostris.

poses the name *Trilophodon*.　In the *Mastodon Ohioticus* the
lower jaw has two tusks in the young of both sexes ; these

* Owen's " Odontography," 4to, 1845, p. 617, pl. 144.

are soon shed in the female, but one of them is retained by the male (fig. 112, *B*). The upper tusks are long and retained in both sexes (fig. 112, *A*).*

An almost entire skeleton of a Mastodon (*M. turicencis*) has been discovered in the pliocene deposits of Asté, Piedmont, and has been described and figured by Professor Sismonda,† from whose beautiful Memoir fig. 112 is taken. The total length from the tail to the end of the tusks is 17 feet. The teeth have the same narrow shape and multi-mammillate structure as in *M. arvernensis*, but in the numerical character of transverse divisions of the crown this species agrees with *M. Ohioticus.*

The Mastodons were elephants with the grinding teeth less complex in structure, and adapted for bruising coarser vegetable substances. The grinding surface of the molars (fig. 113), instead of being cleft into numerous thin plates, was divided into wedge-shaped transverse ridges, and the summits of these were subdivided into smaller cones, more or less resembling the teats of a cow, whence the generic name.‡ A more important modification appeared to distinguish the extinct genus, in respect of the structure of the molar teeth ; the dentine, or principal substance of the crown of the tooth (fig. 113, *d*) is covered by a thick coat of dense and brittle enamel (*e*) ; a thin coat of cement is continued from the

Fig. 113.

Upper molar of Mastodon.

* Owen's "Odontography," p. 618.

† Osteografia di un Mastodonte augustidente, 4to, 1851.

‡ μαστος, a nipple ; οδους, a tooth.

fangs upon the crown of the tooth, but this substance does not fill up the interspaces of the divisions of the crown, as in the elephant's grinder (fig. 118, c). Such at least is the character of the molar teeth of the two species of Mastodon, which Cuvier has termed *Mastodon giganteus* and *Mastodon angustidens* (fig. 113). Fossil remains of proboscidians have subsequently been found principally in the tertiary deposits of tropical Asia, in which the number and depth of the clefts of the crown of the molar teeth, and the thickness of the intervening cement, are so much increased as to establish transitional characters between the lamello-tuberculate teeth of the elephants and the mammilated molars of the typical Mastodons, showing that the characters deducible from the molar teeth are rather the distinguishing marks of species than of genera in the present family of mammalian quadrupeds.

The dentition of this family may be expressed by the formula—

$$d \ i \ \frac{1.1}{} ; \ i \ \frac{1.1}{1.1} ; \ c \ \frac{0.0}{0.0} ; \ d \ m \ \frac{3.3}{3.3} ; \ p \ \frac{1.1}{1.1} ; \ m \ \frac{3.3}{3.3} = 34 ;$$

that is to say, in the Proboscidians in which the dentition most nearly approached to the typical one, thirty-four teeth were developed, as follows :—in the upper jaw, two deciduous

Fig. 114.
Deciduous dentition, young *Mastodon longirostris.*

incisors, followed by two permanent incisors developed as tusks ; six deciduous molars (three on each side, d 2, 3, 4, fig. 114); two premolars (one on each side, p 3, fig. 114), and six true molars (three on each side, m 1, 2, 3, figs. 114 and 115) ;—in the lower jaw, two incisors as tusks (uncertain whether preceded by deciduous tusks), deciduous molars, premolars, and molars, as in the upper jaw.

The elephantoid animal (*Mastodon longirostris*, Kaup ; *Mastodon angustidens*, in part, Cuvier) which exhibited the above

instructive dentition of the proboscidian family, once roamed
over the part of the earth now forming England, France, Italy,
and Germany. The first steps in our knowledge of its dentition
were made by Cuvier, who called it the narrow-toothed Mastodon " Mastodon à dents étroites," or *Mastodon angustidens*. This
name was suggested by the less breadth of the grinding surface
of the teeth as compared with those of a previously described
species of Mastodon from North America, called the *Mast.*

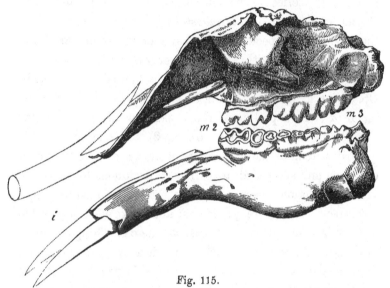

Fig. 115.

Dentition of old *Mastodon longirostris.*

giganteus or *M. Ohioticus*. Cuvier describes and figures a last
molar, upper jaw, from Trévoux, consisting, as in the specimen
from Norfolk Crag (fig. 113), and as in that from Eppelsheim
(fig. 115, *m* 3), of five transverse ridges, with a front and back
talon or subsidiary ridge. The latter is the largest, and subdivided into teat-shaped tubercles, so as almost to merit the
name of a sixth division of the tooth. The principal ridges are
divided into two chief or primary tubercles, with secondary
tubercles in the interspace ; the chief tubercles are more or

less deeply grooved lengthwise, or cleft at top, so that mastication wore them down to small circles of dentine surrounded
by a thick border of enamel, and further attrition reduced
these to a trilobed or trefoil form.

The last lower molar of the *Mast. angustidens* from La
Rochetta di Tanaro,* exhibits the same five principal transverse ridges and the hinder one, as in the corresponding tooth
in the Eppelsheim Mastodon (fig. 115, *m* 3), and being the
first of the series of narrow mastodontoid teeth to which
Cuvier applied the name *angustidens*, it may be regarded as
the type of that species. The characteristic premolar of the
Mast. angustidens, with a quadrate crown of two ridges, each
cleft into two tubercles (fig. 114, *p* 3), is figured by Cuvier, in
Op. cit. pl. i., fig. 3, and again, *in situ*, with the last deciduous
molar (*d* 4) in a portion of the upper jaw of the *Mastodon
angustidens* from Dax (ib. pl. iii., fig. 2). The nature of this
quadrituberculate tooth as a premolar, *i.e.*, as a tooth which
displaced and succeeded an earlier or deciduous tooth in the
vertical direction, was recognized by Cuvier. "Je crois encore
qu'on peut en conclure que la dent antérieure était susceptible
de remplacement de haut ou bas, comme dans l'hippopotame :
ma raison est, que cette petite dent de Dax n'est pas encore
usée et qu'il faut qu'elle soit venue après la grande qui l'est."†
Dr. Kaup has described and figured the same premolar in the
upper jaw of a younger individual of the *Mastodon angustidens*
(his *longirostris*), in which it is still concealed in its formative
cavity above the three-lobed deciduous tooth which it has
replaced in Cuvier's specimen. The tooth next behind, which
is the homologue of the last milk-molar (*d* 4, fig. 114) of the
typical series, consists, in both the Dax and Eppelsheim specimens, of three principal ridges and a posterior bituberculate
talon ; this accessory portion being more developed in the

---

* Cuvier, Op. cit. Divers Mastodontes, pl. iv., fig. 1, top view, fig. 2, side view.
† Ossemens Fossiles, 4to, 1821, tom. i., p. 256.

Eppelsheim specimen. Whether such difference be valid for a specific distinction may be doubted ; but that Cuvier assigned the name *angustidens* to a *Mastodon* with narrower molars than the *M. giganteus*, which had a quadri-tuberculate premolar, a penultimate molar of four principal divisions and a talon, and a last molar with five principal divisions and a talon, is certain. To that Mastodon, therefore, which has the same shaped and sized ultimate and penultimate true molars and premolar, the same name is here assigned.

The antepenultimate molar (fig. 114, *m* 1) consists of four ridges and a talon.

Three molars are developed anterior to this tooth ; the first (fig. 114, *d* 2) is the smallest, with a subquadrate crown of two transverse ridges. The second molar (ib., *d* 3), of twice the size of the first, has three ridges. The third molar (ib., *d* 4), with an increase of one-third the bulk of the former, has three ridges and a bituberculate talon, which in some specimens might almost be reckoned as a fourth ridge. The two-ridged premolar (ib., *p* 3) above described, takes the place of the second of the above molars, after the first and second are shed. The above definition of the molar series applies to both upper and lower jaws, the cut (fig. 114), and the symbolic letters and numbers, preclude the necessity of verbal description.

From the analogy of the existing elephants, it may be in- ferred that the long tusks (fig. 115, *i*) supported by the pre- maxillaries, were preceded by a pair of small deciduous incisors. There is not such ground for concluding their existence in the mandible ; but this jaw, in the male *Mastodon longirostris* (fig. 115), supported two incisive tusks, shorter and straighter than those above.

In the proboscidian quadrupeds the molar teeth, progres- sively increasing in size, and most of them in complexity, follow each other from before backwards, at longer intervals than in other quadrupeds, and are never simultaneously in place. Not more than three are in use at any period on one side of either jaw ; all the molars, save the penultimate (fig. 115, *m* 2) are shed by the time the crown of the last molar has cut the gum, and the dentition is finally reduced to *m* 3 on each side of both jaws, with commonly the loss of the inferior tusks, as in the old *Mastodon turicencis,* from the tertiary deposits of the Po, described and figured by Sismonda.*

The genus was represented by species ranging, in time, from the miocene to the upper pliocene deposits, and in space, cosmopolitan with tropical and temperate latitudes. The transition from the mastodontal to the elephantine type of dentition is very gradual.

Genus ELEPHAS, L.—The latest form of true elephant which obtained its sustenance in temperate latitudes is that which Blumenbach called *primigenius,* the "Mammoth" of the Siberian collectors of its tusks (fig. 119). Its remains occur chiefly, if not exclusively, in pleistocene deposits, and have

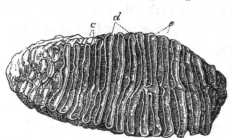

Fig. 116.

Upper grinder of the Mammoth (*Elephas primi-genius*).

even been found in turbary near Holyhead. Its grinders (fig. 116) are broader, and have narrower and more numerous and close-set transverse plates and ridges, than in other elephants. In the existing Indian species, *e. g.,* (fig. 117), the molars are relatively narrower, the plates (*d d*) are less numerous, and their enamelled border (*e e*) is festooned. In the African elephant (fig. 118) the plates are still fewer, are relatively larger, and so expanded at the middle as to present a lozenge shape. The *Elephas priscus,* Gdf., of European

* Osteografia di un Mastodonte Angustidente, 4to, Turin, 1851.

pliocene beds, has molars most like those of the present African species. The tusks of the elephant, like those of the Mastodon,

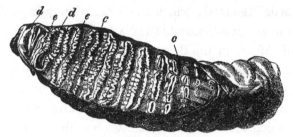

Fig. 117.
Upper molar, Asiatic Elephant.

consist of true ivory, which shows, in transverse fractures or sections, striæ proceeding in the arc of a circle from the

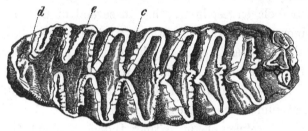

Fig. 118.
Upper molar, African Elephant.

centre to the circumference in opposite directions, and forming, by their decussations, curvilinear lozenges. This character is a valuable one in the determination of fragments of fossil tusks.

The tusks of the extinct *Elephas primigenius*, or mammoth, have a bolder and more extensive curvature than those of the *Elephas indicus;* some have been found which describe a circle, but the curve being oblique, they thus clear the head, and point outward, downward, and backward. The numerous fossil tusks of the Mammoth which have been discovered and recorded, may be ranged under two averages of size—the larger ones at nine feet and a half, the smaller at five feet and a half in length. The writer has elsewhere assigned reasons for the probability of the latter belonging to the female Mammoth, which must accordingly have differed from the existing elephant of India, and have more resembled that of Africa, in the development of her tusks, yet manifesting an intermediate character by their smaller size. Of the tusks which are referable to the male Mammoth, one from the newer tertiary deposits in Essex measured nine feet ten inches along the outer curve, and two feet five inches in circumference at its thickest part; another from Eschscholtz Bay was nine feet two inches in length, and two feet one and a half inches in circumference, and weighed one hundred and sixty pounds. A specimen, dredged up off Dungeness, measured eleven feet in length. In several of the instances of Mammoth's tusks from British strata, the ivory has been so little altered as to be fit for the purposes of manufacture; and the tusks of the Mammoth, which are still better preserved in the frozen drift of Siberia, have long been collected in great numbers as articles of commerce.

In a specimen of the extinct Indian elephant (*Elephas ganesa*, Fr. and C.) preserved in the British Museum, the tusks are ten feet six inches in length, and in consequence of their small amount of curvature, they project eight feet five inches in front of the head. Their apparent disproportion to the size of the skull is truly extraordinary, and exemplifies the maximization of dental development.

The mammoth is more completely known than most other extinct animals by reason of the discovery of an entire specimen preserved in the frozen soil of a cliff at the mouth of the river Lena in Siberia. The skin was clothed with a reddish wool, and with long black hairs. It is now preserved at St. Petersburg, together with the skeleton (fig. 119). This measures, from the fore part of the skull to the end of the mutilated tail, 16 feet 4 inches; the height, to the top of the dorsal spines, is 9 feet 4 inches ; the length of the tusks, along the curve, is 9 feet 6 inches. Parts of the skin of the head, the eye-ball, and of the strong ligament of the nape which helped

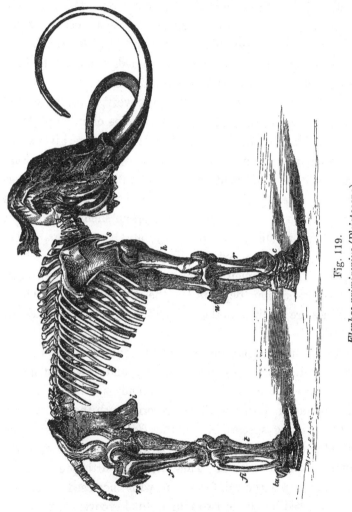

Fig. 119.

Elephas primigenius (Pleistocene).

to sustain the heavy head and teeth, together with the hoofs, remain attached to the skeleton. These huge elephants, adapted by their clothing to endure a cold climate, subsisted on the branches and foliage of the northern pines, birches,

willows, etc.; and during the short summer they probably
migrated northward, like their contemporary the musk-buffalo
which still lingers on, to the 70th degree of N. latitude, retreating during the winter to more temperate quarters.   The
mammoth was preceded in Europe by other species of elephant—*e.g.*, *Elephas priscus*, Goldfuss, and *Elephas meridionalis*,
Nesti, which, during the pliocene period, seem not to have
gone northward beyond temperate latitudes.

Genus HIPPOPOTAMUS, L.—The discovery, in lacustrine and
fluviatile deposits of Europe, of the remains of an amphibious
genus of Mammal now restricted to African rivers, gives scope
for speculating on the nature of the land which, uniting England with the Continent, was excavated by lakes and intersected by rivers, with a somewhat warmer temperature than
at present, to judge by a few S. European shells which occur
in the fresh-water formations—*e. g.*, at Grays, Essex, where
remains of the large extinct *Hippopotamus major* have been
found.   The specimen of the lower jaw (fig. 121) was discovered in similar deposits on the Norfolk coasts.   Other
localities are specified in the writer's "History of British Fossil
Mammals."

The first premolar has a simple subcompressed conical
crown, and a single root ; it rises early, and at some distance
in advance of the second premolar, and is soon shed ; the
other premolars form a continuous series with the true molars
in the existing species, but in the *Hippopotamus major* the
second premolar is in advance of the third by an interval
equal to its own breadth.   This and the fourth premolar retain the simple conical form, but with increased size, and
are impressed by one or two longitudinal grooves on the outer
surface, which, when the crown is much worn, give a lobate
character to the grinding surface.   The true molars are pri-
marily divided into two lobes or cones by a wide transverse
valley, and each lobe is subdivided by a narrow antero-pos-

terior cleft into two half cones, with their flat sides next each other ; the convex side of each half cone is indented by two angular vertical notches, bounding a strong intermediate prominence.   When their summits begin to be abraded, each lobe,

or pair of demicones, presents a double trefoil of enamel on the grinding surface, as shown in fig. 120 ; when attrition has proceeded to the base of the half cones, then the grinding surfaces of each lobe presents a quadrilobate figure.   The crown of the last molar tooth

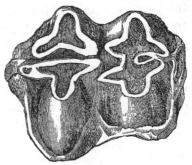

Fig. 120.
Molar tooth, Hippopotamus.

of the lower jaw is lengthened out by a fifth cone, developed behind the two normal pairs of half cones, and smaller in all its dimensions.

The hippopotamus is first met with in pliocene strata. The remains of *H. major* have hitherto been found only in Europe ; they are common along the Mediterranean shore, and do not occur north of the temperate zone.   In Asia this

Fig. 121.
Lower jaw of *Hippopotamus major* (fresh-water Pliocene, Cromer, Norfolk).

form of Pachyderm was represented, perhaps at an earlier period, by the genus *Hexaprotodon*—essentially a hippopotamus, with six incisor teeth, instead of four, in each jaw.

Genus RHINOCEROS, L.—The rhinoceros, like the elephant, was represented in pliocene and pleistocene times, in temperate and northern latitudes of Asia and Europe, by extinct species. One (*Rhinoceros leptorhinus*) associated with the *Hippopotamus major* in fresh-water pliocene deposits ; another (*R. tichorrhinus*) with the mammoth in pleistocene beds and drift. The discovery of the carcase of the tichorrine rhinoceros in frozen soil, recorded by Pallas in his "Voyages dans l'Asie Septentrionale,"* showed the same adaptation of this, at present tropical, form of quadruped to a cold climate, by a twofold covering of wool and hair, as was subsequently demonstrated to be the case with the mammoth. Both the above-named fossil rhinoceroses were two-horned ; but they were preceded, in the pliocene and miocene periods, by species devoid of horns, yet a rhinoceros in all other essentials (*Acerotherium*, Kaup).

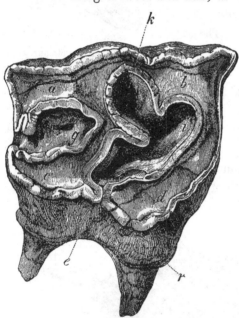

Fig. 122.
Upper molar, Rhinoceros. Nat. size.

The modifications which the upper molars of the rhinoceros present as compared with those of its antetype, the Palæotherium, will be readily understood by comparing fig. 99 with fig. 122, and are as follows :—The concavities (*ff*) on the outer side of the crown, in fig. 99, are almost levelled, and from one of them a

* 4to, 1793, pp. 130-132.

slight convexity projects, in some species of Rhinoceros, giving a gently undulated surface to that side of the tooth. The valley (e) is more expanded at its termination (i), in the Rhinoceros ; and, in some species, it bifurcates and deepens, so that one branch may form an insulated circle of enamel when the crown is worn. The posterior valley (g) is usually deeper and more extended. The ordinary lobes (a, b, c, d) are very similar, and produce, by the confluence of a with c, and of b with d, the two oblique tracts of dentine which are more decidedly established as transverse ridges in the Lophiodont or Tapiroid group. A basal ridge (r) girts more or less completely the inner and the fore and hind parts of the base of the crown. Not fewer than twenty species of extinct rhinoceroses are entered in Palæontological catalogues.

The extinct *Chœropotamus, Anthracotherium, Hyopotamus,* and *Hippohyus,* had the typical dental formula, and this is preserved in the existing representative of the same section of non-ruminant Artiodactyles, the hog. The first true molar when the permanent dentition is completed, exhibits the effects of its early development in a more marked degree than in most other Mammalia, and in the Wild Boar has its tubercles worn down and a smooth field of dentine exposed by the time the last molar has come into place ; it originally bears four primary cones, with smaller sub-divisions formed by the wrinkled enamel, and an interior and posterior ridge. The four cones produced by the crucial impression, of which the transverse part is the deepest, are repeated on the second true molar with more complex shallow divisions, and a larger tuberculate posterior ridge. The greater extent of the last molar is chiefly produced by the development of the back ridge into a cluster of tubercles ; the four primary cones being distinguishable on the anterior main body of the tooth. The crowns of the lower molars are very similar to those above, but are rather narrower, and the outer and inner basal tubercles are

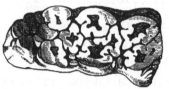

much smaller, or are wanting ; the grinding surface of the last is shown in fig. 123.

Extinct species of hog have been found in miocene beds at Eppelsheim (*Sus palæochœrus*, Kp.), and at Simorre (*S. simor-*

Fig. 123.

Last lower molar, Hog. Nat. size.

rensis, Lt.) ; in pliocene beds (*S. arvernensis*, Crt.), and in pleistocene and later deposits, where the species (*S. scrofa fossilis*) is indistinguishable from the present wild boar.

Order RUMINANTIA.

Of other forms of beasts subsisting on the vegetable productions of the earth, and more akin to actual European Herbivora, there co-existed, in Europe, with the now exotic genera *Elephas*, *Rhinoceros*, *Hippopotamus*, etc., a vast assemblage of species, nearly all of which have passed away. The quadrupeds called " Ruminants," from the characteristic second mastication of the partly-digested food by the act called " rumination" or " chewing the cud," constitute at the present period a circumscribed group of Mammalia, which Cuvier believed to be " the most natural and best-defined order of the class."* He characterized it as having incisive teeth only in the lower jaw (fig. 128, *c*), which were replaced in the upper jaw by a callous gum.   Between the incisors and molars is a diastema, in which, in certain genera only, may be found one or two canines.   The molars (fig. 128), *h*, almost always six on each side of both jaws, have their crown marked by two double crescents, with the convexity turned inwards in the upper set, outwards in the lower.   The four legs are terminated by two toes and two hoofs, flattened at the contiguous

* Règne Animal, tom. i., p. 254.

sides, so as to look like a single hoof cloven ; whence the name
"cloven-footed," also given to these animals. The perfect circumscription and definition of this order, so desirable by the
systematic zoologist, is indeed invaded, in the actual *Ruminantia*, by certain peculiarities of the camel tribe.

In entering upon the evidences of the first appearance in
this planet of the order of animals, which now are the most
valuable to man, it may be well to call to mind the characters
of the *Anoplotherium*. The upper true molars have two
double crescents, convex inwards, one of the inner ones being
encroached on by a large tubercle, the reduced homologue of
which may be seen in the internal inner space of the crescents
in the ox and some other Ruminants. The lower true molars
also, at one stage of attrition, form crescentic islands of enamel,
with the convexity turned outwards, as in Ruminants, the last
molar having the accessory crescent behind. The functional
hoofs were two in number on each foot, but must have resembled those of the camel tribe in shape ; the scaphoid and
cuboid of the tarsus were distinct also, as in the *Camelidæ;*
and the metacarpal and metatarsal bones were divided, as in
the water musk-deer (*Moschus aquaticus*), and in the embryos
of all Ruminants. The dentition of the extinct Dichodon (figs.
102, 103) made a still nearer approach to that of the Ruminants. The chief distinction of this and other extinct Herbivores with double crescentic molars is the completion of the
upper series of teeth by well-developed incisors. But the premaxillaries in the new-born camel contain each three incisors,
one of which becomes fully developed. The *Camelidæ* are hornless, like the Anoplotherioids and Dichodonts ; and with one
exception—the giraffe—all Ruminants are born without horns.

Thus the *Anoplotherium*, in several important characters,
resembled the embryo Ruminant, but retained throughout life
those marks of adhesion to a more generalized mammalian
type. The more modified or specialized form of hoofed animal,

with cloven feet and ruminating stomach, appears at a later period in the tertiary series.

The modification of the upper molars of the existing Ruminant quadrupeds consists in the lower and less pointed lobes of

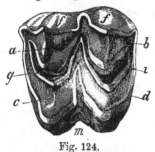

Fig. 124.

Upper molar of Megaceros.

the crown, the unworn summits of which are at first rather trenchant, like curved blades, than piercing. They are soon abraded by mastication, and present the crescentic lobes of dentine (a, b, c, d) shown in fig. 124. The transverse double-crescentic valley (g, i) contains a thicker layer of cement, and forms two detached crescents in worn teeth. The premolars resemble in structure one half of the true molars.

FAMILY I.—BOVIDÆ.

Fossil molars of the ruminant type and bovine character have hitherto been found, with unequivocal evidence, to the writer's knowledge, only in beds or breccias of pliocene and pleistocene age. At those periods in Britain there existed a very large species of bison (*Bison priscus*), and a larger species of ox (*Bos antiquus*), from pliocene fresh-water beds ; whilst a somewhat smaller but still stupendous wild ox (*B. primigenius*) has left its remains in pleistocene marls of England and Scotland. With this was associated an aboriginal British ox of much smaller stature and with short horns (*B. longifrons*), which continued to exist until the historical period, and was probably the source of the domesticated cattle of the Celtic races before the Roman invasion. A buffalo, not distinguishable from the musk kind (*Bubalus moschatus*), now confined to the northern latitudes of North America, roamed over similar latitudes of Europe and Asia in company with the hair-clad elephants and rhinoceroses.

FAMILY II.—CERVIDÆ.

Cuvier[*] first made known the fact of teeth with the character of ruminant molars, and of portions of antlers, being associated with remains of *Lophiodon* and *Mastodon* in the freshwater miocene beds of Montabusard, department of the Loiret. These early ruminant fossils agreed in size with the roebuck; but there were characters showing that they differed almost generically from all known deer. In 1834 Professor Kaup received from the miocene strata near Eppelsheim, Darmstadt, the entire cranium of a small Ruminant, the teeth of which were identical with those described and figured by Cuvier; but the series being complete, showed that the animal had long procumbent canines, as in the *Moschus moschiferus;* in some secondary characters of the teeth, however, as in the proportions of the premolars, and especially the presence of the first of that series, at least in the lower jaw, it was generically distinct from *Moschus* or *Tragulus.* Moreover, the animal had possessed, like the males of the small deer of India called "Muntjac," antlers as well as long canine teeth. Both in the miocene beds of Ingré and Eppelsheim, antlers have been found which were supported on long pedicles, as in the muntjac, and simply bifurcate near their end. It is probable that these horns, which have been referred to the nominal species *Cervus anocerus,* may belong to the *Dorcatherium* of Kaup.

Other species of *Cervidæ* were, however, associated with that remarkable form in the miocene period. Dr. Kaup ascribes some more or less mutilated antlers, which had been shed, to a species he calls *C. dicranocerus.* The beam rises from one to two inches without sending off any branch or brow-antler; it then sends off a branch so large and so oblique that the beam seems here to bifurcate; the anterior prong is, however, the smallest and shortest. The writer has received similar shed and mutilated antlers from the red crag of Sussex, which

[*] Ossemens Fossiles, 4to, tom. iv., p. 104, pl. viii., figs. 5 and 6.

seems to contain a melange of broken-up beds of eocene, miocene, and pliocene age.*

The cervine Ruminants have been divided into groups
according to the forms of the antlers. Of the group with

Fig. 125.
Megaceros Hibernicus (Pleistocene marl).

antlers expanded and flattened at top, of which the fallow-deer
(*Dama*) is the type, no fossil examples have been found in
Britain. Cuvier has described and figured antlers of great size
from the pliocene deposits of the valley of the Somme, near

* Quarterly Jour. of the Geol. Soc., vol. xii., 1856, p. 217, figs. 14-16.

Abbeville, which, from the relative position and direction of the brow-snag and mid-snag, and from the terminal palm, he regards as a large extinct species of fallow-deer ; the name *Cervus Somonensis* has since been attached to this species. But there once existed a group (*Megaceros*, fig. 125) characterized by a form of antler at present unknown amongst existing species of deer. With a beam (*b*) expanding and flattening towards the summit, and a brow-snag (*p*), as in the *Dama* tribe, this antler shows a back-snag (*bz*). Moreover, in antlers, showing an expanse of ten feet in a straight line from tip to tip, and which, from their size and form, seem to have been developed by the deer at its prime, the brow-snag expands and sometimes bifurcates—a variety never seen in the fallow-deer, but which becomes exaggerated in the rein-deer group. The representative of the subgenus *Megaceros* is an extinct species (*M. Hibernicus*, fig. 125), remarkable for its great size, and especially for the great relative magnitude and noble form of its antlers : it is the species commonly but erroneously called the "Irish elk;" for it is a true deer, intermediate between the fallow and rein-deer ; and though most abundant in, it is not peculiar to, Ireland. In that country it occurs in the shell-marl underlying the extensive turbaries. In England its remains have been found in lacustrine beds, brick-earth, red crag, and ossiferous caves.*

The rein-deer (*Cervus Tarandus*) has peculiar antlers (fig. 126), and proportionably the largest of any of existing species. The beam is somewhat flattened throughout, but expands only and suddenly at its extremity, a similar expansion characterizing the brow-snag (*br*) and mid-snag (*bz*), two, three, or more points being developed from all these expansions in fully-developed antlers. The brow-snag is remarkable for its length. There is also frequently a short back-snag. It is plain, therefore, from the presence of this snag, from the great rela-

* Owen, History of British Fossil Mammals, p. 444.

tive size of the antlers, from the complex brow-snag, and the terminal expansion of the beam, that we have in the rein-deer the nearest of kin to the extinct *Megaceros.*

The existing species (*Tarandus*) is restricted to northern latitudes, ranging to extreme ones in Europe, and in America from the Arctic Circle southward to the latitude of Newfoundland, where the large variety called "Carabou" still exists. Rein-deer of similar size, ranged over continental Europe, appear to have been seen by Cæsar in Germany, and have left good

Fig. 126.
Skull and antlers of *Cervus Tarandus.*

evidence of their existence in many parts of England. The specimen figured (fig. 126) was found in pleistocene "till" at Bilney Moor, East Dereham.

A large deer, with subcompressed ramified antlers, slightly expanding at the base of the terminal divisions, but differing from the rein-deer in the absence of the brow-snag, has left its remains in the pleistocene sands of Riège, near Pézenas, France. It is the *Cervus martialis* of Gervais; and seems to have been an intermediate form between the rein-deer (*Tarandus*) and the elk (*Alces*). There is no existing representative of this interesting annectant form of deer.

In formations of corresponding age in France, called "alluvions volcaniques" by Gervais,* fossil antlers of two other extinct species of deer have been discovered, in which, as in *Alces*, the brow-antler is absent, but in which the beam does not expand into a palm.

In North America remains of a large deer (*Cervus americanus fossilis*, Harlan), much resembling the *Wapiti* (*Cervus canadensis*) have been found in pleistocene deposits on the banks of the Ohio. In South America Dr. Lund discovered fossil antlers of two species in bone-caves in Brazil: they were associated with remains of an antelope (*Antilope maquinensis*, Lund) of which genus no living representative is now known in South America.

Of deer with antlers of the type of the existing red-deer (*C. elaphus*), a species is indicated in pleistocene beds and bone caves which rivalled the *Megaceros* in bulk (*Strongyloceros spelæus*); and with this are found, in similar places of deposit, remains of a red-deer with antlers

Fig. 127.

Antler of Red-deer, from alluvium, Ireland.

* Zoologie et Paléontologie Française, 4to, p. 82.

equalling or surpassing the finest that have been observed within the historical period.

Fig. 127 represents one of a pair of antlers from the bed of the Boyne at Drogheda, now in the museum of Sir Philip Egerton, Bart., which measures 30 inches in length, and sends off not fewer than fifteen branches or " snags." *a* is the " brow-snag," which rises immediately above the " burr ;" *b* the second, *c* the third, and *d* the " crown" or terminal cluster of snags, which gave to the deer developing them at the period of his full perfection the title of " crowned hart."

The little roebuck, like the red-deer, appears from its fossil remains to have continued to exist from the prehistorical pleistocene times to the present period.

Order CARNIVORA.

The quadrupeds which subsist by preying upon others co-existed under corresponding varieties of form, and in adequate numbers, with the numerous and various *Herbivora* of the newer tertiary periods. A brief description has already been made of some of the singular forms, the genera of which are extinct, that lived in eocene and miocene times.

Genus GALECYNUS, Ow.—In 1829 the fossil skeleton of a Carnivore, of the size of a fox, was discovered by Sir Roderick I. Murchison in the pliocene schist of Œningen. On a close comparison of this specimen, the writer finds that the first premolar is smaller, and the third and fourth larger than in the fox, and all the teeth are more close-set and occupy a smaller space than in the genus *Canis;* the bones of the feet are more robust ; and these, with other characters, indicate an extinct genus intermediate between *Canis* and *Viverra.** The unique specimen is now in the British Museum.

Genus FELIS, L.—As it is by this form of perfect Carnivore that Cuvier chiefly illustrated his principle of the correlation

* See Quarterly Journal of the Geological Society, vol. iii., 1847, p. 55.

of animal structures, it will be exemplified more particularly in this place, and by the aid of the subjoined cut (fig. 128). The founder of palæontology thus enunciates the law which he believed to guide effectively his labours of reconstructing extinct species :—

" Every organized being forms a whole, a single circumscribed system, the parts of which mutually correspond and concur to the same definitive action by a reciprocal re-action. None of these parts can change without the others also changing, and consequently each part, taken separately, indicates and gives all the others." *

Cuvier did not predicate that law by an à priori method, by any of those supposed short cuts to knowledge, the fallacy of which Bacon so well exposes ; he arrived at the law inductively, and after many dissections had revealed to him the facts—of the jaw of the Carnivore being strong by virtue of certain proportions ; of its having a peculiarly shaped and articulated condyle, with a plate of bone of breadth and height adequate for the implantation of muscles, with power to inflict a deadly bite—a process grasped by muscles of such magnitude as necessitated a certain extent of surface for their origin from the cranium, with concomitant strength and curvature of the zygomatic arch ; the facts of the modified occiput and dorsal spines in relation to vigorous uplifting and retraction of the head when the prey had been griped ; the size and shape of the piercing, lacerating, and trenchant teeth ; the mechanism of the retractile claws, and of the joints of the limb that wielded them ;—it was not until after Cuvier had recognized these facts, and studied them and their correlations in a certain number of typical *Carnivora*, that he felt justified in asserting that " the form of the tooth gives that of the condyle, of the blade-bone (s), and of the claws, just as the equation of a curve evolves all its properties ; and exactly as, in

* Ossemens Fossiles, 4to, tom. i. (1812), p. 58.

taking each property by itself as the base of a particular equation, one discovers both the ordinary equation and all its properties, so the claw, the blade-bone, the condyle, the femur, and all the other bones individually, give the teeth, or are

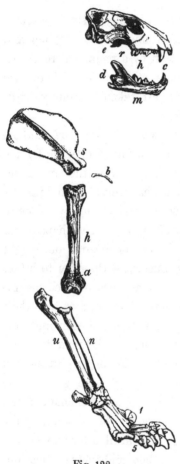

given thereby reciprocally; and in commencing by any of these, whoever possesses rationally the laws of the organic economy will be able to reconstruct the entire animal." The principle is so evident, that the non-anatomical reader will have little difficulty in satisfactorily comprehending it by the aid of the subjoined diagram.

In the jaws of the lion (fig. 128, *h*, *m*), there are large pointed teeth (laniaries or canines, *c*) which pierce, lacerate, and retain its prey. There are also compressed trenchant teeth (*h*), which play upon each other like scissor-blades in the movement of the lower upon the upper jaw. The lower jaw (*m*) is short and strong; it articulates to the skull by a transversely extended convexity or condyle (*d*), received into a corresponding

Fig. 128.
Palæontological characters of a Feline Carnivore.

concavity (*e*), forming a close-fitting joint, which gives a firm attachment to the jaw, but almost restricts its movements to one plane, as in opening and closing the mouth. The plate of bone, called coronoid process (*r*), which gives the surface of attachment to the chief biting muscle (crotaphyte or temporal) is broad and high ; the surface on the side of the skull (temporal fossa, *t*) from which that muscle arises is correspondingly large and deep, and is augmented by the extension of ridges of bone from its upper and hinder periphery.

The bar or bridge of bone (zygomatic arch) which spans across the muscle, bends strongly outwards to augment the space for its passage ; and as it gives origin to another powerful biting muscle (masseter), the arch is also bent upwards to form the stronger point of resistance during the gripe of that muscle. From almost all the periphery of the back surface of the skull there is a strong pitted ridge, affording extensive attachment to powerful muscles which raise the head, together with the animal's body which the lion may have seized with his jaws ; this beast of prey being able to drag along the carcase of a buffalo, and with ease to raise and bear off the body of a man. If we next examine the framework of the fore limb, which is associated with the above-defined structure of the skull, we find that the fore paw consists of five digits (1-5) ; the innermost and shortest (1) answering to our thumb, and having two bones ; the other four digits having each three bones or "phalanges." All those digits enjoy a certain freedom of motion and power of reciprocal approximation for grasping ; but their chief feature is the modification of the terminal phalanx, which is enlarged, compressed, subtriangular, and more or less bent ; with a plate of bone, as it were, reflected forwards from the base, from which the pointed termination of the phalanx projects like a peg from a sheath. A powerful, compressed, incurved, sharp-pointed, hard, horny claw is fixed upon that peg, its base being firmly wedged into the interspace between the peg and the sheath. The toe-joint so armed is retractile. This complex, prehensile, and destructive paw is articulated to the two bones of the fore leg (radius, n, and ulna, u) ; they are both strong, are both distinct, are firmly articulated to the arm-bone (h) by a joint, which, although well knit, allows great extent and freedom of motion in bending and extension ; and, besides this, the two bones are reciprocally joined so as to rotate on each other, or rather the radius upon the ulna, carrying with it, by the greater expansion of its lower end, the whole paw, which can thus be turned " prone" or " supine ;" whereby its application as an instrument for seizing and tearing is greatly advantaged. The humerus or arm-bone (h) is remarkable for the extension of strong ridges from the outer and inner sides, just above the elbow-joint. These ridges indicate the size and force of the supinator, pronator, flexor, and extensor muscles of the paw. To defend the main artery of the fore leg from compression during the action of these muscles, a bridge of bone (a) spans across it as it passes near their origin. The upper end of the arm-bone is equally well marked by powerful ridges for muscular implantation, especially for the deltoid ; but these ridges do not project beyond the round " head" of the bone, so as to impede its movements in the socket.

The blade-bone (scapula s) is of great breadth, with well-developed processes (spine, acromion, and coracoid) for muscular attachments ; the size and shape of this bone relate closely to the volume of the muscles which operate upon the arm-bone and fore limb. A small clavicular bone (b) is interposed between a muscle of the head and one of the arm, giving additional force and determination of action reciprocally to both muscles.

Such are some of the modifications of the teeth and framework of a beast of prey, which concur, and were deemed by Cuvier to be correlated, in the organization of such animals.

Let us compare them with those of the corresponding parts in an ox (fig. 129). The teeth answering to the great laniaries in the lion are absent ; at most, one recognizes the homologues of the lower canines, reduced in size and altered in shape, so as to form the outer teeth (*c*) of a bent row of incisors terminating the lower jaw.   The back teeth (*h*)

instead of being trenchant, have broad and flat crowns, roughened with hard ridges, opposing each other with a grinding action, like mill-stones. The lower jaw is long and slender ; it articulates to the skull by a flat condyle (*d*), admitting of rotatory movements upon a flattened articular surface on the skull, and limiting the extent of opening and shutting the mouth. The coronoid process (*r*) is very slender, and the fossa which marks the size of the temporal muscle (*t*) is correspondingly small. The zygomatic arch (*o*) is short and feeble, and its span is narrow ; it is almost straight, or with a slight bend downwards. The parts of the skull (pterygoid plates) which afford attachment to the rotating muscles of the jaw, and the (angular) part (*f*) of the jaw into which they are inserted, are of great extent.

Fig. 129.
Palæontological characters of a Ruminant (*Bos*).

The ox masticates grass with great efficiency ; it inflicts no injury to other animals with its teeth. The horns are its weapons, and they are chiefly defensive.

The fore foot of the ox is reduced to two principal toes, with two rudimentary ones dangling behind. Each of these has its extremity enveloped by a thick horny case, or hoof; this modification is accompanied by a junction or coalescence of the radius (n) and ulna (u), preventing reciprocal rotation or movement of those bones on each other; by a joint restricting the movement of the fore arm (antibrachium) upon the arm (brachium or humerus, h) to one plane; by a long and narrow blade-bone (s), with a stunted coracoid and no clavicle; in short, by modifications adapting the limb to perform the movements required for locomotion, and almost restricting it to such. This type of fore limb is always associated with broad grinding teeth, and with the modifications of jaw and skull above defined. The due amount of observation assured Cuvier that these several modifications, like the contrasted ones in the Carnivora, were correlated, and he enumerates the physiological grounds of that correlation.

These grounds may be traced to a certain degree in the secondary modifications of the carnivorous order. If the retractibility of the claw be suppressed, the carnassiality of the teeth is reciprocally modified. If the unguiculate foot is reduced from the digitigrade to the plantigrade type, the dentition is still more altered, and made more subservient to a mixed diet.

By the application of the correlative principle to the fossil mammalian remains of pliocene and later deposits, the Herbivora have been distinguished from the Carnivora; and out of the latter have been reconstructed extinct species of the feline, viverrine, ursine, and other families of the order. In England and continental Europe a peculiarly destructive feline quadruped existed, with the upper canines much elongated, trenchant, sharp-pointed, sabre-shaped, whence the name *Machairodus* proposed for this feline sub-genus. It was represented by species as large as a lion (*M. cultridens* and

M. latidens); and by others of the size of a leopard (*M. palmidens* and *M. megantereon*). This form is first found in the miocene of Auvergne and of Eppelsheim; next in the pliocene of the Val d'Arno; and finally in cave breccia in Devonshire.

The penultimate tooth in the upper jaw and the last tooth in the lower jaw were denominated by Cuvier "dents carnassières." The carnassial or sectorial is a very characteristic tooth in the carnivorous order, but undergoes many modifications, and preserves its typical form, as represented in figures 130 and 131, only in the most strictly flesh-feeding species. In it may be

Fig. 130.

Working surface of the upper sectorial tooth, Hyæna. Nat. size.

distinguished the part called the " blade" (fig. 130, *b, b*), and the part called the " tubercule" (fig. 130, *t*). The lower sectorial in the genus *Felis* consists exclusively of the blade (fig. 131), which is pretty equally divided into two lobes. The blade of the upper sectorial always plays upon the outside, and a little in advance of the lower sectorial.

The upper sectorial succeeds and displaces a deciduous tubercular molar in all Carnivora, and is, therefore, essentially a premolar tooth; the lower sectorial comes up behind the deciduous series and has no immediate predecessor; it is, therefore, a true molar, and the first of that class. The sectorial teeth present gradational varieties of form in the carnivorous series, from *Machairodus*, in which the crown consists exclusively of the " blade" in both jaws, to *Ursus* (fig. 132, *m* ɪ), in which it is totally tubercular;

Fig. 131.

Side view of lower sectorial tooth, Lion. Nat. size.

the development of the tubercle bearing an inverse relation to the carnivorous propensities of the species.

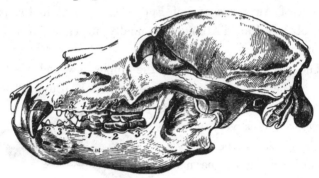

Fig. 132.
Dentition of the Bear (*Ursus*).

The finest examples of the large pleistocene lion (*Felis spelœa*) have been discovered in bone-caves — *e. g.*, in those of Banwell, Somersetshire, and of Belgium. The production of the apex of the nasal process of the maxillary, as far back as that of the nasal bone, proves this species to have been a lion, not a tiger. It roamed over pliocene and pleistocene Europe, and has left its remains in many stratified deposits of the former period in Britain.

Under similar circumstances have been found, more abundantly in Germany, the remains of the gigantic bear (*Ursus spelœus*), and more abundantly in England those of the great hyæna (*Hyæna spelœa*), probably a spotted one, like the fierce " Crocuta" of the Cape. Wolves, foxes, badgers, otters, wolverines, and martin-cats, foumarts and weasels, have left their remains in the newer tertiary deposits and bone caves. Bats, moles, and shrews, were then, as now, the forms that preyed upon the insect world in Europe. The majority of these Carnivora, like the hares, rabbits, voles, and other Rodents, are not distinguishable from the species which still exist. These smaller unguiculate Mammals, like the smaller pleistocene Ruminants, seem to have survived those changes during which the larger species perished. It is probable that the horse and ass are descendants of species of pleistocene antiquity. At the pliocene period there existed a species similar in size to a zebra. There is no certain character by which the present wild boar can be distinguished specifically from the *Sus*, which was contemporary with the mammoth.

Order RODENTIA.

This order includes an extensive series of small Mammals in which a single pair of large, curved, ever-growing incisors in each jaw is associated with many other peculiarities of structure. These incisors (fig. 133, *i*), separated by a wide interval from a short series of molars, characterize the whole order of Rodents ; the single exceptional family, *Leporidæ*, including hares, rabbits, and Picas or tailless hares of

Fig. 133.
Skull and teeth of a Porcupine.

Siberia, retaining a second minute incisor behind each of the larger ones in the upper jaw.

The small size of the great majority of the species of this order leads to the neglect or the oversight of their fossil remains by the labourers in quarries and other deposits of stone, to whom the palæontologist is usually indebted for his first acquaintance with characteristic fossils of such formations. No evidence has yet been obtained of any unequivocal remains of a rodent animal in strata more ancient than the eocene tertiary deposits. Cuvier detected remains of Rodents allied to the dormouse (*Myoxus*) and squirrel (*Sciurus*) in the eocene building-stone of the Montmartre quarries near Paris. The lacustrine marls of the middle (miocene) tertiary period have yielded evidences of not fewer than eleven genera of Rodentia distinct from any now known to exist. The deposits at Eppelsheim, near Darmstadt, of the same miocene age, have given evidence of Rodents akin to the marmot and the beaver. The more recent tertiary formations and the bone-caves in England have furnished fossil remains not distinguishable from the existing beaver, hare, and rabbit, water-vole and field-vole, as well as remains of a Pica, or tailless hare, belonging to the genus *Lagomys*, now confined as an existing species to Asia; and of a very large Rodent, akin to the beaver, called *Trogontherium*. Similar fossil remains have been abundantly found in the pliocene and pleistocene formations of continental Europe, including representatives of the genus *Hystrix*, or fossil porcupines (*H. refossa*, Ger.), from the pliocene of Issoire (fig. 133). The coeval deposits of America have yielded fossil remains of extinct species belonging to genera—*e. g., Lagostomus, Echimys, Ctenomys, Cœlogenys*, and other Cavies— now restricted to South America. In North America, fossil remains of a Rodent of comparatively gigantic size have recently been discovered. Some parts of the skeleton, and more especially the dentition of the rodent order, are highly characteristic—the form of the articular surface for the lower jaw, which is a longitudinal groove, the molars, especially of the phytiphagous kinds, crossed by enamel plates more or less transverse—these, with the long, curved, chisel-shaped incisors, two in each jaw, suffice to determine the ordinal relations of the fossil. The incisors alone would not be always so safe a guide, for the rodent modification of these teeth is repeated in the marsupial wombat and the lemurine aye-aye.

With regard to the Rodentia, the great beaver (*Trogontherium*) seems to have become extinct in England and the Europæo-Asiatic continent before the historical period; whilst the smaller pliocene beaver continued to exist with us, like

the wolf, until hunted down by man.   It still survives in a
few of the great continental rivers.   Of the little Lagomys of
our ossiferous caves no living example remains in either
England or Europe.   The species, indeed, may be extinct :
its genus is now limited to Central and Southern Asia.

GEOGRAPHICAL DISTRIBUTION OF PLEISTOCENE MAMMALS.

A most interesting generalization has been educed from
the mass of facts relating to the fossil Mammals of the later
tertiaries—viz., the close correspondence between the fauna
of those and of the present periods in the Europæo-Asiatic
expanse of dry land.   For here species continue to exist of
nearly all those genera which are represented by pliocene and
post-pliocene mammalian fossils of the same natural continent,
and of the immediately adjacent island of Great Britain.

The bear has its haunts in both Europe and Asia ; the
beaver of the Rhone and Danube represents the great
Trogontherium ; the Lagomys and the tiger exist on both
sides of the Himalayan mountain chain ; the hyæna ranges
through Syria and Hindostan ; the Bactrian camel typifies
the huge *Merycotherium* of the Siberian drift ; the elephant
and rhinoceros are still represented in Asia, though now confined to the south of the Himalayas.   The true macacques
are peculiar to Asia, and though most abundant in the
southern parts of the continent and the Indian Archipelago, also exist in Japan ; a closely-allied sub-genus
(*Innus*) is naturalized on the rock of Gibraltar at the present day.   A fossil species of Macacus was associated with
the elephant and rhinoceros in England during the period of
the deposition of the newer pliocene fresh water beds.   The
more extraordinary extinct forms of Mammalia, called *Elas-
motherium* and *Sivatherium,* have their nearest existing pachy-
dermal and ruminant analogues in the same continent to

which those fossils are peculiar. Cuvier places the Elasmo-
there between the horse and rhinoceros. The existing four-
horned antelopes, like their gigantic extinct analogues, the
Sivathere and Bramathere, are peculiar to India. It may be
regarded as part of the same general concordance of geographical distribution, that the genus *Hippopotamus*, extinct in
England, in Europe, and in Asia, should continue to be represented in Africa, and in none of the remoter continents of the
earth—Africa also having its hyæna, its elephant, its rhinoceros, and its great feline Carnivores. The discovery of
extinct species of *Camelopardalis* in both Europe and Asia,
of which genus the sole existing representative is now, like
the hippopotamus, confined to Africa, adds to the propriety
of regarding the three continuous continental divisions of the
Old World as forming, in respect to the geographical distribution of pliocene, post-pliocene, and recent mammalian genera,
one great natural province. The only large edentate animal
(*Pangolin gigantesque*, Cuvier ; *Macrotherium*, Lartet) hitherto
found in the tertiary deposits of Europe, manifests its nearest
affinities to the genus *Manis*, which is exclusively Asiatic
and African.

Extending the comparison between the existing and the
latest of the extinct series of Mammalia to the continent of
South America, it may be first remarked that, with the
exception of some carnivorous and cervine species, no representatives of the above-cited mammalian genera of the Old
World of the geographer have yet been found in South
America. Buffon* long since enunciated a similar generalization with regard to the existing species and genera of
Mammalia ; it is almost equally true in respect of the fossil.
Not a relic of an elephant, a rhinoceros, a hippopotamus, a
bison, a hyæna, or a lagomys, has yet been detected in the
caves or the more recent tertiary deposits of South America.

* Histoire Naturelle, tom. ix., p. 13, 4to, 1758.

On the contrary, most of the fossil Mammalia from those formations are as distinct from the Europæo-Asiatic forms as they are closely allied to the peculiarly South American existing genera of Mammalia.

The genera *Equus*, *Tapirus*, and the still more ubiquitous *Mastodon*, form the chief, if not sole exceptions. The representation of *Equus* during the pliocene period by distinct species in Asia (*E. primigenius*) and in South America (*E. curvidens*), is analogous to the geographical distribution of the species of *Tapirus* at the present day.

South America alone is now inhabited by species of sloth, of armadillo, of cavy, aguti, ctenomys, and platyrrhine monkey; but no fossil remains of a quadruped referable to any of these genera have yet been discovered in Europe, Asia, or Africa. The types of *Bradypus* and *Dasypus* were, however, richly represented by diversified and gigantic specific

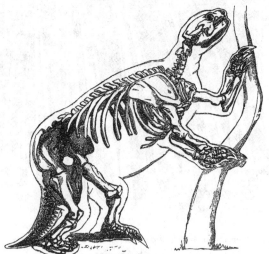

Fig. 134.
Extinct Terrestrial Sloth, *Mylodon robustus* (Pleistocene, S. America).

forms in South America during the geological periods immediately preceding the present. The skeleton of one of these forms of the sloth tribe is represented in fig. 134; it measures from the fore part of the skull to the end of the tail, 11 feet. It was discovered buried 12 feet deep in the fluviatile deposits seven leagues north of the city of Buenos Ayres in the year 1841. It forms the subject of a work entitled, Description of the Skeleton of an Extinct Gigantic Sloth (*Mylodon robustus*),[*] in which are set forth in detail the grounds for regarding it as a member of the same natural family as the present small arboreal sloth, and as being modified to obtain its leafy food by uprooting and prostrating trees.

A still larger species of terrestrial sloth (*Megatherium*) coexisted with the Mylodon in South America. Its skeleton, now complete in the British Museum, measures 18 feet; its dentition

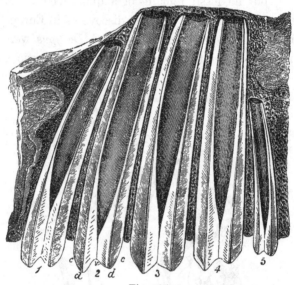

Fig. 135.

Section of upper molar teeth, *Megatherium* (one-third nat. size), Pleistocene, South America.

agrees as to number and kind of teeth with that of the sloths (*Bradypus*). But the molars (fig. 135) are longer, more deeply implanted, of more complex structure, and with grinding

[*] 4to, 1842, Van Voorst.

surfaces of the bilophodont type. The elephants, which subsist on similar food to that of the Megatherium, had their
grinding machinery maintained by a numerous succession of
teeth : the same end was attained in the Megatherium, by a
constant growth and renovation of the same teeth. The formative pulps were lodged in the deep basal cavities, exposed
in the section figured (fig. 135). The molar teeth were five
in number on each side of the upper jaw, and four in number
on each side of the lower jaw (fig. 136).
In this bone the fore part is much prolonged, and grooved above, to support
a long, cylindrical, powerfully muscular
tongue, by which the Megatherium,
like the giraffe, stripped off the small
branches of the trees its colossal strength
enabled it to prostrate. The dentition
of *Mylodon* differed only from that of
Megatherium in the shape of the teeth.
The same may be said of the allied
genera *Megalonyx* and *Scelidotherium*.
They were contemporary and geographically associated genera of the same, now
quite extinct, family of great terrestrial
sloths.

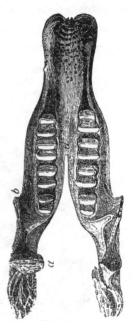

Fig. 136.
Lower jaw and teeth of
Megatherium (Pleistocene,
South America).

In like manner, the small loricated
and banded quadrupeds of South
America called armadillos were represented in pleistocene times in that
continent by as well-defended species, rivalling the Megatherioids in bulk. The specimen of the almost entire skeleton and
bony armour (fig. 137) is one of the smaller species of these
great extinct non-banded armadillos ; yet it measures from the
snout to the end of the tail, following the curve of the back, 9
feet ; the tesselated trunk-armour being 5 feet in length and 7

feet across, following the curve at the middle of the back.
These large extinct species differ from the modern armadillos,
in having no bands or joints in their coat of mail, for the
purpose of contracting or bending the body into the form of a
ball.   They also differ in the fluted form of the teeth (fig. 138) ;
whence the generic name (*Glyptodon*) assigned to them.   The
species are distinguished, like their present puny representatives (*Dasypus*), by peculiar patterns of the outer surface of the
constituent ossicles of the tesselated mail.   In the species figured
(*G. clavipes*), a large raised central circular plate is surrounded
by smaller portions.   The species named *G. reticulatus*, *G. tuber-*

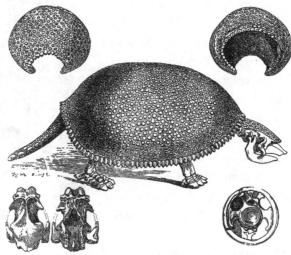

Fig. 137.
Extinct gigantic Armadillo (*Glyptodon clavipes*).

culatus, *G. ornatus*, etc., have their names from other modifications of the sculptured surface of their armour.   Above
the principal figure in cut 137 are shown the front and back
margins of the body-armour ; below it, opposite the left hand,
are upper and under views of the cranium, which was defended
by a tesselated bony casque.   The tail also had its independent osseous sheath, supported by the vertebræ within, as
shown in the figure opposite the right hand.

*Toxodon,** *Macrauchemia,*† and *Protopithecus,*‡ are additional evidences of extinct South American Mammals, matched only by species now peculiar to that continent.

Australia in like manner yields evidence of an analogous correspondence between its last extinct and its present aboriginal mammalian fauna, which is the more interesting on account of the very peculiar organization of most of the native quadrupeds of that division of the globe. That the Marsupialia form one great natural group, is now generally admitted by zoologists ; the representatives in that group of many of the orders of the more extensive placental sub-class of the Mammalia of the larger continents have also been recognized in the existing genera and species : the dasyures, for example, play the parts of the *Carnivora,* the bandicoots (*Perameles*) of the *Insectivora,* the phalangers of the *Quadrumana,* the wombat of the *Rodentia,* and the kangaroos, in a remoter degree, that of the *Ruminantia.* The first collection of mammalian fossils from the ossiferous caves of Australia§ brought to light the former existence on that continent of larger species of the same peculiar marsupial genera : some,

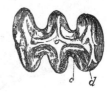

Fig. 138.

Teeth of great extinct Armadillo (*Glyptodon clavipes*), Pleistocene, South America.

as the *Thylacine,* and the dasyurine sub-genus represented by the *D. ursinus,* are now extinct on the Australian continent, but one species of each still exists on the adjacent island of Tasmania ; the rest were extinct wombats, phalangers, poto-

* Owen, "Fossil Mammalia of the Voyage of the Beagle," 4to, 1839.

† Ib.   ‡ Lund. Annales des Sciences Nat., 2d series, tom. xiii., p. 313.

§ Mitchell's (Sir Thos.) Three Expeditions into the Interior of Australia, 8vo, 1838, vol. ii., p. 359.

roos, and kangaroos—some of the latter (*Macropus Atlas, M. Titan*) being of great stature. A single tooth, in the same collection of fossils, gave the first indication of the former existence of a type of the marsupial group, which represented the Pachyderms of the larger continents, and which seems now to have disappeared from the face of the Australian earth. Of the great quadruped, so indicated under the name *Diprotodon* in 1838, successive subsequent acquisitions have established the true marsupial character and the near affinities of the genus to the kangaroo (*Macropus*), but with an osculant relationship with the herbivorous wombat. The

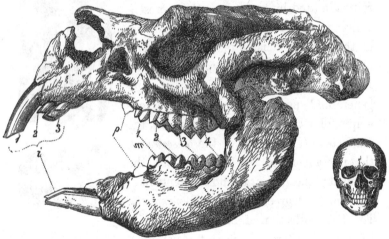

Fig. 139.

Skull, gigantic Pachydermoid Kangaroo (*Diprotodon Australis*) Pleistocene Australia.

entire skull of the *Diprotodon Australis* (fig. 139) has lately been acquired by the British Museum, showing *in situ* the tooth (*i*) on which the genus was founded. This skull measures 3 feet in length ; that of a man is inserted in the cut to exemplify the huge dimensions of the primeval kangaroo. Like the contemporary gigantic sloth in South America, the Diprotodon of Australia, while retaining the dental formula of its living homologue, shows great and remarkable modifications of its limbs. The hind pair were much shortened and strengthened, compared with those of the kangaroo ; the fore pair were lengthened as well as strengthened ; yet, as in the case of the Megatherium, the ulna and radius were maintained free, and so articulated as to give the fore paw the rotatory actions. These in *Diprotodon*, would be needed, as in the herbivorous kangaroo, by the economy of the marsupial pouch. The dental formula of *Diprotodon* was $i \frac{3-3}{1-1}$, $c \frac{0}{0}$, $p \frac{1-1}{1-1}$, $m \frac{4-4}{4-4} = 28$,* and, as in *Macropus major*, the first of the grinding series (p) was soon shed; but the other four two-ridged teeth were longer retained, and the front upper incisor (i, 1) was very large and scalpriform, as in the wombat. The zygomatic arch sent down a process for augmenting the origin of the masseter muscle, as in the kangaroo. The foregoing skull, with parts of the skeleton, of the *Diprotodon Australis*, were discovered in a lacustrine deposit, probably pleistocene, intersected by creeks, in the plains of Darling Downs, Australia.

The same formation has yielded evidence of a somewhat smaller extinct herbivorous genus (*Nototherium*), combining, with essential affinities to *Macropus*, some of the characters of the Koala (*Phascolarctus*).† The writer has recently communicated descriptions and figures of the entire skull of the *Nototherium Mitchelli* to the Geological Society of London.‡ The genus *Phascolomys* was at the same period

Fig. 140.
Grinding surface of molar of *Phascolomys gigas* (nat. size), Pleistocene, Australia.

represented by a wombat (*P. gigas*) of the magnitude of a tapir, one of the grinding teeth of which is represented, of the natural size, in fig. 140.

* See that of *Macropus*, explained in Ency. Brit., article Odontology, p. 449.
† "Report on the extinct Mammals of Australia," Trans. of Brit. Assoc., 1844.
‡ Quarterly Journal of the Geol. Soc., pt. iv., 1858.

The pleistocene marsupial *Carnivora* presented the usual relations of size and power to the *Herbivora,* whose undue increase they had to check. Fig. 141 represents an almost entire skull, with part of the lower jaw of an animal (*Thylacoleo*)

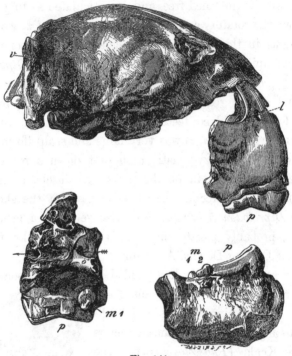

Fig. 141.
Skull of a large extinct Marsupial Carnivore (*Thylacoleo carnifex*),
Pleistocene, Australia.

rivalling the lion in size, the marsupiality of which is demonstrated by the position of the lacrymal foramen (*l*) in front of the orbit; by the palatal vacuity (*o*), by the loose tympanic bone, the development of the tympanic bulla in the alisphenoid, by the very small relative size of the brain, and other characters detailed in the "Philosophical Transactions"* for 1859. The carnassial tooth (*p*) is 2 inches 3 lines in longitudinal extent, or

* "On the Fossil Mammals of Australia. Part I. Description of the *Thylacoleo carnifex.*" By Prof. Owen, etc.

nearly double the size of that in the lion. The upper tubercular tooth (*m*, 1) resembles in its smallness and position that
in the placental Felines. But in the lower jaw the carnassial
(*p*) is succeeded by two very small tubercular teeth (*m*, 1 and 2),
as in *Plagiaulax* (fig. 93, p. 320) ; and there is a socket close
to the symphysis of the lower jaw of *Thylacoleo* which indicates
that the canine may have terminated the dental series there,
and have afforded an additional feature of resemblance to the
Plagiaulax.

The foregoing are some of the more interesting illustrations
of the law, that " with extinct as with existing Mammalia, particular forms were assigned to particular provinces, and that
the same forms were restricted to the same provinces at a
former geological period as they are at the present day."*
That period, however, was the more recent tertiary one.

In carrying back the retrospective comparison of existing
and extinct Mammals with those of the eocene and oolitic
strata, in relation to their local distribution, we obtain indications of extensive changes in the relative position of sea and
land during these epochs, in the degree of incongruity between
the generic forms of the Mammalia which then existed in
Europe and any that actually exist on the great natural continent of which Europe now forms part. It would seem, indeed,
that the further we penetrate into time for the recovery of
extinct Mammalia, the further we must go into space to find
their existing analogues. To match the eocene Palæotheres
and Lophiodons, we must bring Tapirs from Sumatra or South
America, and we have to travel to the antipodes for Myrmeco-
bians, the nearest living analogues to the Amphitheres of our
oolitic strata.

On the problem of the extinction of species, little, demon-
stratively, can be said ; and on the more mysterious subject

* Report on the Extinct Mammals of Australia, 1844.

of their coming into being, no light has yet been thrown by experiment or observation. As a cause of extinction in times anterior to man, it is most reasonable to assign the chief weight to those gradual changes in the conditions affecting a due supply of sustenance to animals in a state of nature which must have accompanied the slow alternations of land and sea brought about in the æons of geological time. Yet this reasoning is applicable only to land-animals; for it is scarcely conceivable that such operations can have affected sea-fishes. There are characters in land-animals rendering them more obnoxious to extirpating influences, which may explain why so many of the larger species of particular groups have become extinct, whilst smaller species of equal antiquity have survived. In proportion to its bulk is the difficulty of the contest which, as a living organism, the individual of such species has to maintain against the surrounding agencies that are ever tending to dissolve the vital bond, and subjugate the living matter to the ordinary chemical and physical forces. Any changes, therefore, in such external agencies as a species may have been originally adapted to exist in, will militate against that existence in a degree proportionate to the bulk of the species. If a dry season be gradually prolonged, the large Mammal will suffer from the drought sooner than the small one; if such alteration of climate affect the quantity of vegetable food, the bulky Herbivore will first feel the effects of stinted nourishment; if new enemies be introduced, the large and conspicuous animal will fall a prey, while the smaller kinds conceal themselves and escape. Small quadrupeds are more prolific than large ones. Those of the bulk of the Mastodons, Megatheria, Glyptodons, and Diprotodons are uniparous. The actual presence, therefore, of small species of animals in countries where larger species of the same natural families formerly existed, is not the consequence of degeneration—of any gradual diminution of the size of such

species—but is the result of circumstances which may be illustrated by the fable of the "oak and the reed;" the smaller
and feebler animals have bent and accommodated themselves
to changes to which the larger species have succumbed.

That species, or forms so recognized by their distinctive
characters and the power of propagating them, have ceased to
exist, and have successively passed away, is a fact no longer
questioned. That they have been exterminated by exceptional
cataclysmal changes of the earth's surface has not been proved.
That their limitation in time, in some instances or in some
measure, may be due to constitutional changes accumulating
by slow degrees in the long course of generations, is possible.
But all hitherto observed causes of extirpation point either
to continuous slowly operating geological changes, or to no
greater sudden cause than the, so to speak, spectral appearance of mankind on a limited tract of land not before inhabited.
It is most probable, therefore, that the extinction of species,
prior to man's presence or existence, has been due to ordinary
causes—ordinary in the sense of agreement with the laws of
organization and of the never-ending mutation of the geographical and climatal conditions on the earth's surface. The
species, and individuals of species, least adapted to bear such
influences, and incapable of modifying their organization in
agreement therewith, have perished. Extinction, therefore, on
this hypothesis, implies the want of self-adjusting power in
the individuals of the species subject thereto.

But admitting extinction as a natural law, which has
operated from the beginning of life under specific forms of
plants and animals, it might be expected that some evidence
of it should occur in our own time, or within the historical
period. Reference has been made to several instances of the
extirpation of species, certainly, probably, or possibly, due to
the direct agency of man. The hook-billed parrot (*Nestor
productus*) of Philip's Island, west of New Zealand, is, perhaps,

the latest instance of this kind. But this cause avails not in the question of the extinction of species at periods prior to any evidence of human existence; it does not help us in the explanation of the majority of extinctions, as of the races of aquatic Invertebrata and Vertebrata which have successively passed away.

The Great Auk (*Alca impennis*, L.) seems to be rapidly verging to extinction. It has not been specially hunted down, like the dodo and dinornis, but by degrees has become more scarce. Some of the geological changes affecting circumstances favourable to the well-being of the *Alca impennis*, have been matters of observation. The last great auks, known with anything like certainty to have been seen living, were two which were taken in 1844 during a visit made to the high rock, called "Eldey," or "Meelsoekten," lying off Cape Reykianes, the S. W. point of Iceland. This is one of three principal rocky islets formerly existing in that direction, of which the one specially named from this rare bird " Geirfugla Sker" sank to the level of the surface of the sea during a volcanic disturbance in or about the year 1830. Such disappearance of the fit and favourable breeding-places of the *Alca impennis* must form an important element in its decline towards extinction. The numbers of the bones of *Alca impennis* on the shores of Iceland, Greenland, and Denmark, attest the abundance of the bird in former times.

Within the last century, academicians of Petersburg and good naturalists described and gave figures of the bony and the perishable parts, including the alimentary canal, of a large and peculiar fucivorous Sirenian—an amphibious animal like the Manatee, which Cuvier classified with his herbivorous Cetacea, and called *Rytina Stelleri*, after its discoverer. This animal inhabited the Siberian shores and the mouths of the great rivers there disemboguing. It is now believed to be extinct, and this extinction appears not to have been due to any special

quest and persecution by man. We may discern in this fact
the operation of changes in physical geography, which have at
length so affected the conditions of existence of the Siberian
manatee as to have caused its extinction. Such changes had
operated, at an earlier period, to the extinction of the elephant
and rhinoceros of the same region and latitudes : a future
generation of zoologists may have to record the final disappearance of the Arctic buffalo (*Ovibos moschatus*). Remains of
Ovibos and *Rytina* show that they were contemporaries of
Elephas primigenius and *Rhinoceros tichorrhinus.*

But recent discoveries indicate that, in the case of the
last two extinct quadrupeds, a rude primitive human race
may have finished the work of extermination, begun by antecedent and more general causes.

Flint weapons called " celts," unquestionably fashioned
by human hands, have been discovered in stratified gravel,
containing remains of the mammoth, in the valley of the
Somme near Abbeville and Amiens, at different periods,
from the year 1847 (Boucher de Perthes, "Antiquités cel-
tiques et antediluviennes," Paris, 1847) to the present time.

These evidences of the human species have been extracted
from the deposit in question, by Mr. Prestwich ("17 feet from
the surface in undisturbed ground," "Proceedings of the Royal
Society," May 26, 1859) ; by Mr. Flower, ("20 feet from the
surface, in a compact mass of gravel," "Times," November
18, 1859) ; by M. Gaudry ("L'Institut," October 5, 1859) ; and
by M. Geo. Pouchet,—all with their own hands in the course
of the year 1859. Besides the *Elephas primigenius*, remains
of *Rhinoceros tichorhinus, Cervus somonensis, Ursus spelœus,* and
of a large extinct Bovine animal, have been found in the
same bed of gravel.

Mr. Prestwich, F.G.S., after a careful study of the geolo-
logical relations of this bed, refers it to the post-pliocene

2 D

age; and to a period "anterior to the surface assuming its present outline, so far as some of its minor features are concerned."

Similar flint weapons had been discovered by Mr. John Frere, F.R.S. (" Archeologia," vol. xiii., " An account of flint weapons discovered at Hoxne in Suffolk," 1800) in a bed of flint gravel, 16 feet below the surface, of the same geological age as that in the valley of the Somme.

Flint weapons have been discovered mixed indiscriminately with the bones of the extinct cave-bear and rhinoceros. One in particular was met with beneath a fine antler of a rein-deer, and a bone of the cave-bear, imbedded in the superficial stalagmite in the bone-cave at Brixham, Devonshire, during the careful exploration of that cave conducted by a committee of the Geological Society of London in 1858 and 1859.

Dr. Falconer, F.G.S., has communicated (" Proceedings of the Geological Society," June 22, 1859) the results of his examination of ossiferous caves in Palermo; and in respect to the " Maccagnone cave," he draws the following inferences :—That, " it was filled up to the roof within the human period, so that a thick layer of bone splinters, teeth, landshells, coprolites of hyæna, and human objects, was agglutinated to the roof by the infiltration of water holding lime in solution; that subsequently and within the human period, such a great amount of change took place in the physical configuration of the district as to have caused the cave to be washed out, and emptied of its contents, excepting the floor-breccia and the patches of material cemented to the roof, and since coated with additional stalagmite." (P. 136.)

Sir Charles Lyell believes " the antiquity of the Abbeville and Amiens flint instruments to be great indeed, if compared to the times of history or tradition. . . ." " It must have required a long period for the wearing down of the chalk which supplied the broken flints for the formation of so much gravel at various heights, sometimes 100 feet above the present level of the Somme, for the deposition of fine sediment including entire shells both terrestrial and aquatic, and also for the denudation which the entire mass of stratified drift has undergone, portions having been swept away, so that what remains of it often terminates abruptly in old river cliffs, besides being covered by a newer unstratified drift. To explain these changes, I should infer considerable oscillations in the level of the land in that part of France; slow movements of upheaval and subsidence, deranging but not wholly displacing the course of ancient rivers. Lastly, the disappearance of the elephant, rhinoceros, and other genera of quadrupeds, now foreign to Europe, implies in like manner a vast lapse of ages, separating the era in which the fossil implements were framed, and that of the invasion of Gaul by the Romans."*

As to the successive appearance of new species in the course of geological time, it is first requisite to avoid the common mistake of confounding the propositions, of species being the result of a continuously operating secondary cause, and of the mode of operation of such creative cause. Biologists may entertain the first without accepting any current hypothesis as to the second.

That the species of the mineralogist and the botanist should be owing to influences so different as is implied by the operation of a second cause, and the direct interference of a first cause, is not probable. The nature of the forces operating in the production of a lichen may not be so clearly understood as those which arranged the atoms of the crystal on which the lichen spreads. Pouchet has contributed the most valuable evidence as to the fact and mode of the production by external

---

* Address, on opening the Section of Geology, at the Meeting of the British Association at Aberdeen, September 15, 1859.

influences of species of Protozoa.* With regard to the species of higher organisms, distinguishable as plants and animals, their origin is as yet only matter of speculation.

Buffon† regarded varieties as particular alterations of species, which illustrated the mutability of species themselves. The so-called varieties of a species, species of a genus, genera of an order, etc., were with him but so many evidences of the progressive degrees of change, which had been superinduced by time and successive generations, and chiefly by degradation from a primordial type. Applying this principle to the species of which he had given the history in his great work, he believed himself able to reduce them to a very small number of primitive stocks, of which he enumerates "fifteen."

Lamarck,‡ adverting to observed ranges of variation in certain species, affirmed that such variations would proceed and keep pace with the continued operation of the causes producing them ; that such changes of form and structure would induce corresponding changes in actions, and that a change of actions, when habitual, became another cause of altered structure ; that the more frequent employment of certain parts or organs leads to a proportional increase of development of such parts ; and that as the increased exercise of one part is usually accompanied by a corresponding disuse of another part, this very disuse, by inducing a proportional degree of atrophy, becomes another element in the progressive mutation of organic forms.

A third theorist§ calls to mind the instances of sudden departure from the specific type, manifested by a malformed or monstrous offspring, and quotes the instances in which such malformations have lived and propagated the deviating structure. He notes also the extreme degrees of change and

* Heterogenie, 8vo, 1859.
† Histoire Naturelle, Dégénération des Animaux, tom. iv., p. 311.
‡ Philosophie Zoologique, 8vo, 1809, tom. i., chs. 3 and 7.
§ "Vestiges of the Natural History of Creation."

of complexity of structure undergone by the germ and embryo
of a highly organized animal in its progress to maturity.   He
speculates on the influence of premature birth, or on a somewhat prolonged fœtation, in establishing the beginning of a
specific form different from that of the parent.

Mr. Wallace,* assuming that varieties may arise in a wild
species, shows how such deviations from type may tend to
adapt a variety to some changes in surrounding conditions,
under which it is better calculated to exist, than the type-
form from which it deviated.

No doubt the type-form of any species is that which is
best adapted to the conditions under which such species at
the time exists ; and as long as those conditions remain
unchanged, so long will the type remain ; all varieties
departing therefrom being in the same ratio less adapted to
the environing conditions of existence.   But, if those conditions change, then the variety of the species at an antecedent date and state of things will become the type-form of the
species at a later date, and in an altered state of things.

Mr. Charles Darwin had, previously to Mr. Wallace,
pondered over and worked at this principle, which he
illustrates by ingenious suppositions, of which I select the
following :—" To give an imaginary example from changes
in progress on an island :—let the organization of a canine
animal which preyed chiefly on rabbits, but sometimes
on hares, become slightly plastic ; let these same changes
cause the number of rabbits very slowly to decrease, and the
number of hares to increase ; the effect of this would be that
the fox or dog would be driven to try to catch more hares ;
his organization, however, being slightly plastic, those individuals with the lightest forms, longest limbs, and best eye-
sight, let the difference be ever so small, would be slightly
favoured, and would tend to live longer, and to survive during

* Proceedings of the Linnæan Society, August 1858, p. 57.

that time of the year when food was scarcest ; they would also rear more young, which would tend to inherit these slight peculiarities. The less fleet ones would be rigidly destroyed. I can see no more· reason to doubt that these causes in a thousand generations would produce a marked effect, and adapt the form of the fox or dog to the catching of hares instead of rabbits, than that greyhounds can be improved by selection and careful breeding.*

Observation of animals in a state of nature, however, is still required to show their degree of plasticity, or the extent to which varieties do arise ; whereby grounds may be had for judging of the probability of the elastic ligaments and joint-structures of a feline foot, for example, being superinduced upon the more simple structure of the toe with the non-retractile claw, according to the principle of a succession of varieties in time.

Farther discoveries of fossil remains are also needed to make known the antetypes, in which varieties, analogous to the observed ones in existing species, might have occurred, seriatim, so as to give rise ultimately to such extreme forms as the Giraffe.

This application of palæontology has always been felt by myself to be so important that I have never omitted a proper opportunity for impressing the results of observations showing the "more generalized structures" of extinct, as compared with the "more specialized forms" of recent animals.

But observation of the effects of any of the above hypo-thetical transmuting influences in changing any known species into another has not yet been recorded. And past experience of the chance aims of human fancy, unchecked and unguided by observed facts, shows how widely they have ever glanced away from the gold centre of truth.

* Proceedings of the Linnæan Society, August 1858, p. 49. The principle is more fully illustrated in the work "On the Origin of Species," 8vo, 1859.

The principles, based on rigorous and extensive observation of facts, which have thus been inductively established, and have tended to impress upon the minds of the closest reasoners in Biology a conviction of a continuously operative secondary creational law, are the following :—the law of irrelative or vegetative repetition : the law of unity of plan or relations to an archetype : the phenomena of parthenogenesis : the progressive departure from general type as exemplified in the series of species from their first introduction to the present time.

TABLE OF GEOLOGICAL DISTRIBUTION OF MAMMALS.

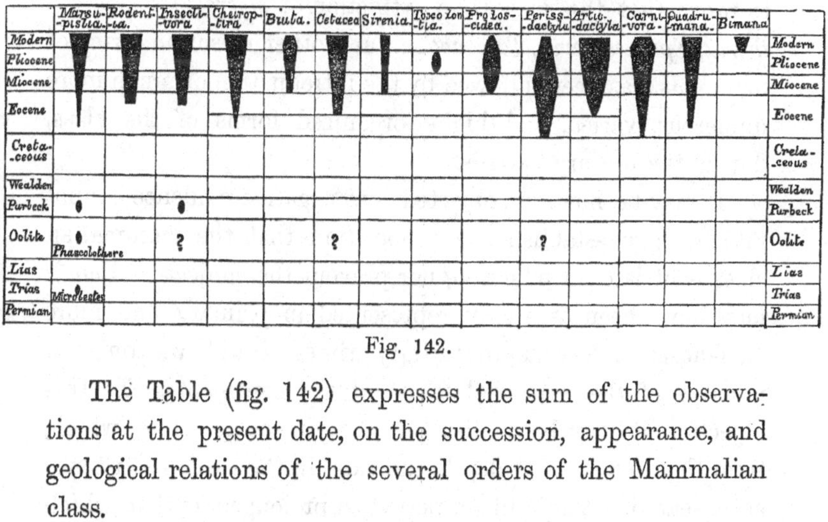

	Marsu-pialia.	Rodent-ia.	Insecti-vora	Chirop-tera	Bruta.	Cetacea	Sirenia.	Toxo-dontia.	Pro Los-cidea.	Periss-dactyla	Artio-dactyla	Carni-vora.	Quadru-mana.	Bimana	
Modern															Modern
Pliocene															Pliocene
Miocene															Miocene
Eocene															Eocene
Creta-ceous															Creta-ceous
Wealden															Wealden
Purbeck	•		•												Purbeck
Oolite	• Phascolotherium		?			?				?					Oolite
Lias															Lias
Trias	Microlestes														Trias
Permian															Permian

Fig. 142.

The Table (fig. 142) expresses the sum of the observations at the present date, on the succession, appearance, and geological relations of the several orders of the Mammalian class.

The earliest evidences are of small species, which, whenever they have presented grounds for ordinal determination, have proved to belong to the low organized Marsupialia. The doubt, when it has existed, lies between this and the Insectivorous order, also low in the class according to cerebral characters.* One example only, from Stonesfield oolite, the *Stereognathus*, may prove to be a minute Ungulate, as is indicated by the note of interrogation under *Perissodactyla*. The

* Owen, " On the Classification and Geographical Distribution of the Mammalia," 8vo, 1859.

similar mark, under *Cetacea*, refers to the fossil, probably washed out of an Upper Oolitic bed, referred to at p. 321. The *Marsupialia* recur, under distinct generic forms, in the eocene strata, and, according to actual knowledge, present their fullest development in pliocene and modern times, more especially in Australia. The orders *Bruta*, *Perissodactyla*, and *Carnivora*, have become reduced in numbers ; the *Proboscidia* still more so ; the representatives of the singular group *Toxodontia* have wholly disappeared.

The sum of the evidence which has been obtained seems to prove that the successive extinction of *Microlestes*, *Amphitheria*, *Spalacotheria*, *Triconodons*, and other mesozoic forms of mammals, has been followed by the introduction of much more numerous, varied, and higher-organized forms of the class, during the tertiary periods.

It may be, however, objected that negative evidence cannot satisfactorily establish the proposition that the mammalian class is of late introduction, nor prevent the conjecture that it may have been as richly represented in primary and more ancient secondary as in tertiary times, could we but get remains of the terrestrial fauna of the continents. To this objection it may be replied: in the palæozoic strata, which, from their extent and depth, indicate, in the earth's existence as a seat of organic life, a period as prolonged as that which has followed their deposition, no trace of mammals has been observed. Were mammals peculiar to dry land, such negative evidence would weigh less in producing conviction of their non-existence during the Silurian and Devonian æons, because the explored parts of such strata have been deposited from an ocean, and the chance of finding a terrestrial and air-breathing creature's remains in oceanic deposits is very remote. But in the present state of the warm-blooded, air-breathing, viviparous class, no genera and species are represented by such numerous and widely dispersed individuals, as those of the order *Cetacea*, which, under the guise of fishes, dwell, and can only live, in the ocean.

In all *Cetacea* the skeleton is well ossified, and the vertebræ are very numerous: the smallest Cetacean would be deemed large amongst land-mammals; the largest surpass in bulk any creatures of which we have yet gained cognizance: the hugest ichthyosaur, iguanodon, megalosaur, mammoth, or megathere, is a dwarf in comparison with the modern whale of a hundred feet in length.

During the period in which we have proof that *Cetacea* have existed, the evidence in the shape of bones and teeth, which latter enduring characteristics in most of the species are peculiar for their great number in the same individual, must have been abundantly deposited at the bottom of the sea; and as cachalots, grampuses, dolphins, and porpoises are seen gambolling in shoals in deep oceans, far from land, their remains will form the most characteristic evidences of verte-brate life in the strata now in course of formation at the bottom of such oceans. Accordingly, it consists with the known characteristics of the cetacean class to find the marine deposits which fell from seas tenanted, as now, with vertebrates of that high grade, containing the fossil evidences of the order in vast abundance.

The red crag of Suffolk and Essex contains petrified frag-ments of the skeletons and teeth of various *Cetacea*, in such quantities as to constitute a great part of that source of phos-phate of lime for which the red crag is worked for the manu-facture of artificial manure. The scanty and dubious evidence of *Cetacea* in secondary beds seems to indicate a similar period for their beginning as for the soft-scaled cycloid and ctenoid fishes which have superseded the ganoid orders of mesozoic times.

We cannot doubt but that had the genera *Ichthyosaurus*, *Pliosaurus*, or *Plesiosaurus*, been represented by species in the

same ocean that was tempested by the Balænodons and Dio-
plodons of the miocene age, the bones and teeth of those
marine reptiles would have testified to their existence as abundantly as they do at a previous epoch in the earth's history.
But no fossil relic of an enaliosaur has been found in tertiary
strata, and no living enaliosaur has been detected in the
present seas: and they are consequently held by competent
naturalists to be extinct.

In like manner does such negative evidence testify to the
non-existence of marine mammals in the liassic and oolitic
times. In the marine deposits of those secondary or mesozoic
epochs, the evidence of vertebrates governing the ocean, and
preying on inferior marine vertebrates, is as abundant as that
of air-breathing vertebrates in the tertiary strata ; but in the
one the fossils are exclusively of the cold-blooded reptilian
class, in the other, of the warm-blooded mammalian class. The
Enaliosauria, Cetiosauria, and *Crocodilia,* played the same part
and fulfilled similar offices in the seas from which the lias
and oolites were precipitated, as the *Delphinidæ* and *Balænidæ*
did in the tertiary seas, and still do in the present ocean.
The unbiassed conclusion from both negative and positive
evidence in this matter is, that the *Cetacea* succeeded and
superseded the *Enaliosauria.* To the mind that will not
accept such conclusion, the stratified oolitic rocks must cease
to be trustworthy records of the condition of life on the earth
at that period.

So far, however, as any general conclusion can be deduced
from the large sum of evidence above referred to, and contrasted, it is against the doctrine of the Uniformitarian.
Organic remains, traced from their earliest known graves, are
succeeded, one series by another, to the present period, and
never re-appear when once lost sight of in the ascending
search. As well might we expect a living Ichthyosaur in the
Pacific, as a fossil whale in the Lias: the rule governs as

strongly in the retrospect as the prospect. And not only as respects the *Vertebrata*, but the sum of the animal species at each successive geological period has been distinct and peculiar to such period.

Not that the extinction of such forms or species was sudden or simultaneous: the evidences so interpreted have been but local. Over the wider field of life at any given epoch, the change has been gradual; and, as it would seem, obedient to some general, continuously operative, but as yet, ill-comprehended, law. In regard to animal life, and its assigned work on this planet, there has, however, plainly been "an ascent and progress in the main."

Although the mammalia, in regard to the plenary development of the characteristic orders, belong to the Tertiary division of geological time, just as " *Echini* are most common in the superior strata; *Ammonites* in those beneath, and *Producti* with numerous *Encrini* in the lowest"* of the secondary strata, yet the beginnings of the class manifest themselves in the formations of the earlier preceding division of geological time.

We are not entitled to infer from the *Lucina* of the permian, and the *Opis* of the trias, that the Lamellibranchiate Mollusks existed in the same rich variety of development at those periods as during the tertiary and present times; and no prepossession can close the eyes to the fact that the Lamellibranchiate have superseded the Palliobranchiate bivalves.

On negative evidence, *Orthisina, Theca, Producta,* or *Spirifer* are believed not to exist in the present seas: on negative evidence the existing genera of siphonated bivalves and univalves are deemed to have been very rare in permian, triassic, or oolitic times. To suspect that they may have then abundantly existed, but have hitherto escaped observation, because certain Lamellibranchs with an open mantle, and some holostomatous

---

* A generalization of William Smith's.

and asiphonate Gastropods, have left their remains in secondary strata, is not more reasonable, as it seems to me, than to conclude that the proportion of mammalian life may have been as great in secondary as in tertiary strata, because a few small-forms of the lowest orders have made their appearance in triassic and oolitic beds.

Turning from a retrospect into past time for the prospect of time to come, I may crave indulgence for a few words, of more sound, perhaps, than significance, relative to the amount of prophetic insight imparted by Palæontology. But the reflective mind cannot evade or resist the tendency to speculate on the future course and ultimate fate of vital phenomena in this planet. There seems to have been a time when life was not; there may, therefore, be a period when it will cease to be.

Our most soaring speculations still show a kinship to our nature: we see the element of finality in so much that we have cognizance of, that it must needs mingle with our thoughts, and bias our conclusions on many things.

The end of the world has been presented to man's mind under divers aspects : as a general conflagration; as the same, preceded by a millennial exaltation of the world to a paradisiacal state,—the abode of a higher race of intelligences.

If the guide-post of Palæontology may seem to point to a course ascending to the condition of the latter speculation, it points but a very short way, and in leaving it we find ourselves in a wilderness of conjecture, where to try to advance is to find ourselves "in wandering mazes lost."

With much more satisfaction do I return to the legitimate deductions from the phenomena which have been under review.

In the survey which has been taken of the various forms of life that have passed away—of their characters, succession, geological position, and geographical distribution—if I have succeeded in demonstrating the adaptation of any structure to the exigencies, habits, and well-being of the species, I have fulfilled one object which I had in view, viz., to set forth the beneficence and intelligence of the Creative Power.

If, in all the striking changes of form and proportion which have passed under review, we could discern only the results of minor modifications of a few essential elements, we must be the more strikingly impressed with the unity of that Cause, and with the wisdom and power, which could produce so much variety, and at the same time such perfect adaptations and endowments, out of means so simple. For, in what have those contrasted limbs, hoofs, paws, fins, and wings, so variously formed to obey the behests of volition in denizens of different elements, differed from the mechanical instruments which we ourselves plan with foresight and calculation for analogous uses, save in their greater complexity, in their perfection, and in the unity and simplicity of the elements which are modified to constitute these several locomotive organs ?

Everywhere in organic nature we see the means not only subservient to an end, but that end accomplished by the simplest means. Hence we are compelled to regard the Great Cause of all, not like certain philosophic ancients, as a uniform and quiescent mind, as an all-pervading *anima mundi*, but as an active and anticipating intelligence.

By applying the laws of comparative anatomy to the relics of extinct races of animals contained in and characterizing the different strata of the earth's crust, and corresponding with as many epochs in the earth's history, we make an important step in advance of all preceding philosophies, and are able to demonstrate that the same pervading, active, and beneficent intelligence which manifests His power in our times, has also manifested His power in times long anterior to the records of our existence.

But we likewise, by these investigations, gain a still more

important truth, viz., that the phenomena of the world do not
succeed each other with the mechanical sameness attributed
to them in the cycles of the epicurean philosophy; for we are
able to demonstrate that the different epochs of the earth were
attended with corresponding changes of organic structure; and
that, in all these instances of change, the organs, still illustrating the unchanging fundamental types, were, as far as we could
comprehend their use, exactly those best suited to the functions
of the being.   Hence we not only show intelligence evoking
means adapted to the end; but, at successive times and
periods, producing a change of mechanism adapted to a change
in external conditions.   Thus the highest generalizations in
the science of organic bodies, like the Newtonian laws of
universal matter, lead to the unequivocal conviction of a great
First Cause, which is certainly not mechanical.

INDEX.

Printed in the United States
By Bookmasters